Seeking News, Making China

Seeking News, Making China

INFORMATION, TECHNOLOGY, AND THE EMERGENCE OF MASS SOCIETY

John Alekna

STANFORD UNIVERSITY PRESS
Stanford, California

Stanford University Press
Stanford, California

© 2024 by John Alekna. All rights reserved.

No part of this book may be reproduced or transmitted in any form or by any means, electronic or mechanical, including photocopying and recording, or in any information storage or retrieval system, without the prior written permission of Stanford University Press.

Printed in the United States of America on acid-free, archival-quality paper

Library of Congress Cataloging-in-Publication Data
Names: Alekna, John, author.
Title: Seeking news, making China : information, technology, and the emergence of mass society / John Alekna.
Description: Stanford, California : Stanford University Press, 2024. | Includes bibliographical references and index.
Identifiers: LCCN 2023039425 (print) | LCCN 2023039426 (ebook) | ISBN 9781503636675 (cloth) | ISBN 9781503638570 (paperback) | ISBN 9781503638587 (ebook)
Subjects: LCSH: Press and politics—China—History—20th century. | Radio journalism—Political aspects—China—History—20th century. | Radio broadcasting—Political aspects—China—History—20th century. | China—Politics and government—20th century.
Classification: LCC PN5367.P6 A44 2024 (print) | LCC PN5367.P6 (ebook) | DDC 070.4/4932095105—dc23/eng/20230912
LC record available at https://lccn.loc.gov/2023039425
LC ebook record available at https://lccn.loc.gov/2023039426

Cover design: Gabriele Wilson
Cover art: *Wuxiandian* [Radio], no. 1, 1956

For my parents

Contents

	List of Figures	ix
	Acknowledgments	xi
	A Note on Names and Transliterations	xiii
INTRODUCTION	The Question of the Sparrows	1
1	The Newsscape of 1919	25
2	Sun Yat-sen, Shanghai, and the Technopolitics of Semicolonial China, 1922–1925	43
3	The Manchurian State Constructs a Newsscape, 1922–1931	61
4	Reading the Radio, Listening in the Streets, 1927–1937	84
5	The Occupation of the Mind, 1937–1945	110
6	Red News and Red Women, 1937–1949	155
7	Socialized Media, 1949–1958	190
8	The Technopolitics of Disorder	238
CONCLUSION	Desire and the Transformation of the Newsscape	258
	Notes	269
	Bibliography	323
	Index	341

List of Figures

4.1. A man in rural Hebei reading a blackboard of transcribed news during the War of Resistance against Japan, 1938 — 85

6.1. A group of radio monitors studying radio repair during the Korean War — 156

6.2. A group of radio monitors practicing the transcription of news from a simulated broadcast — 175

7.1. A crowd of men, women, and children gather to listen to a radio receiver in a rural area — 191

7.2. A utopian image of progress under the worker's state features advanced technologies like vacuum tubes and radio broadcast towers — 193

7.3. Rockets soar and Sputnik flies, while Tiananmen's radio waves echo around the world — 197

7.4. The installation of wired radios signals the prosperity of a happy rural home — 201

8.1. A cadre uses a telephone handset and amplifier to exhort hundreds of people hoeing the earth — 239

8.2. A broadcast worker adjusts a loudspeaker as people labor in teams to pull carts of earth uphill — 243

8.3. People laboring at night gather to hear encouraging news from the radio loudspeaker — 246

Acknowledgments

The world in which this book began seems to hardly exist anymore. When I wrote an early sketch of the project in Janet Chen's modern China seminar at Princeton ten years ago, much more felt certain. The West was full of stable democracies. China was open and optimistic. Around the world, the path to a more mobile, diverse, and accepting society seemed clear. Critical studies of news in the construction and deconstruction of society seemed an artifact for historians and historians alone. Sadly, the cascading crises of the last decade have made clear the urgency of understanding the historical and contemporary interactions of information, technology, and politics. If this book can contribute in some small way to that project, then it will have repaid the scholarly debts accrued with more than mere gratitude.

Until then, I am grateful to the many people who made this work possible. First, I must thank the two individuals who have done more than any others to bring this book into the world. Janet Chen plowed through hundreds of pages of doglegged storytelling to find kernels of historiographical import. She shepherded this project through six years of work and steered it away from many shoals. The book would not be here without her advice and keen insight. Zhang Li helped turn a rather aimless endeavor into an archivally rich story of experience. I would not have been able to write this story without her advice, assistance, and sponsorship across more than seven years, first as a student and then as a colleague.

The teachers and institutions at Princeton shaped how I think about history and this project. Two generous scholars, Emily Thompson and Benjamin Elman, developed my understanding of technology and inspired much of the theoretical and historiographical context of this book. The work and teaching of Stephen Kotkin influenced my framing of the story within the search for a mass society. Sheldon Garon made sure I never forgot Japan. The History of Science Program Seminar read and greatly improved my writing on communications and society in the early People's Republic. Later, Jacob Eyferth's thorough reading of an unformed dissertation helped push this research to become more theoretical and substantive.

Several institutions and programs supported related research and training. The Princeton University History Department, the Blakemore Foundation, the Fulbright Program, Peking University, and the Institute for Advanced Interdisciplinary Study have made substantial investments of time and financial assistance. My colleagues in the History of Science, Technology, & Medicine Department have been unendingly supportive. I would not have been able to reach this point without the patient friendship and assistance of Zhu Yanmei. Our chairman emeritus, Professor Han Qide, took a chance on this project and has proved a most welcoming scholar.

Over the years, chapters have appeared in various states of undress across the Sinological landscape. I presented chapter 1 at the Association for Asian Studies as part of a panel on the history of news. I previously published parts of chapter 2 in *Technology & Culture*. I delivered a version of chapter 8 as a lecture at Peking University. I am grateful to Dayton Lekner for organizing a wonderfully active and inspiring sound studies working group, who reviewed two sections of this text. The anonymous readers for Stanford University Press provided thoughtful critiques that helped improve the historiographic structure and content of the book. To my editor, Dylan Kyung-lim White, who saw potential in this project, thank you. To Noemi Swierski, whose fine ear and careful proofreading were invaluable, thank you. To every friend that provided feedback on my writing and argumentation, thank you. To my family, without whose support none of this would have been possible, thank you most of all.

A Note on Names and Transliterations

Most Chinese proper names and all textual transliterations in this book follow the pinyin system of romanization. I have made certain exceptions to improve readability and accessibility for a nonspecialist audience. Two personal names use nonstandard romanization: Chiang Kai-shek (Jiang Jieshi) and Sun Yat-sen (Sun Zhongshan) retain the foreign-language renderings that they chose in life. Geographical terms are more complicated. Many Chinese places changed names over the course of the twentieth century. Others cycled through multiple foreign-language transliteration systems. In this text, cities and provinces are rendered mostly as Mandarin pinyin, with the exception of the cities of Canton and Hong Kong, which are well-known to English-language readers by their Cantonese-derived names. In quotations from English-language sources, I retain the historical names of some major cities (e.g., Peking) to preserve readability. More minor cities I render in brackets with the modern pinyin transliteration (e.g., Dalian instead of Ta'lien or Dairen). In the case of Fengtian, I have chosen to use the historical Mandarin pinyin name, not its modern name Shenyang or its Manchu name, Mukden, or its older romanization, Fengtien. I use the traditional English names for the two great rivers of China—the Yangtze (Changjiang) and the Yellow (Huanghe).

Mandarin Pinyin	Alternate Names or Romanizations
Beijing	Peking, Pekin, Beiping, Peiping
Changjiang	Yangtze River
Chongqing	Chungking
Guangzhou	Canton
Huanghe	Yellow River
Nanjing	Nanking
Shenyang	Fengtian, Fengtien, Mukden
Tianjin	Tientsin
Xianggang	Hong Kong, Hongkong
Wuhan	Hankow

INTRODUCTION

The Question of the Sparrows

On April 19, 1958, the alleyways and courtyards of Beijing stirred far earlier than usual. Shuffling and whispers disturbed the predawn darkness as people lumbered toward their combat posts—awake but in near silence. Weapons lay at the ready: a housewife with her pots, an auntie with her washbasin. Groggy students emerged from their dormitories with gongs and bugles and horns, flags and bamboo poles. At 4:30 a.m., the stage was set. Standing at the municipal radio station, Wang Kunlun, acting as chairman of the Encircle and Suppress Sparrows Headquarters Committee, gave the signal to attack. Two hundred ninety wired broadcast networks, spread like a web across the capital, shouted as one, bringing Wang's instructions to key junctions of the city, each garrisoned by speakers. "Attack!" they barked. Three million Beijingers began to shake the sky with a thunderous clatter.

The biology of a small bird, the common sparrow, lay at the root of this curious event. These birds startle instinctually at noise or when faced with brightly colored objects, and so it was with streamers and sound, poison and shot, that a vast enterprise mobilized to kill them. Noise washed over the birds from all directions. For four hours that morning, they were kept in the air by sound, flying about without a place of refuge in the vast city. Loudspeakers urged on the hunt, reporting success stories from all corners of the metropolis. Hunters pursued the prey so diligently that they remembered, decades later, having no time to eat. People chased the birds across gray-tiled rooftops and up park trees, through temples and down winding lanes. The sparrows flew over a city alive with racket until, at last, their tiny hearts gave out. They fell from the sky by the hundreds and thousands, dead. Smiling young soldiers shoveled their carcasses onto trucks and strung dozens of birds into garlands, talismans of the people's organized power, to decorate the sides of the flatbed. A loudspeaker hanging on the tailgate broadcast news of the day's victories, attracting a cortege of bicyclists to pedal alongside.[1]

Coordination carried this campaign to eliminate pests to a successful conclusion. The people, mobilized in unison, could achieve great leaps forward in production, development, and sanitation. The people, mobilized in unison, could wipe out the birds that, it was said, stole ripe grain from the fields and from the bellies of China's millions and ranked among the worst threats to public health. Twice a day, at dawn and dusk, the broadcast network led the people in a campaign to eradicate the animals. Even after the evening noisemaking ended, it urged brigades of Beijingers to collect the avian dead, brush away nests, clean houses, and sweep the streets. For three days and three nights, commands passed through the radio to awaken a mighty din across the capital. In the end, a broadcaster announced the margin of victory: 401,160 sparrows dead.[2]

Such statistics echoed across the country as provinces and districts organized similar campaigns. In Shandong, the sparrow war began when six hundred thousand people sprang into action at the command of the Jinan radio station on May 11 at 4:30 a.m. Old residents remember how the sound of hundreds of thousands of people shouting at once rent the sky with deafening noise.[3] A national mass education campaign had been preparing the way for the sanitation movement since the previous December. In Shanxi Province, programs on the dangers of sparrows and the other "Four Pests" (rats, flies, mosquitos) blared from more than seven hundred thousand loudspeakers and more than ten thousand radios in sixty counties and cities. Close to a million civilians and cadres listened in.[4] A *People's Daily* editorial, transmitted from the Beijing radio station, urged the mass sanitation drive forward, pressing the people to overturn that old imperialist label "the sick man of Asia." Petty officials across the country transcribed the editorial as it was broadcast. They copied it onto village blackboards, pasted it on news walls, and distributed it in mimeographed newspapers.[5] Spurred by the arrival of this propaganda and these commands, rural counties throughout China replicated the experience of Beijing. In Hunan Province, Lingling County established an Eliminate Sparrows Headquarters and Committee, as did every village, town, and cooperative in its jurisdiction. On the evening of June 29, the county chairman, speaking from the broadcast post, instructed the village and cooperative farm committees to prepare for battle. On the cooperatives, production team leaders dutifully divided residents into sparrow-capturing brigades. Then, at 5:30 a.m. the next day, a party official, communicating through the dozens of miles of wire strung across the county, through the loudspeakers set up in village squares, communal dormitories, cafeterias, and streets, ordered the attack. That morning, 384,693 Lingling residents spread out to hunt sparrows over the fields and through the mountains. The sound of horns, guns, and hoarse voices chased the birds all day. After dusk, farmers lit lamps so that the villagers could continue the chase, as loudspeakers urged them on.[6]

The War against the Sparrows is remembered as a lesson in the folly of human interventions against nature. Sparrows eat insects. Eradicating the birds, therefore, increased the number of destructive locust-like pests. The war thus contributed to a vast famine that killed more people than any peacetime disaster in human history. Nevertheless, the persecution of the sparrows was not a wholly irrational action. In fact, many societies have carried out misguided eradication campaigns to refashion the environment into a more utilitarian form; think, for instance, of the extinct American carrier pigeon or long-suffering plains bison. What made the War against the Sparrows seem outlandish and memorable to contemporaries was its synchronicity and scale—the fact that tens of millions of people could be mobilized at once toward a common task.[7] The question posed by the sparrow campaign, therefore, is not necessarily why it happened but rather *how* it happened. How did a small clique of men at the top of the Communist Party—Mao Zedong and a rotating cast of adjutants—gain mastery of the birds of the air and the pests of the ground? How did they compel tens of millions of Chinese to go out, before dawn, on a spring day in 1958, and shout into the sky?

To ask the question of the sparrows is to ask how a mass society emerged. Only a mass society, with the expectation of universal connection, consumption, and participation, could carry out such a feat. Until now, attempts to answer this question have centered on the strength of party organization, the persuasive power of propaganda, and the pervasive militarization of society. The nested hierarchies of Leninist organization, almost infinitely replicable both vertically and horizontally, administered and organized action from Beijing to the smallest village. Responsible committees like the Sparrow Elimination Headquarters could be set up at all levels of government across the entire nation. Propaganda, like that of the *People's Daily* editorial, convinced people of the policy's necessity by striking themes like national pride (overturn those foreign imperialists' idea of China as a "sick man"!) and hope (ridding ourselves of pests will improve harvests), bestirring the "affective revolution."[8] Finally, a society militarized by decades of war was conditioned to accept the call to "battle stations," to hunt in "brigades," in the "war" on the sparrows.[9] Though essentially correct, these explanations do not stand on their own, missing as they do the fundamental connective tissue that ties them together—that is, the communication infrastructures and practices that allowed the War against the Sparrows to succeed. Overlooking the *ways* that news flowed, they miss the dynamic power of information, how it can illuminate the relationship between urban and rural, elite and nonelite, literate and illiterate. In short, the rise of mass political action in China cannot be understood as arising through war, nationalism, and the introduction of party organization alone. It must also be understood through the package of techniques, processes, and behaviors we call technology.

At every stage of the sparrow campaign, broadcasting played a central role, pacing and encouraging the fight, carrying propaganda and instructions from Beijing to the provinces, from the provinces to the counties, from the counties to every resident. Broadcasting was the technological infrastructure through which the decision-makers at the top of the Communist Party enacted their agenda. Yet this book is not just a story about radio or the

Communist Party. Broadcasting was just one technological practice within a larger communications ecosystem, which I call the *newsscape*. This newsscape, which eventually allowed a small group of men to mobilize an entire nation, developed largely outside the party. They merely deployed and extended it toward its logical conclusion. Consequently, this book seeks the origins of the infrastructural substratum of Chinese mass society across a fifty-year span of history, integrating actors and spaces not often considered in the usual Communist-Nationalist duality. Indeed, using a technological lens to understand the organization and mobilization of Chinese politics after decades of internal decay throws into relief the complicated evolution and politically problematic inheritances of modern China.[10] The dream of Chinese radio originated in semicolonial Shanghai but was pioneered in warlord Manchuria. Japanese-collaborationist regimes promoted the idea of communication's centrality to the nation-state and built much of China's broadcasting and listening infrastructure. The Nationalist government invented a system to bring news, information, and administration into the countryside through radio, but the Communists carried out the scheme. This book is a story about the technological construction of a Chinese mass society that evolved through many iterations and contained many possible expressions. It is a polyphonic story, where the Chinese Communist Party plays but one part. In fact, this book highlights the ways in which a mass society was constructed outside of *any* single government, nation, or ideology, through the reordering of everyday news practices. Though states intentionally developed the infrastructures that reorganized quotidian experience, the individual and collective desire for information also drove what I call the technopolitical process.

The Technopolitical Process

Most people who have lived through the first decades of the twenty-first century will understand that new communications technologies can dramatically transform society, producing shifts in both large-scale phenomena, like national politics, and intimate, microscale experiences, like personal relationships and daily behaviors. New ways to send, receive, and interact with information have overturned key aspects of our political system, reshaped our media landscape, and sent large segments of society marching in different directions. Yet as we contemplate the social and political phenomena that shifts in communications technology have enabled in our lifetimes, we would do well to remember that the contemporary West is hardly the first society to experience such upheavals. Other people have been here before. Interwoven technological, informational, and political revolutions have occurred many times in the past, most famously in the rise of newspapers and mass politics in the modern period—the subject of a vast literature encapsulated by Anderson's *Imagined Communities*.[11] But as this book will show, the story of newspapers (and other print media) and the creation of nineteenth-century mass politics is hardly a universal one. Other types of information revolution have occurred and, indeed, are still possible. Studying the many ways these transformations have unfolded, appreciating how contemporaries understood them, and recognizing their varied outcomes will do much to illuminate the present, shedding light on both our own experiences and the historical evolution of this technopolitical moment. This book therefore aims to explore the way that individuals, especially the nonelite, experienced one such interlocked technological, informational, and political upheaval in the twentieth century.

After the First World War, most of humanity was still marginally literate, living in rural areas where news arrived slowly, unreliably, and usually by word of mouth. China was no exception. Like other agrarian societies, its people desired technologies and social formations that could promise greater information flows. At the same time, they faced an existential crisis intimately connected to the old information order —namely, the challenge of responding to the era of mass politics that had emerged in the previous decades of imperialism and war.[12] In fact, around the world, a reorientation toward universal consumption (of politics, goods, and culture) was coming to redefine or challenge most aspects of society—"mass" was the calling card of the age. Michel Foucault has argued that a certain set of conditions, which he calls an *episteme*, defines the possibilities of knowledge and organization in a given period.[13] Drawing from this thought, we might say that a mass orientation—this insistence on production for universal consumption—was the political, technological, and philosophical episteme of the epoch. In Europe and North America, which the Chinese of the early twentieth century called the West, mass media practices designed for consumption by a theoretical everyman had already emerged. Point-to-point communication made room for the diffuse transmission of information. Thus newspapers displaced personal letters as the primary source of news. Soon, broadcasting would displace (though not replace) wireless telegrams.[14] Similarly, mass politics, which held out the possibility of universal participation, was fast displacing other forms of social organization. Though the mass episteme was invisible to most people who lived in it, it defined the bounds of intellectual legitimacy, of economic, political, and sometimes even physical survival.

Chinese history shows just how imperative adjusting to this new environment became. In 1895, Imperial Chinese forces were defeated in a decisive if contingent campaign by a Japanese military reorganized under a new mass-political and mass-media system. The mass societies of Western Europe and North America, developed over the course of the nineteenth century through what Eric Hobsbawm calls the "dual revolutions" and hardened by the recent experience of total war, pressured China from their treaty port bases in the East.[15] As recently as 1900, they had invaded and soundly swept aside Chinese armies. And now a new form of mass organization had arisen in the former Russian Empire to threaten from the north. Still, the Bolshevik Revolution and its Leninist mass-membership revolutionary party pointed toward new ways of responding to the challenge of mass organization, especially for what their stadial Marxist theory termed underdeveloped nations.

Mass politics could be created through war, revolution, and political parties, but widespread political mobilization also required technological infrastructures to coordinate geographically and socially disparate bodies—provincial and local governments, civil organizations, and individuals. In fact, mass politics and the technologies of mass media should not be imagined as two distinct phenomena. They are, instead, facets of the same process; one is prerequisite of the other. A mass-political state or party can only be sustained if there is mass communication, existing insofar as they can transmit information. The reverse is also true. Mass communication can only exist in a society with a large-scale production-and-distribution infrastructure maintained and protected by a universal (though not necessarily strong) state. Thus, I argue, mass politics and mass media are two aspects of a single *technopolitical process* that advance and grow together, interwoven so deeply that they are largely indistinguishable. This idea will not sound completely foreign to scholars in the field of science, technology, and society (STS). I see this technopolitical process as a particular facet of the concept of *coproduction* developed by scholars of that field over the last half-century. Coproduction argues, in the words of Sheila Jasanoff, that "we gain explanatory power by thinking of natural and social orders as being produced together."[16] The concept of the technopolitical process takes aspects of this framework (selecting technologies from within the wider scope of "the natural" and politics from within the all-encompassing frame of "the social") and demonstrates how these coproduced phenomena change over time.

Exploring China's technopolitical process from 1919 to 1968, this book joins many classic and contemporary works that examine infrastructural technology's relationship to the political, economic, and social construction of society. Research has shown, for instance, how the development of the United States Post Office shaped the new republic—with some scholars going so far as to argue that its network, along with improved transportation, resulted in an explosion of newspapers and allowed for the emergence of Jacksonian democracy.[17] The study of the role of pamphlets, books, newspapers, and rumor (that is, print technologies and their associated behaviors) in the French Revolution is now several decades old and nearly a field unto itself.[18] More recently, scholars have begun to address how technology, energy, and politics are coproduced.[19] Just as relevant to our study, scholars have demonstrated that "domestic" technologies shape everyday life, culture, and social relations—even our ideas of gender.[20] Adding to this significant body, this book argues that news and its associated technologies were just as important as coal, railroads, or the home in coproducing the experience of twentieth-century life.

The Newsscape

Like the integral technopolitical process, technologies themselves elude discrete identification. They are rarely, if ever, one thing. "Radio" is not a single reified entity. Rather it is a set of practices, a bundle of techniques, which interacts with other "technologies" like the web of human relations (called "culture" or "society"), the human body (especially its labor and its voice), as well as the natural and built environments. Interpreting China's twentieth century through the technopolitical process, through techniques of power and information, this study seeks to understand how a mass media and mass political society emerged and to convey how people experienced these enormous shifts. Such an understanding requires a holistic heuristic that encompasses interlocking technological practices and social patterns, natural geography and built infrastructure, that can comprehend all aspects of the way information moves through a social and physical landscape. This is what I call the *newsscape*.

Developed in the next chapter and expanded on throughout the rest of the book, the concept of newsscape identifies technology as merely one aspect of an ecosystem of news-bearing social behaviors. By integrating space and society, information and culture, the newsscape reflects the state of the technopolitical process at a certain point in time, making it legible to historians. The newsscape's dynamic evolution over the fifty-odd years between 1919 and 1968 will illustrate not only the social, geographical, and infrastructural context of news in China but also the many, sometimes hidden, ways that news influences all human experience.

The act of coining new terminology like *newsscape* cannot be taken lightly. It must be demonstrably useful, rooted first in the unsuitability of other terms like *media landscape, communication ecosystem,* or *soundscape*. Though scholars know media as any intermediary form, practice, or substance, contemporary usage reifies media as discrete corporate entities like Fox News, Twitter, or the *Washington Post*. The phrase "media landscape" often refers to this constellation of large, urban corporations.[21] The newsscape does not refer to a series of institutions that control information or even to how a few forms of media interact. It encompasses the patterns and structures of behavior in a geographical and social landscape. Religion, wealth, literacy, gender (all markers of identity) influenced how news traveled, as did economic structures like market days, and physical constructions like teahouses and crossroads. Daily circulation through the natural and built environment mattered more than corporate media. Even when understood in a scholarly way, the word *media* can emphasize discrete substances and materials, whereas I wish to contextualize objects and underline social practice.

Similarly, an ecosystem usually refers to a closed system of functionalities, sometimes but not necessarily across different media forms. For instance, a media-, communications-, or information- ecosystem might signify the interaction of Facebook and the *New York Times* digital platform, or it may refer to the fact that Wechat embraces different functionalities (text, call, payment, social media) hosted within a single network.[22] Sometimes an ecosystem means a political echo chamber—a self-enclosed, self-referential information bubble.[23] This sense that an information ecosystem is closed (though organic and holistic) contradicts my intended meaning of newsscape. Even when scholars use *ecosystem* to signify a collection of media forms, they do not necessarily notice the dynamic interaction of those forms.[24] In contrast, this book highlights how information moves through multiple forms, thus placing news and its experience at the center.

The various permutations of the word *communication* offer a different set of challenges for scholars. In his classic essay "Mass Communication and Cultural Studies," James Carey addresses critics of *communications* who believe that "the term generally overlooks the fact that communication is first of all a set of practices, conventions, and forms." These critics point out that it is usually conceived as isolated "from the expressive and ritual forms of everyday life."[25] Carey answers by embracing a totalizing, cultural definition of communication (taking after Clifford Geertz) and by arguing that communication is "the maintenance of a society in time . . . not the act of imparting information or influence but the creation, representation, and celebration of shared even if illusory beliefs."[26] While I welcome Carey's contention that information and ritual practice combine to construct the form of society, I still find the early critiques of the use of *communications* to label this phenomenon valid. In contemporary discourse, *communications* too frequently means infrastructure and systems, not the comprehensive cultural definition for which Carey so eloquently argued. Indeed, Carey himself encounters this problem in his influential essay on the telegraph. In it, he demonstrates that technology and ideology are profoundly linked, but at the same time his usage partially reproduces the definition of communications as an infrastructural (and not cultural) system. Like Carey, I also problematize *communication*, echoing his observation of its bifurcation into two meanings (transmission and transportation). But unlike him, I continue to use the word primarily to refer to general systems and the act of transmission, without the time-specificity of news or the experiential focus of the newsscape.

Turning from communications studies, the concept of newsscape draws on the pioneering theory of soundscape, which interprets sound as a social phenomenon within the built and natural environments.[27] I try to make a similar jump, understanding the concept of news as a social practice, with socially constructed meanings, but inscribed within the built and natural environments. As my work makes clear, no single mode of reception (whether listening or reading) was ever dominant or even distinct. Therefore, a focus on the experience of sound, or the use of the term *soundscape*, is inadequate to holistically describe the circulation of news.

Still, the heuristic derived from it, *newsscape*, has positive value. Rooted in news, it is by definition intangible, dynamic, and ephemeral. While news can color the physical world, its hues quickly fade. While news can shake society, the patterns of its movement are quickly forgotten. It is fleeting, in contrast to the greater permanence of the organizations and materials. So, though I sometimes emphasize materiality, the function of information (its forms, movement, and social significance) matters more. Placing news at the heart of our story decenters objects and institutions, while foregrounding experience. This emphasis encourages new perspectives on change, helping remove the tyranny of certain narratives about the nation, infrastructure, or broadcasting.[28] Tracing patterns of behavior outside of all these established categories, newsscape analysis shows us not a teleology of the rise of the state, or even the rise of a set of technologies that we call radio, but rather the rise of a mass society through a reorganization of social practices motivated by the desire for information. This "mass society" can allow for the emergence of nationalism or an intensely administered state, but it is not necessarily the same as either. You can have a mass society with multiple nationalisms or one with a barely functioning central government. Reifying and cohering the experience of news into a legible form, the newsscape therefore confirms information practices as a fundamental axis of change in the modern period outside of these more familiar concepts.

In addition, the idea of the newsscape allows this study to move across political regimes, geographical regions, and the rural-urban divide so prevalent in studies of what were once called "peasant societies." Moving between the village and the city, to take one example of the heuristic's analytical power, allows this book to address the issue of space. All agrarian societies transitioning into a mass orientation through the technopolitical process faced the challenge of collapsing space into manageable units, of finding a way to effectively control territory. Reviel Netz has written eloquently about how nineteenth-century American farmers, ranchers, and railroads confronted this problem using an agricultural technology—barbed wire.[29] Stringing it up across the prairies, they lay claim to the landscape of a continent. Those seeking to administer China faced a similar conundrum. What technological practices could lay claim to the vast area formerly administered by the loosely governed Qing Empire? Certainly, railroads and steamships were one method. But as Elizabeth Köll has shown, the railroads faced serious political headwinds before the 1950s and, at the time of the advent of radio broadcasting, were immensely inadequate to the task of integrating the country.[30] Riverine shipping could only do so much in the face of unnavigable rapids and shallow headwaters.[31] Wireless broadcasting, by contrast, connected previously isolated districts to urban centers, increasing the speed and volume of information. When understood within wider social practices of news—as a newsscape—such changes in communications seem revolutionary. Indeed, by bringing central government movements and directives to rural areas where 90 percent of the population lived, I show that radio technology quickly and inexpensively reordered the political geography of China, creating the space for (but not the certainty of) a reorganized and powerful Chinese state in the latter half of the twentieth century.

Similarly, the newsscape's holistic analysis allows us to move across sensory experience (whether auditory or visual) and social class. Taking up Lydia Liu's proposition, the newsscape views writing as a technology, thus allowing us to break down the distinctions between sound reproduction, orality, and script.[32] I demonstrate the frequency with which the written word became speech and sound became transcript. This is far from a mere theoretical intervention. Historically, the vast majority of people in China were illiterate, so the complicated and ill-defined boundary between sound and script is immensely important for the history of information—and thus all sociopolitical history. The shifts in the newsscape that loudspeakers made possible incorporated large numbers of illiterate individuals into the mass information age for the first time. The advent of sound media thus began to destabilize the privilege of the written word, even while information continued to weave in and out of both forms of being.

Lastly, the newsscape also breaks down the barriers of what is considered "news." In this book, *news* is simply defined as time-sensitive information. Indeed, there is no fixed way to delineate news, information, propaganda, or—as I will show—government orders and state administration. "We have begun a campaign against the sparrows" or "the Soviets have launched the first satellite" arrived through the same channels, intermediaries, and practices as "arrest these men for treason," "rebuild that school," or "register all the young men for the draft." Orders and directives are just as time sensitive as the range of information categories we in the contemporary West consider "news." Thus, this book uses *news*, despite its anachronism and foreign-ness, to subsume the multiple vocabularies Chinese people used to refer to the practice of the transmission of information—that is, the practices variously called news (*xinwen*), information (*xiaoxi*), education (*jiaoyu*), and communications (*jiaotong, tongxun, chuanbo*). The most commonly used term for news or information in the period of study, *xuanchuan* (propaganda or, as a verb, to propagandize), is so freighted with derogatory connotations in English that its use often distorts the original meaning. In Chinese, as in other languages, the term is neutral. The word demonstrates, however, just how intertwined news and governance were—how, as the works of many historians have implied, the reach of news is coextensive with the reach of the state.[33]

Perhaps unsurprisingly, the history of news has been the subject of much fruitful academic work in recent years. Heidi Tworek has elucidated the political significance of news in twentieth-century Germany, placing its wireless network and news agencies in geopolitical context. Arthur Asseraf has delivered a groundbreaking opus on the political and social behaviors of news in colonial Algeria. Julia Guarneri has described the newsscape *avant la lettre* in her work on the newspapers of the nineteenth- and twentieth-century United States, analyzing in productive detail the buildings, squares, regional networks, and social behaviors associated with news in the cities of that period.[34] All these studies have, to greater and lesser extents, highlighted the intermediality of news—that is, the movement of information across media like memory, print, song, telegraph, and letter—a fact that the newsscape heuristic used in this book only highlights further. In China, for instance, a medium like radio was consumed through sound but also through newspapers and public blackboards that contained transcribed broadcasts. Following this pervasive thread of intermediality allows this study to provide a corrective to a Chinese media historiography still largely focused on print, urban areas, and intellectuals while emphasizing how deeply technologies are bound in contextual practice.

Radio and the Historiography of China's Distinctive Newsscape

The conventional narrative of modern communications history is largely based on a narrow set of industrialized Western exemplars. China's experience differed significantly. In the early years, radio in China was read more than it was heard. It arrived secondhand as often as it was directly experienced. It was participatory rather than passive. For decades, it took the form of a massive wired loudspeaker network rather than private receivers. It was an indispensable tool of administration as much as mass entertainment. Women were the voice of broadcast authority rather than men. Radio transmitted orders, guidelines, economic plans, and social policies to rural officials. It was the technological means by which the state enacted its program and the people expanded the political possibilities of their society. Broadcasting allowed nascent, resource-constrained regimes to bring central government campaigns to the countryside, and it allowed the population who lived there to mobilize themselves to comply. A comparative historiography of the technopolitical process in China further illuminates these essential differences in context and narrative, emphasizing, in particular, the distinctive place of wireless technologies within the history of its newsscape.

When radio first arrived in China in the winter of 1922–23, it burst into a world where mass communication was slow, disjointed, or nonexistent. In this way, the country resembled the other, mostly illiterate, agrarian societies that would become known as the "Third World" later in the twentieth century. Less than 10 percent of the population ever read newspapers, which often took days or weeks to reach even relatively proximate locations. On the eve of the Second World War, China's largest newspaper had a circulation of just 150,000 in a country with more than 400 million residents.[35] That just twenty years later, at the beginning of the Great Leap Forward, radio broadcasting reached hundreds of millions of people instantaneously, every day, hints at the radical changes that took place. The mass episteme in China was inseparable from wireless broadcasting; elsewhere, it largely grew with print.

In fact, the political economy of news in China differed substantially from the West. For scholars like Richard John, Jonathan Silberstein-Loeb, or Jurgen Habermas, the expansion of news and associated behaviors reflected capitalistic economic organization, both within the news industry and the emerging commercial-bourgeois classes.[36] In contrast, I show that although economic concerns were frequent, the emergence of a mass society through the technopolitical process was, in this instance, strikingly lacking in capitalist concerns and structures. The profit motive, wage labor, and concentrated industry make remarkably few appearances here—and not just within the sections dealing with communist ideology and news practice. Yes, the news was always circumscribed by material and economic concerns, just as Michael Stamm describes in *Dead Tree Media*, but for many people in much of this story, news is not the corporate business it is elsewhere.[37] News often arrives through voluntary effort or government sponsorship, through a reorganization of behaviors in which no one maximized profit. This was not, of course, universal. Urban newspapers were certainly capitalist in nature; imperial period Peking gazettes were already publishing businesses, as Emily Mokros shows.[38] But I demonstrate that the vast majority of people did not interact with those urban papers on a regular basis. For them, the lived experience of this aspect of the technopolitical process existed outside of a capitalist logic.

Still, the regular readers of the commercial press remain historically and historiographically significant. The rise of newspapers and other publications, even on a small scale, was critical for China's intellectual and political trajectory; the urban, intellectual "reading public" comprised the most influential members of society. But they remained a distinct minority. In her landmark study of "public sympathy" in the 1930s, Eugenia Lean acknowledges that this reading public was limited to urban residents, usually the upper or middle classes, thus revealing an essential contradiction in this and other important studies of the media and politics of Republican China.[39] The importance of print technology and culture in the rise of "mass media" relies on a narrow definition of *mass*. Rural residents (some 90 percent of the population) and the urban working class are excluded. If we seek the origins of a truly mass media and mass political engagement, we must look elsewhere.

The fact that China's print market remained remarkably small in relation to the population may seem puzzling. Chinese cultures, after all, invented printing more than one thousand years ago.[40] China's first modern newspaper, *Shanghai Xinbao*, appeared in 1861, though there was already a rich history of court gazettes based on official reports from the capital dating back to the Song dynasty (960–1279).[41] Newspapers and magazines multiplied in the early twentieth century, becoming influential among intellectual and bourgeois classes. These print items became especially important for defining the ideological trajectory of China as they spread ideas like liberalism, communism, nationalism, and a general reform-minded spirit.[42] Nonetheless, the size of the print "mass media" remained limited.[43] Low absolute circulation numbers point, in part, to a low level of literacy, generally pegged at below 10 percent of the population. The print numbers also point to an extensive secondhand market for papers: once read, a paper could be resold or even rented. Reports indicate that a single paper might be passed a dozen times until the ink was illegible. But the numbers also indicate a poorly functioning marketplace. Henrietta Harrison has demonstrated that newspapers did indeed reach the outlying villages of major cities by the late 1920s, but they continued to arrive days or weeks late.[44] Villages relied on travelers to bring news, or letters from friends and family, or on itinerant merchants to sell old scraps of newspaper. Under these conditions, the arrival of a radio receiver represented a sea change in the newsscape.

Distribution constituted the primary chokepoint in newspaper circulation. By the 1920s, print technology in and of itself could have enabled the growth of a large national market. As the case of Japan proves, as early as the 1890s, industrial print technology could easily accommodate East Asia's logographic writing systems.[45] In the 1920s, several Japanese newspapers had daily print runs exceeding one million copies. Thus, the true challenges for the creation of a Chinese mass media were political; media practices could not intensify without the development of the entire technopolitical process.[46] The collapse of the national government after 1911 and the resulting decades of civil war and maladministration caused roads to decay, railroad construction to idle, and postal efficiency to stall. The print media therefore appeared only sporadically outside major cities. Telegraphy, which first arrived in China in the 1870s, provides another illustration of the political circumscription of communications technology.[47] While telegraphed "circulars"—widely distributed messages intended for publication in newspapers—became a common way to propagandize (indeed, circular telegrams became the primary method of coordinating civilian political action in the period following 1919), the medium remained unreliable. The wires stood in constant danger of being torn down by warlords on military campaigns or by bandits in search of a quick profit at the scrapyard. In addition, because messages had to be transcribed and retransmitted in relays, telegraphy was slow. It could take days for a telegram to reach far corners of the country, if they arrived at all; telegrams were susceptible to interception by political opponents at relay stations.[48] Since political conditions and consequent infrastructural weaknesses conspired to impede the implementation of communications technologies like the telegraph and newspapers, a nationwide mass media market emerged only haltingly.

Radio, by contrast, allowed those who controlled broadcasting facilities to overcome the political obstruction of borders, the geographical obstruction of distance, and the social obstruction of literacy. Radio waves did not respect the boundaries between warlord fiefdoms; information could travel nationwide with fewer impediments. Wireless technology shortened the chain of communication between information producer and consumer, crossing distances that had yet to be bridged by decent roads, telephone wires, or railroads. In rural areas, civilians and government officials transcribed broadcast content for further distribution. Private listeners eagerly discussed broadcast reports with family, friends, and acquaintances, broadening the reach of information. In urban areas, radio bridged the divide between literate and nonliterate individuals. The working class, more likely to be unable to read or afford a newspaper, listened to loudspeakers on the streets, becoming direct consumers of regular news. A common medium for urban and rural areas, rich and poor, north and south, radio expanded the organizational possibilities of Chinese politics.

It is significant, then, that radio arrived as China reached a nadir of its political existence. By 1923, the year of China's first broadcast, a dozen major warlord polities had dissected the country and vied among themselves for dominance. A consortium of imperial powers exercised control over the financial levers of government, as well as many of the largest businesses. Foreign powers occupied the most prosperous cities through a system of concessions. The year-old Communist Party had fewer than a hundred members, as much a reading club as a revolutionary organization. The Nationalist Party claimed a few thousand scattered and loosely organized adherents.[49] These were the two largest political parties in the country. Political mobilization remained limited to urban students, intellectuals, bourgeois businessmen, and some working-class unions. Political action was organized through a corporatist system overseen, generally, by municipal-level chambers of commerce and ad hoc committees. These could turn out tens or hundreds of thousands of people for protests—impressive and potent numbers but only a fraction of a percent of the population.[50] Five decades later, a coherent national leadership could mobilize hundreds of millions of Chinese toward any goal. The coevolution of Chinese politics and Chinese information technologies over these fifty years is the very essence of the technopolitical process.

What changed? One central shift began the same week as China's first radio broadcast: the reform of both Nationalist and Communist parties along Leninist lines. Hierarchies of committees administered under the principle of democratic centralism, the embedding of committed party members in the state bureaucracy and the army, and the concept of elite membership proved a highly effective method of political organization. The Leninist party-state system allowed both parties to dramatically expand their membership and efficiently command action.[51] However important the introduction of a party-state system, the Sino-Japanese War of 1937–45 was the signal event in China's technopolitical process. The cliché that the state makes war and war makes the state rings true.[52] The last twenty years have seen a renaissance in wartime studies illustrating the point.[53] The war cohered extant nationalist (small *n*) feelings into a formidable ideological force while increasing state involvement in arenas as varied as welfare, agriculture, and sanitation; war became the most extreme avatar of mass politics.[54] At the same time, historians like Rana Mitter have emphasized that Japan began to wear out the Nationalist state late in the war, destroying rather than fortifying its institutions.[55] Considering this decline, did the War of Resistance "make" the Chinese state? This study answers with an uncomfortable and much-overlooked fact: Japanese-led collaborationist governments helped forge a Chinese mass society (if not a Chinese state) by building much of the technological infrastructure necessary to enact mass mobilization. Just as important, the war also radically expanded the demand for information. As this book demonstrates, people sought ways to participate in a mass information system, and they reorganized their behavior accordingly.

Though the developmental link connecting war, media, and politics echoes similar evolutions in other societies, the relationship between communications technology and nationalism in China is less clear. In his *Imagined Communities*, Benedict Anderson famously draws a link between the rise of nationalism since the Gutenberg Revolution and what he identifies as print capitalism.[56] How print capitalism, which historians have demonstrated existed by the late Qing period, if not far before, contributed to the development of identification with the "nation" is debatable and beyond the scope of this book.[57] But what is clear is that some form of national identity existed before the advent of regular newspaper readership.[58] In China, modern mass media was not a prerequisite for ethnogenesis. Still, the presence of radio receivers did substantially change the newsscapes of rural districts, bringing news about China's conflict with Japan (for instance) to a wide audience much more quickly. Regardless of the content of the news or propaganda, a larger percentage of the population could follow national events. Thus, radios and the news-sheets they produced integrated many areas into wider political currents for the first time, laying the groundwork for a mass society that could later embrace a modern nationalism distinct from early modern ethnic identity. McLuhan's famous dictum "the medium is the message" has strong echoes here.[59]

China was not unusual in that radio increased identification with larger political trends. Rebecca Scales has provided a portrait of an imagined "radio nation" called into being in France in the interwar period.[60] To the extent that people could hear radio live, the themes she draws of broadcasting contributing to a national imaginary ring true for China, particularly in later years with their accelerated information flow and synchronous listening. But neither radio nor any other modern media constructed nationalism on its own. Whatever aspects of modern nationalism did not already exist in imperial times, the experience of wars with Japan and struggles with imperialist powers, especially Britain, supplied. Furthermore, in many other respects, the experience of radio in China differed fundamentally from countries like Scales's France. The absolute number of radio receivers remained small; consequently, for most people, radio was not a direct or even exclusively aural phenomenon. In this it resembled colonial India or Soviet Russia, where rumor and transcription also played outsized roles.[61] In rural China, the information radio conveyed was often experienced intermedially, through blackboards, radio newspapers, and word of mouth—all of which collated strands from multiple sources into a new product that sometimes differed strikingly from the original. This indirectness meant that language and content played a less direct role in strengthening national identity. I argue, then, that accelerated information flows were not as significant for their stimulation of an imagined belonging (as Anderson theorized) so much as for their encouragement of physical and social organization. These acts—the gathering of people, the building of news spaces, the receiving of information, the maintaining of infrastructure—eventually coalesced through repetition into a mass society that could take many forms. At various points in twentieth-century China, it was imperial, transnational, class-based, or commercially oriented. Nationalism was only one consequence of the transforming newsscape.

Recently, a number of scholars have begun to describe the experience of sound technologies, including broadcasting, in the early People's Republic, bringing much-needed attention to the particularities of Chinese acoustic history. In particular, Jie Li has outlined how radios and loudspeakers contributed in heterogeneous ways to the Chinese revolution, aiming to provide "a soundtrack to a previously silent historiography."[62] Wei Lei has offered a study of how radio helped transform social life in China, valuable especially for its analysis of the post-Maoist era.[63] Dayton Lekner has described the acoustic and affective revolution as a source of political mobilization in the Hundred Flowers Campaign of 1956. Paulina Hartono has offered a creative analysis of the voices of Communist broadcasters and the cultural-political significance of their *sound*.[64] This book agrees wholeheartedly with their work and seeks to complement them by placing acoustic history within the greater context of the newsscape. Sound did not exist in isolation but was part of a social and technological ecosystem that interacted with the printed word, the natural landscape, and built environments in complex ways. This context existed outside the Chinese Communist Party (CCP) and stretched back more than thirty years before 1949.[65] To understand how sound reproduction, wireless practices, and communications technology shaped the People's Republic, we must examine the material, social, and geographical contexts of their development.

Sources and Structure

The complicated history of China's engagement with the technopolitical process, of the inextricable growth of its mass media and mass politics, is, for perhaps obvious reasons, considered "sensitive" (*mingan*) by the government of the People's Republic. The story is not necessarily always a direct or flattering one. Much of the putatively extant archive, for instance the records of the Republican period (1911–49) Central Broadcasting Administration, is kept under lock and key at the Second Historical Archive in Nanjing, unavailable to either Chinese or foreign researchers. Records of national and most provincial broadcasters from after the founding of the People's Republic in 1949 are similarly off-limits. The topic, like much of twentieth-century Chinese history, must be researched obliquely. The historical materials on which this study is based were therefore collected from a number of municipal and provincial archives across China, as well as British and American repositories. Throughout, technology shaped the possibilities of research, allowing for a new and different kind of historical work that operates as far as possible at the level of experience, as told through small actors in diverse sources. Keyword searches for calling up files and fonds uncovered documents that would not be revealed by looking through the records of a radio station or propaganda agency—police investigations of impoverished radio thieves, colorful children's books portraying the power of broadcasting, letters begging for help getting radio news to wartime refugees. Likewise, I also make use of a widespread genre of memoir, collected and published by local governments, which tell the story of China's twentieth century "from below," outside of national elites and centralized institutions. I have mined examples of how the newsscape functioned in practice from these "cultural-historical materials" (*wenshi ziliao*), reading them as far as possible against the grain with an awareness of their "official" origins and the possibility of confirmation bias. Regardless of these caveats, they proved to be a rich source whose utilization would not have been possible without widespread indexing and database search capabilities. Any innovations of perspective in this book must be largely attributed to the innovations of methodology that technology has allowed. Any missteps are my own.

The book opens with an examination of the newsscape of China before the arrival of radio broadcasting. Taking the protests of May 1919 as a case study, the first chapter shows how infrastructure, the natural landscape, and the built environment interacted to shape the flow of news. Developing the concept of the newsscape through particular instances of news transmission, the chapter illustrates the degree to which Chinese politics and media were constrained in the immediate postwar period. Chapter 2 discusses the advent of radio within the context of China's technopolitical process. Examining the micropolitics of amateur radio building and customs regulations, as well as the macropolitics of Sun Yat-sen's turn toward broadcasting and Leninism, the chapter shows that the mass politics and mass media of the technopolitical process were inseparable from birth. In chapter 3, we turn north to Manchuria, where, for the first time, a Chinese state attempted to use wireless technologies to self-consciously construct a newsscape. Marshal Zhang Zuolin and his son Zhang Xueliang built a world-class radio system, for a time the most powerful in Asia. They pioneered the use of radio as a tool for both international relations and domestic propaganda, cementing the Northeast's status as a heartland of China's wireless modernity.

The book next turns to examine the impact of radio on the newsscape as experienced by nonelite individuals. Chapter 4 argues that the spread of radio technology in the 1930s marked the beginning of a newsscape revolution in China. Some rural areas experienced daily news for the first time. Illiterate workers could listen in for free on the streets and from public loudspeakers; communal listening culture pervaded both rural and urban areas. Through radio-listening posts and the radio-news fliers that they published, information could be heard or read contemporaneously throughout the country as the crises of the late 1930s mounted. Indeed, in chapter 5, I contend that the drive for information lay near the heart of the Chinese wartime experience (1937–45), as broadcasting became a tool for both conquest and resistance, and civilians desperately sought news. Though radio equipment quickly deteriorated in their territory, the Nationalist Chinese government in Chongqing communicated with a network of thousands of small newspapers behind enemy lines, keeping communication with occupied areas open throughout the struggle. Here the book also demonstrates how Japan, counterintuitively, constructed large parts of China's radio broadcasting and listening infrastructure, irrevocably altering the Chinese newsscape.

The final three chapters address the newsscape of CCP-held territories from 1937 to 1968. Chapter 6 examines the distinctive gendering of the CCP's newsscape, exploring the context in which voices of news and authority became female. At war's end, Communist women led the charge to take over Japanese-built broadcasting stations, co-opt Japanese technicians, and insert themselves into the existing newsscape throughout the country. The chapter thus also highlights the Communist Party's dependence on established infrastructures in years before its seizure of power. Chapter 7 follows the establishment of the People's Republic in 1949 and examines the government's use of radio broadcasting to reinforce its propaganda goals, direct campaigns, and implement policy. In the CCP's "socialized media" environment, radio listening became required and participatory; citizens listened hours a day to broadcasts about social campaigns, national development, and international affairs. Through this sometimes forced, but often enthusiastic, embrace of radio, people gave the center the power to synchronize the nation, keeping everyone apace of each major and minor campaign, of every political advance and reversal. It enabled society to move as one body and radically expanded the possibilities of national political life. This phenomenon intensified from 1954 on, when the government began constructing a wired broadcasting system. Loudspeakers installed in houses, shops, dining halls, and factories connected an eager populace to a central radio. Through the analysis of communications networks in the 1950s, this book illustrates the importance of using experiential and technological lenses to investigate the construction of Chinese state power. Chapter 8 uses the Cultural Revolution to consider the changes in the Chinese newsscape over the previous fifty years and to meditate on the nonteleological nature of the technopolitical process. A conclusion offers my thoughts on the implications of interpreting the emergence of a mass society from below. It concludes by reflecting on the role that the desire for information has played in human sociopolitical evolution and speculates about lessons for our future.

ONE

The Newsscape of 1919

Early on the evening of May 4, 1919, an injured university student named Xiang Shixiang hurried into a China post office near Tiananmen Square, took a paper slip marked Express Mail Telegram (*kuai you dai dian*), and began to write a plea for help to his elder brother in a small coastal town in Zhejiang. Nearby, a twenty-five-year-old university auditor and sometime newspaper correspondent named Wang Guangqi, fresh from a scene of arson, rushed into a China telegraph office to dash off a report to his friend at a radical newspaper in Chengdu. Just a few dozen characters, it would have to do until the longer letter he would write that evening could wend its way south to Shanghai and then west, up the Yangtze to Sichuan. Meanwhile, at least two individuals made long-distance telephone calls across the eighty-odd miles to Tianjin, Beijing's port city, so that the events of the day would make the morning papers. Across the city, in telegraph installations, post offices, and phone booths, dozens of their compatriots rushed to the same urgent task: calling the nation to action with news of a student uprising before the minister of communications, whose home they had just sacked and whose colleague they had just beaten with bars from an iron bed frame, could attempt to cut off the rebellious capital.[1]

The student protest and riot that had occurred that day would reverberate through twentieth-century Chinese history. The country faced the loss of parts of Shandong Province, the homeland of Confucius, to the avaricious Japanese Empire, despite the fact that it had contributed hundreds of thousands of men to the Allied war effort.[2] But students had stood up to a seemingly supine government in the process of being forced to sign away integral territories and valuable rights both in Paris and in the proverbial back rooms of Beijing ministries. The story of how the students' violent reaction to China's defeat at the negotiating tables in Versailles galvanized an inchoate nationalism has been retold many times, to the point of cliché. This chapter does not aim to retell the events of May 4 and the weeks that followed. Rather, I aim to illustrate that *how* the news of these events traveled mattered just as much as the events themselves. What information was transferred and through what media? How did the forms of technology, the dictates of geography, the configurations of the built environment, and the social rhythms of "traditional" society shape what was heard and read? What can this event tell us about the history of news? More important, what can the history of news tell us about the society in which the event occurred?

Scrutinizing the flow of news can reveal much about society and politics—illustrating who has power and why, as well as uncovering the connections between different regions, peoples, and social classes. But the paths of information flow are rarely ever linear. Instead, they exist in a complex matrix of technological, material, cultural, and political contexts across multiple media. I use the word *newsscape* to refer to this intermedial information environment. Both infrastructure and practice, it acknowledges the materiality of news—the paper, the walls, the wires, roads, and rails through which information travels—as well as the social contexts that shape it. Customs of gathering and dispersal, conventions of conversation, impulses of transcription and reproduction, cultures of literacy and politics—all these are just as integral to the experience of news as any print or wireless technology.

The words of the young activists of Beijing sketched in this chapter's opening paragraph entered a newsscape that would, at first glance, have seemed unrecognizable to their fathers. Information sped along wires at seemingly instantaneous speeds. Newspapers transformed this information into a public commodity like never before. As measured by the number of interested readers and print circulation, access to news had increased by scales of magnitude over the previous generation. But a closer look at how information moved reveals that these new technologies constrained news while also liberating it. An appreciation of scale shows that readerships were exceedingly small. An examination of reception timelines demonstrates that the "instantaneous" was in fact sometimes only as fast as people or paper could travel.

An earlier generation of scholars concerned with information and empire shared an unwritten assumption that technologies like the telegraph or newspaper were perfect machines for transmitting information, that news, especially, would flow from source to subject in a relatively smooth, unmediated process.[3] A single communications technology could come in and sweep away what had come before and replace it with something new and revolutionary for society, economics, and politics. A close, holistic analysis of the historical development of news, not just its component technologies, has challenged this view. The most essential insight the field of the history of news offers is the *intermediality* of information—the idea that no item of news or information exists in isolation. Historically, a newspaper might nearly always be read aloud and exist alongside other forms of writing like manuscript newsletters and personal missives. All of these would be contained within oral news networks that interpreted, verified, accreted, and deleted information. Developed by media studies in the second half of the twentieth century to critique views that were seen as overly "deterministic," the idea of *intermediality* reacted strongly against the likes of Marshall McLuhan, Elizabeth Eisenstein, Jurgen Habermas, and Benedict Anderson, arguing that there were no discrete eras or ages of communication technology.[4] While this critique oversimplifies the arguments of those foundational scholars, the heuristic has proven influential—though not until much more recently in the field of history per se, which has enthusiastically adopted the idea in practice if not in name. The adoption of an intermedial view of news began with critiques of European print history that pointed out the centuries-long survival of manuscript writing in the post-Gutenberg era. People continued to convey news by handwritten letters even in the age of the newspaper.[5] According to this account, in our excitement to trace the development of print, we have overlooked the fact that newspapers, magazines, and books constituted only a small fraction of the totality of human communication.[6]

I wish to push this established idea of technological intermediality further by integrating an understanding of how natural, social, and built geography also shaped the flow of news, thus forming the newsscape. As Robert Darnton shows in his study of news in late *ancien régime* France, human places and behaviors influence how news is spread and experienced. French people heard news under a certain tree in the park where they knew to gather, in aristocratic salons, and through handwritten newsletters.[7] Similarly, Chinese people heard news in markets, along streets, in teahouses, and at parks through what Henrietta Harrison calls the oral news network.[8] They read telegrams in newspapers and letters from friends and family. All these, the people and the paper, traveled along the rail lines that were beginning to cross the country but also along old roads, canals and rivers. The flow of news continued to be affected by weather and tide, the rhythms of festivals and market days, the shape of temple walls and theaters, and the human impulse to investigate noise and commotion. A spectrum of technologies, from copper wire to the human voice, carried information, interacting in multitudinous and unexpected ways. Taking the events surrounding May 4, 1919, as a case study, this chapter uncovers the forms and functions of the Chinese newsscape at the beginning of the twentieth century.

The Limitations of Telegraph and Newsprint: The Case of Chengdu

The sometime-correspondent Wang Guangqi had moved to Beijing from Sichuan several years before these events unfolded. A frequent auditor of classes at Peking University (many like Wang and Mao Zedong, who could not qualify or pay for official enrollment, hung around to listen in on lectures), he also maintained close contact with his hometown of Chengdu. Wang made sure to send his friend Li Jieren, then editor of a small progressive newspaper, frequent telegrams—necessarily short messages of around twenty to thirty characters updating his friend on events in Beijing and overseas. Around the period of May 4, Wang dispatched nearly fifty of these messages, an almost daily stream of updates, including on the events of the day itself, in which he participated.[9] The fact that Wang had gone from sacking the communications minister's house directly to a telegraph office did not mean that the news appeared in the Chengdu papers the next day, however. Far from it. The constraints of the telegraph, of geography, and of the postal system meant that the news would take the better part of two weeks to arrive in the capital of Sichuan in any meaningful sense.

In China in the spring of 1919, communication (*jiaotong*) still functionally and semantically meant transportation, though the word had begun to bifurcate in other languages. In the nineteenth and early twentieth centuries, the English word *communication* and its variants referred to two ideas: the transmission of information and the movement of people across space. Railroads were a *communications* network. In fact, the sense of transportation was perhaps the primary connotation because the two concepts were then inseparable. Until the arrival of the telegraph, transmitting information nearly always meant someone traveling from one place to another.[10] Though the telegraph began to shift this truism, static information-sharing remained restricted to either end of the wire—a point-to-point communication. Significantly, it was not until after the advent of wireless, with its pretensions to universal "broadcasting," that the word *communication* finally split in two, becoming "communication" (the transmission of information) and "transportation" (to move from one place to another). Indeed, we might say that modern information orders are characterized by the divergence of communication and transportation or that modernity itself can be defined by such a division.

In Chinese, the word *jiaotong* (communications) bifurcated much later and much less definitively than in the English vocabulary—the *Jiaotong* Ministry, controlled by the hated Cao Rulin, supervised railroads, highways, telegraphs, mail, and radio. In 1919, communication in China (in the unified nineteenth-century sense) still relied to a remarkable degree on movement, largely because of limitations to the functionality of the country's telegraph system. Wang Guangqi's message from Beijing to Sichuan illustrates this point. The process began in the telegraph office, where Wang would have paid per word to have his handwritten telegram slip accepted by the office worker. Despite the seeming urgency of the news, Wang's slip went to the back of the out-tray pile. The recipient, Li, recalled that "in those days news telegrams were slower than official or business telegrams." Because the telegraph office generated less revenue from telegrams categorized as "news" compared to private business telegrams, and because they were legally required to prioritize official messages, Wang's message would have to wait.[11] Nor was this sluggishness a problem limited to the relatively distant Sichuan. "While the telegraph service is modern and generally gives satisfaction," wrote an observer describing the districts around Beijing and Tianjin, "there is public complaint that the time taken for transmission of a telegram is often greater than is required for a letter sent by train or steamer and that the tariff is too high. Both of these grievances grow out of the fact that official telegrams go free of charge and take precedence over all others."[12]

When its turn finally came, Wang's telegram would have to be encoded into a series of four-digit numbers, representing the characters of the message, set according to the official telegraph codebook. The translation of character to code required a great deal of training, a feat of memory difficult to grasp today. Though there are many accounts of operators (plausibly) claiming to have memorized the numerical digits for all seven thousand odd characters, it is equally clear that it remained a hurdle for the telegraphic system for decades, as Thomas Mullaney and others have discussed at length.[13] As late as the Second World War, a communications expert embedded with the Chinese Communists noted continuing issues with the telegraph code:

> Really skilled operators know nearly all the nine thousand odd groups [four numeral combinations] for the characters in standard telegraph code; I have seen operators listening to four figure groups and writing down characters. However, ordinary army operators would have had to spend hours with the telegraph code book for even basic service messages and learning a little English was far more convenient. I was told that it took at least five years to train an expert operator to go between four figure groups and characters without continual reference to the code book.[14]

The telegraphic system in China, in other words, could not have been more different than the Boy Scout–level simplicity of English Morse.

Having been translated into four-digit numerical codes, Wang's message was then converted into dashes and dots—long and short pulses of electricity—according to international Morse code and transmitted over the wire west from Beijing, over Shanxi, Shaanxi, and finally into Chengdu. As the sounds of the dashes and dots arrived at the receiving office, coming one at a time across the wire, a telegraph operator listened, transcribing them first into numbers and then into characters. The handwritten transcription-cum-translation could then be picked up or delivered by courier to the intended recipient. As one might imagine, this process was prone to mistakes. Characters were often transposed or incorrectly translated. The per-character pricing encouraged economies of phrase, which led to staccato, almost cryptic messages. These two factors meant that the full meaning of dispatches could sometimes be lost in the long process of transmission.[15] In any event, Li remembered, "This important and simple news [of May 4] did not appear in big characters in [Chengdu's] *Chuan Bao* until May 7," a full three days after it had occurred. Li no doubt expected an explosive reaction, at least among the intellectual and student classes of the city.[16]

Yet the news fell flat. "At that time in Chengdu," Li says "the average person did not pay much attention [to this news]. Only we editors were different."[17] Reports in other papers indicate that for days, even weeks, well-informed people in Sichuan had no inkling of the events in Beijing or, at least, had no appreciation of their importance. On May 8, a resolution from a meeting in Ba County, near Chongqing, regarding the situation in Paris made no mention whatsoever of student actions in the capital, indicating that the news had not yet reached that county.[18] On May 12, the Sichuan All School Student Association sent a telegram regarding the Qingdao controversy without mentioning imprisoned students or the riots, suggesting that the news had not reached them or, at the very least, that they had not understood its significance.[19] The news had not been received clearly or widely. Wang's short message was carried in Li's small paper, whose circulation figures do not survive but were almost certainly under one thousand—probably on the order of hundreds of copies. *Chuan Bao* resembled a handbill or flier more than it did a modern, industrially produced paper.

Indeed, even when the news from Beijing did appear in Sichuan's largest paper on May 9, it was hidden in a nondescript spot in the "Collection of Important Telegrams" section. The form and restriction of the telegram genre dictated the message's short, abbreviated style—reading like a headline but containing the entire report. At just forty-seven characters, it may have resembled Wang Guangqi's dispatch since its economical phrasing also did not excite a large response among the reading public:

> All Beijing Students Angry at Cao Rulin and other Country-Sellouts; Crowd Went to Traitor Cao, Asked Cao to Answer Questions, Cao Understood Situation Delicate, Fled; Zhang Zongxiang Beaten, Badly Wounded; Student Crowd Burned Down Cao Residence; Several Dozen Arrested.[20]

Though the paper continued to discuss events in France, including a very short editorial on May 11 about the future of Shandong, there was no further mention of what had happened with the student uprising until May 12, when another short update was published:

> On May Fourth, Beijing students from every school gathered at Tiananmen, all carrying small white banners, written in [metaphorical] blood "Return our Qingdao" these four characters. They loudly called Cao and Zhang *National Traitors* and marched, boycotting classes. A spectacular event.[21]

Over the course of eight days following May 4, only these two short notices were published in *Guomin Gongbao*, which, it seems, represented the totality of information about what had occurred in Beijing. Even for those interested, there was nothing further to be had except by private communication from friends or family. *Guomin Gongbao* was seeking further information, along with at least a few individuals in the province. It published rumors from private telegrams on May 15:

> Regarding the Beijing Students Burned Cao Rulin Residence, Beat Zhang Zongxiang Incident, we now hear that certain local gentlemen have received private telegrams from Beijing: Many students have been arrested, more than ten Sichuan students among them.[22]

These spare telegraphic updates did not cause a sensation in Sichuan. There was simply not enough information. The real impact of the news of May 4 would not occur until the first mail arrived from Beijing in mid-May. Li Jieren, the editor, recalled how "on May 16, Wang Guangqi's long letter, that he had written on the very night of May 4, finally arrived. We quickly drew out the important sentences and using size 35 font published it, along with stimulating headlines before and after the letter." This time, Wang's news landed "like a bomb" among Chengdu's intellectuals.[23] A diarist and writer in Chengdu confirms the outlines of this account, recording that in May 1919, certain publications and magazines from Beijing took nineteen days to arrive. A magazine called *Meizhou Pinglun* (Weekly Editorial) was the first print material he recorded that contained detailed eyewitness descriptions of everything that had occurred during the student protest; it did not reach Chengdu until May 30.[24]

For the people of Chengdu, both the quality and quantity of news mattered. The full impact of May 4 was not felt through the snippets of information allowed by the telegraphic form. Having longer tracts of text, which could only be transported by mail, mattered greatly. And that mail was dramatically slower, moving at the rate of a river steamer, or post boat, up the Yangtze and its tributaries. In the gorges, these were pulled by teams of men walking along the shore, dragging boats through the rapids. Having reached major cities and other distribution points, mail was carried by pony packtrains or on the backs of laborers, up to sixty-four pounds per man. "On most routes they travel day and night with first-class mail," wrote one contemporary observer of Sichuan's communications system.[25]

This postal system distributed *Guomin Gongbao*'s approximately four thousand daily copies; it was the only one of the dozen newspapers printed in the province whose output exceeded one thousand copies per issue.[26] These circulation figures underline just how rare newspapers were as a product and just how elite newspaper reading was as a practice. In Sichuan, a province almost the size of France, with a population larger than either France or Germany, had a local newspaper circulation of around ten thousand copies a day, though more editions came from the coastal treaty ports.[27] According to their own statistics, in 1917, the Chinese post office carried 7,706,900 items of "newspaper and printed matter" to Sichuan. This would have included newspapers and magazines from the major eastern cultural and political centers like Beijing, Tianjin, and Shanghai.[28] If, for the sake of experiment, we assume each of these items was a daily newspaper like *Dagongbao*, *Shibao*, or *Shenbao*, we come to a circulation figure of slightly more than twenty-one thousand papers per day for the seventy-eight million people who lived in Sichuan—one newspaper for every 3,714 people.

These circulation figures are not unusual for China, where newspaper readership remained remarkably low. Harrison considers the estimate that 1 percent of the population read papers too high.[29] These lower-end estimates ring true, though very little is firmly reliable with any statistics in China during this period.[30] Newspapers did not release circulation figures. When they did, they were incentivized to exaggerate their figures to attract higher advertising prices. Carl Crow, an advertising pioneer whose job it was to know the Chinese newspaper industry intimately, gave the total circulation of the fourteen largest papers as perhaps 293,000.[31] If we assume that each paper was read by ten individuals through the secondary market and group reading, as contemporaries estimated, we may say that 2,930,000 individuals could read any day's newspapers.[32] (Compare this total to an aggregate newspaper circulation in the United States and Canada of thirty-one million per day in the same year).[33] Taking the Chinese population at 350 million, we may estimate in the roughest sense that eight out of every one thousand people came into direct contact with newspapers regularly. This figure jibes with what we know about the extreme illiteracy of the period. Again, few reliable contemporary statistics exist, only widely varying estimates, but one survey of older generations conducted after 1950 suggests that 5 to 10 percent of the population was literate.[34] Of course, literacy is not a binary but a spectrum—some people could read more than others—and newspapers in this period certainly required a great deal of education.[35] A writer remembered the challenges of reading papers in Beijing in those days: "the form and content of the Chinese newspapers were pitiful. The writing was inferior classical Chinese, neither quaint nor popular. The commentary was fine. The news was fine. But there was no punctuation, no sentences. It just went from beginning to end like a steel cable. The ink smelled, and the characters were misplaced. There was no beauty in its form."[36]

Even where people possessed the education and desire to read newspapers, they were not always readily available. A student in Jinan, Shandong, recalled that, normally, students were not allowed to read the newspapers the teachers subscribed to. They were locked away in the teacher's lounge, and the teachers "never discussed national affairs."[37] Zhou Shizhao, a student from Hunan, recalled that even when newspapers were accessible, their rarity meant that they were always read aloud. On the morning of May 9, after his first classes, he entered the school breakroom and heard someone reading that day's newspaper in a strong voice. The telegrams and express letters reprinted therein described the events in Beijing for the students, who immediately spread the news throughout the school, excitedly discussing the affair in groups of three or five both in and out of class.[38] Newspaper study groups and public bulletin boards also reflected the scarcity of newspapers and the need to pool resources to obtain them. Young people, including the Hunanese student Mao Zedong, organized themselves not so much to study as to share information obtained from various news sources and discuss its implications.[39] They reordered their behavior to maximize information.

The Oral and Aural in Taizhou Prefecture

Given the constraints on the circulation of print, oral and auditory forms of information transmission must retain a central place in any description of the May 4 newsscape. This sets a difficult task for the historian, who must contend with an archive overwhelmingly biased toward elite and print accounts of news transmission. The chain of transmission, from eyewitnesses in the capital to oral recitations in remote villages, is often complicated and opaque. But we possess a number of documents and memoirs recounting the process of oral transmission conducted by student "speech troupes" among urban workers and rural towns and villages. And in the rare instance of Xiang Shixiang, the injured student from Zhejiang Province, we can trace precisely how the news arrived and then spread throughout a local region.

Xiang addressed the Express Mail Telegram slip to his elder brother, the librarian and high school headmaster of the sleepy stone-walled town near the East China Sea where they had grown up. Linhai, located where a mountain river widened into the narrow east Zhejiang plain, was some thousand miles south of Beijing and must have seemed a world away. Still, the younger Xiang would have been able to pay a flat rate, ten cents cash, up front at the cashier window for the two stamps to send his plea home. Chinese employees of the foreign-run China postal service carried the slip to a mail carriage destined for nearby Tianjin, where it was routed to another train that traveled 626 miles south to Pukou, where a ferry bore it across the Yangtze River. Arriving in Shanghai's massive sorting station, the letter again moved south toward Ningbo in Zhejiang, where the railroad ended. From there, it made its way to Taizhou, the small port at the mouth of the mountain river, just a few miles from its destination. Xiang Shixiang's message reached his brother, by some miracle of organization, on May 7, after only three days. The headmaster, reading the note, knew that the news it contained would be explosive. Beijing authorities had arrested three local boys while putting down the protest-turned-riot. Scions of local society, these capital scholars were being held as common criminals by a corrupt administration that had seemingly betrayed the country's interests, not only at the Paris Peace Conference but through numerous treaties, concessions, and loans that had divided the country.[40]

Headmaster Xiang called a joint meeting of the two schools he supervised and conveyed the news in an evidently moving speech. The students soon organized a march down the main commercial street in the town, bearing banners with slogans on white cotton cloth. Shouting at the top of their lungs, they caught the attention of the townsfolk, who would seek to inform themselves about the commotion. These parades, part of a revolutionary repertoire soon embraced across the country, should not be understood the same way we see protests today—as a group of people registering the level of their discontent and agitating for change. This understanding of protest parades is fundamentally shaped by our comprehension of the contemporary newsscape, where information is so pervasive as to be universal, where the intended audience is (to make a broad generalization) previously aware of the situation being protested and is meant to be impressed by the scale and depth of feeling regarding the matter. The students' march in Linhai walled town—the voices, the noise, the visual spectacle—served another, equally important, purpose: to spread the news of what had happened in Beijing to their townsfolk, many of whom were illiterate and all of whom lacked access to daily newspapers. Ears heard what eyes could not read.

In lieu of newspapers, the activists attempted to print fliers, but these were limited in number and usefulness. Xiang Shiyuan published a flier on May 9 with an editorial, "Never Forget National Humiliation," that described China's situation in plain language. The students, however, used classical Chinese in their flier, "Consider [Ming dynasty General] Qi Jiguang." The stilted prose overflowed with historical allusions hardly intelligible to the average person even if they could read most of the characters. Attempting another tact, the students published some doggerel verse and handed it out at the local Buddhist temple, where a festival was being held. The content informs rather than convinces, explaining that "there is a country to the east called Japan, [those] treacherous ghosts more slippery than snakes and scorpions, who desire the rich land of our Shandong, who have taken advantage of our country's troubles. On May 9 they forced us [to sign] the deadly Twenty-One Demand treaty.... Shame! Shame! Shame!... Devotee! Devotee! Pray ardently to the Buddha.... When you return home explain these things to young and old alike."[41] Asking templegoers to carry the news back home from the festival meeting place, the students took advantage of the existing social network revolving around the religious calendar, transmitting the news not only as paper but as memories of things heard and seen to be repeated at home. Despite these fliers, the work to spread the news throughout the prefecture would remain a primarily oral endeavor over the following weeks.[42]

This process of reproduction inevitably produced its own omissions, insertions, and embellishments, as in most instances of oral transmission. Already, the news handed out and told to the festivalgoers did not include information about the protest, only its most general context—the Japanese encroachment on China. Harrison and Bian Donglei have shown just how this information could be garbled using case studies of a Shanxi literatus. The man, Liu Dapeng, lived only a day's journey by foot or cart from the provincial capital of Taiyuan. Though Liu read newspapers, in 1919, these were secondhand and often weeks late. In early May, Liu's son witnessed a student parade protest much like that of Linhai. On May 11, the son returned from Taiyuan and reported to his father that the ambassador to Japan had been shot in Tianjin (rather than beaten in Beijing), prompting the arrest of nineteen students. Though the details were incorrect, Liu's son accurately described the arrests and the grievances of the students—the loss of Qingdao.[43] On May 31, Liu again noted news about student activities witnessed by someone he knew: "A man from my village, Wang Jian, went to the provincial capital and says that in the capital [Taiyuan] students of every school have stopped going to class. Every day they send groups to the market to give speeches saying that all present officials are thieves selling out the country. Already [these officials] have illicitly sold Shandong and Qingdao to Japan. All the schools in Beijing are also like this. The schools in every province are like this."[44] From these diary entries, we can deduce that the messages spread by the students arrived in only the vaguest detail when passed through intermediaries. The student-activists themselves also realized that the oral news network could not be trusted, that a comprehensive understanding of their actions and the international situation would get through only with more direct transmission. They therefore began forming "speech troupes" (*yanjiang tuan*) to spread the news directly.

Chen Jingqiu, a student at Headmaster Xiang's Number Six High School, joined one of these speech troupes. By chance, the May Fourth Movement had broken out just before summer vacation. A native of the mountain district west of Linhai, Chen and his comrades headed up into isolated towns and villages of that region for the break. In his old age, Chen remembered speaking in the busy market streets of these settlements. The troupe carried a large map painted on cloth, showing China with the districts occupied by Japan painted blood red. They spoke in theaters and markets, public spaces where people already gathered habitually to hear news. "I stood on the stage," he remembered, "with two other members of the propaganda team standing behind me with the big map. I bowed to the audience in front of the stage." Chen spoke about the significance of the Beijing students' movement and pointed to the map to describe the actions of the Japanese. "We are like a mulberry leaf. Japan is like a silkworm. Now it has eaten up a piece of Qingdao, Shandong, and is trying in vain to eat up the rest of China," he explained before describing Japanese atrocities in colonial Korea, the Twenty-One Demands, and the crisis at the Paris Peace Conference. "Nearly a thousand villagers, old and young, listened in silence and shared a common hatred for the crimes of traitors and the Japanese devils."[45] Encouraged by the response, they continued to follow the rhythms of rural social life, speaking in teahouses, village markets, and temples. He recalled taking advantage of one village's annual temple fair. Two or three thousand locals came to pray and hear theater, as they did every year. In 1919, these villagers experienced Chen and his classmates as a form of theater—a performance, often on a stage, involving communal listening to "broadcast" sound.

While Chen followed hill paths and socioreligious cycles, speech troupes elsewhere in the country followed other lines of communication: railways, rivers, and canals. In a world offering seemingly instantaneous transmission over electric wire, news most often still traveled only as quickly as a person or slip of paper could move—fastest by train, slowest by foot over a mountain road. In Tianjin, the closest major city to Beijing, the news of May 4 had been received the same day through two separate telephone calls and at least one telegram.[46] But the nonreading urban public did not necessarily experience the news with the same depth of understanding as the newspaper editors or the same excitement as the young activists. With this in mind, students and their sympathizers made public events out of their speeches, with incentives to draw crowds into the entertainment. Speaking in a park in Tianjin's British Concession, an apothecary sent along tea for the listeners, who quickly became swept up in the moment. "Women without any knowledge or learning," that is, those who were illiterate, "having heard the most sorrowful parts [of the speeches], could not help but crying," reported the student paper.[47] Still, though "the knowledge of the people in cities and towns expanded daily, the villages were still cut off (*bisai*)," receiving little news. Students, therefore, began to travel around the Tianjin region's surrounding villages and towns at the end of May, giving speeches so that "the common people can receive a little knowledge of China's present situation." In a region crisscrossed by canals, they often traveled by boat, the most convenient mode of transport.[48] In one such village, "the audience all expressed welcome and wrangled over the printed material with the words *Boycott Japanese Goods*. You can see," the students concluded, "that our fellow Chinese countrymen are hardly unpatriotic, only no one has awakened them yet."[49] Later, a similar troupe from Beiyang University encountered a squall that nearly swamped their small boat, soaking them to the bone. The weather still played a role in the newsscape, having almost stopped the news-bearing speakers. They persevered, however, and spoke to crowds in the markets and in the village theater, where even tea-bearers and the people hired to keep other patrons' seats (the lowest-status individuals) said, according to the students' own flattering report, "What the gentlemen are saying is correct. It would be great if they could come back once a month, so that no one ever forgets."⁵⁰

Was this reportedly enthusiastic response a sign of latent patriotism or a request for more frequent news? Repeated descriptions of the audience as unawakened and needing reminders point toward the latter conclusion. The thousands who came to the fairs and theaters and temples came, like in every other year or market day, to trade, pray, eat, play, and hear the latest information. They organized not because of nationalism but because of news and their other bodily desires. This supposed ignorance made the students all the more eager to reach them. Yet, the challenges were many. Even if thirty or forty small groups left from the universities in Tianjin, they could only cover a small fraction of the region. More important, the troupes only followed existing lines of communication: the canals and railroads. Students from Shanghai's Fudan University worked along the Beijing-Shanghai-Hangzhou railway, putting together their own sort of whistle-stop tour with speeches, fliers, and copperplate photos of the three supposed traitors, including the minister of communications.⁵¹

Still, the traveling students tried to reach all sectors of society through their talks. The future wife of Zhou Enlai remembered giving speeches, though she recalled that women such as her could not speak in the streets because of "feudal" social conventions. Women were limited to the preaching places in markets or education centers on public occasions.⁵² Such "feudal" customs also meant that most of the propagandizing aimed toward women revolved around girls' schools, which acted as nodes of information. One Tianjin woman recalled trying desperately to find more information about events in Beijing. She finally found it on the afternoon of May 6 in the form of a speech being given at her old school, from which she had graduated some years prior.⁵³ For her, as for so many others in the city, the human voice remained central to the newsscape. A young Zhou Enlai used a megaphone (probably just a cardboard tube) to project his voice to a crowd of hundreds of people waiting at the police station for news of the arrested students—a far cry from the electrified radio newsscape that would carry his voice to hundreds of millions of people in future decades.⁵⁴

Other forms of sound also played an important connective role in the May 4 newsscape, alerting listeners to significant events, piquing their interest, and drawing them in to investigate. Certain Beijing students carried a gramophone on muleback to the suburbs west of the city. Entering a village, they would set up their record player under a big tree, beside a small temple, or at the head of the main street to attract an audience. Usually, as one of the speakers recalled, dozens of farmers, men, women, and children would then come to listen to music and hear the news the students had to convey.[55] But sound did not need to be pleasant to function in the newsscape. After midnight on May 6, a nineteen-year-old junior at Fudan University's affiliated high school in Shanghai was awakened in his dormitory by a furiously ringing bell:

> The whole school woke up from their slumber, baffled and confused. Some students thought maybe there was a fire in the cafeteria and ran out of the room. Soon they returned. Someone asked, what is the reason for ringing the bell? They said, "Mr. Shao Lizi, the Chinese teacher, is ringing the bell there. He asked us to call you to the dining room. He has important news to report." The students then rushed to the dining room and saw that Mr. Shao was talking with senior students from the university. The students, some sitting, some standing for want of room, listened to Mr. Shao. Standing on a stool, he gave an almost three hour report on the news from Beijing, which he had just received (late on May 5) via telegram in his capacity as an editor of the *Minguo Ribao*.[56]

Like Alain Corbin's famous French church bells, the sound of the school bell had a social meaning for these students, serving to call together the community and forming them into a node in the newsscape.[57] Like the noise made by the marching students in Linhai and Taiyuan, the sound brought people into the chain of communication that radiated out from Tiananmen on May 4 across the many interconnected media of electricity, manuscript, memory, newsprint, and voice.

Indeed, any one of these news technologies can only exist within a complex matrix of other social techniques—though sometimes these can be as simple as human speech or a piece of chalk. Studying a holistic newsscape reveals this intermediality, whether in 1919, 1949, or later, warning that we should not overlook the centrality of orality, sound, and the experience of illiterate individuals. Similarly, we should not overestimate the influence of the telegraph, which suffered from structural deficiencies, or the role of print in the lived experience of news. When we examine print as a transmitter of news, it must be considered within the context of the postal network and therefore the roads, railways, and watercourses—the natural and built geographies of the newsscape—that constrained it.

But in 1919, the newsscape I have described, like the politics and society of China, stood on the brink of immense change. The speed, frequency, and industrial scale that defines the modern newsscape had already begun with newspapers and magazines; such existing technologies would continue to expand and penetrate ever deeper into rural society and the urban working classes. But for the vast majority of the population, these characteristics of modernity would not emerge until the advent of radio, which obviated the problems inherent in the telegram and newspaper. Messages were no longer translated into numerical code and electric pulses, one at a time, from single point to single point. The use of speech meant that, in theory, anyone could listen to and transcribe news. Radio waves could overcome the primary limitations of newspapers—the fact that they were a material object, printed on a finite natural resource, that had to be physically distributed. Shifting the bounds of these constraints would lay the groundwork for new mass techniques of political and social organization. People would quickly adapt their behavior when presented with practices that allowed higher volumes and frequencies of information. Perhaps unsurprisingly, these fundamental changes in the newsscape began in China's most modern and prosperous city—Shanghai—and with the revolutionaries who lived there.

TWO

Sun Yat-sen, Shanghai, and the Technopolitics of Semicolonial China, 1922–1925

Dr. Sun Yat-sen was playing a dangerous double game in the last week of January 1923. While socialites across Shanghai feted the birth of radio, dancing in hotel ballrooms to the latest New York jazz, the professional revolutionary and inveterate politicker worked to set two Great Powers against one another. He had little choice. Residing in the safety of Shanghai's French Concession, a *de jure* refuge for radicals and dissidents, he lacked an army or any formal diplomatic recognition for his "government." Only his reputation and his small Nationalist Party, as disorganized as it was, remained. He would have to use both that week, along with the newborn power of radio, to salvage the situation and, in the process, drastically alter the course of Chinese history.

January 1923 witnessed both the birth of broadcasting and the formation of two Leninist revolutionary parties in China. These events were neither coincidental nor isolated. They were aspects of a larger global moment in which new mass techniques, emerging from the crucible of the First World War, entered the technopolitical process of many nations in parallel.[1] In the realm of communications, a bundle of techniques called broadcasting evolved from

43

the advances in wireless made during the war. Meanwhile, in the realm of politics, mass revolutionary parties and anticolonial nationalist movements emerged from the social stresses of wartime organization.[2] These new political and communications techniques developed from existing practices and institutions. Broadcasting, for instance, built on newspaper, telegraph, and wireless infrastructure. The newly Leninist Nationalist and Communist parties built on extant but loosely structured organizations like earlier iterations of the Guomindang and Communist study groups. From these bases, the new mass techniques would interact with one another dialectically to accelerate the speed, depth, and reach of the technopolitical process. This fact became apparent almost immediately when China's first radio station broadcast propaganda on behalf of Sun Yat-sen, helping to hasten the Bolshevik embrace of his Nationalist Party. Indeed, as this chapter will demonstrate, communications techniques and practices were experienced as a thoroughly political phenomenon in most contexts, from the macroframeworks of transnational movements and customs administration, to the microperspectives of radio ownership and listening.

This manifestation of the technopolitical process was not merely evident in retrospect. Sun himself was a great exponent and theorist of the conflation of communications and politics, seeing essentially no difference between the two. Indeed, he spent a great deal of his energies elaborating on the need for communications infrastructure (by which he meant both the movement of people and information) as the basis for the political reorganization of China. His 1922 book, *The International Development of China*, expounded on the necessity of railroads, highways, telegraph, and wireless across more than 250 detailed pages.[3] His plan, if implemented, would do nothing less than revolutionize the Chinese newsscape. To him, the political significance of such development was clear:

> The size of a country should not be considered in terms of the area of its territory but rather by the nature of its communications. If communication facilities are lacking, a country, though small in area, is not unlike a large one. If on the other hand, a large country is knitted together by fully developed communication lines, its remote parts seem to be near. Should China to-day continue to carry on the tradition that members of

a community may live an entirely secluded and individualistic life, there could be no good government even in a small area of a hundred square li. But if full use is made of scientific achievements for the promotion of communications, the government can direct the various parts of the Republic even as the brain controls the limbs and digits. Under such conditions, what difference can it make whether the political system is one of centralization or decentralization?[4]

Thus, the events of January 1923 held special significance for Sun. On the evening of Thursday the twenty-fifth, just two days after the country's first broadcast, listeners across China tuned in to its only radio station to hear Sun's "Peaceful Unification Manifesto" (*heping tongyi xuanyan*) proclaimed over the airwaves. Concerned as always with the infrastructural roots of China's disunity, Dr. Sun diagnosed the essential problems facing China in the manifesto's preamble. "The Beijing government has not yet fully implemented the rule of law," he wrote, "so there are still many independent provinces. Beijing's orders (*mingling*) cannot be speedily delivered (*ju ji*), and the cause of reunification is still endless."[5] Sun chose a poetic phrasing to describe the transmission of orders from the capital, invoking the word for imperial horse courier (*ju*). These mounted riders had acted as the emperor's postal service, tying the empire together. He thus emphasized the interrelated problems of discontiguous politics and communications, implying that the fragmentation of the country's newsscape was equally as important as its political fragmentation.

The remainder of the "Peaceful Reunification Manifesto" urged the four major warlord factions to confine themselves to their respective territories, reduce troop levels, and refrain from interfering in each other's affairs.[6] Seemingly contrary to his reputation as a fiery nationalist, Sun also called for a "friendly power" (*you bang*) to assist and oversee the disarmament process. The identity of this foreign power was left purposefully vague, though he was likely playing to both British and American audiences. This was the risky part of his strategy. He intended the appeal for "friendly" intervention to flatter and tempt Britain while frightening Soviet Russia with the prospect of an Anglo American–aligned China on its southern border. He would thus win one or the other's firm backing for a unified China.[7] There was no guarantee that he would not merely alienate both, however, and end up an irrelevancy or, worse, dead. (He had only recently survived an assassination attempt.) Sun, therefore, had to deploy every possible means to shape the political and media environment in his favor.

On Thursday night, the same evening as the broadcast, he invited the editors of the most influential Chinese-language papers, as well as their journalist and wire service colleagues, to his residence on Rue Molière in the French Concession for a banquet. He urged them to make a concerted effort to publicize his plan for peaceful unification over the next three months.[8] Though his party controlled a paper, the *Republican Daily News* (*Minguo Ribao*), its reach was limited, and he needed their help to spread the news. Theoretically, they were well-positioned to assist. Shanghai was home to China's oldest and most substantial news industry, possessing outsized national influence. International wire services like Reuters were headquartered there. Institutions in the city like the *China Press* (an American-owned English-language paper) and *Shenbao* (a Chinese-language paper) were cutting-edge innovators, later helping to found China's earliest radio stations. His twenty or thirty guests, therefore, constituted the most connected individuals in the country, the most prominent nodes in the newsscape. In a speech, he praised the power of information, of "pen and ink," as he put it, to advance the cause of China's unification. In particular, he commended certain political cartoons in Hong Kong. The artists who created them "painted images and sounds that spoke vividly." People throughout that colony used these pictures as advertisements and decorations, causing opinion to harden against the warlords.[9] Sun's vision of the newsscape, then, was many-channeled. It ranged from painting to the printed word of the newspaper, from public speaking to the signals of the telegraph wire. And now, while he addressed the newsmen, it extended to the electromagnetic wave. Shanghai's new radio station broadcast the plan as the editors were meeting, scooping the newspapers. Sun's call for disarmament and international intervention went out across the nation.

Sun delighted in the wireless transmission of his speech. His reaction, first carried in the English-language *China Press*, expressed his optimism about the possibilities of using radio to enact political reform in the country:

> It was my earnest hope that everyone in China should read or hear my manifesto and the fact that it was broadcasted [sic] and heard by hundreds of people with receiving sets—some of them as far away as Tientsin and Hongkong—is a delightful surprise and a great pleasure. We who are working for the reunification of this great country of ours welcome such forward steps as the radio. It will not only closely link China orally with the rest of the world, but it will greatly assist in knitting the various cities and provinces of the country much more closely together.[10]

He did not speak into the microphone himself, being occupied at his banquet, but his words nonetheless reached a broad audience across thousands of square miles, instantaneously. The audience was not the "masses" who would hear Maoist speeches decades later. They were largely foreigners and wealthy, well-educated Chinese, but they were exactly the audience Sun needed to reach to gain support for his Anglo American–supervised unification and scare the Russians into compromise. Listeners were few, but so were the newspapers outside of Shanghai that carried the story of the Peaceful Reunification Manifesto—those that did mention it printed only a pithy summary.[11] The proclamation and movement made only the briefest impression. Today it is almost entirely forgotten, even by historians. Though he seemed to have a great command of the newsscape, its infrastructural limitations meant that, as yet, Sun's message could not get far or last long. In many ways, then, the broadcast was a stunt, serving the new radio station's business as much as the politician.

Nonetheless, it did improve Sun's bargaining position. The morning after the broadcast and banquet, Sun met with the Soviet ambassador, Adolph Joffe, to discuss Soviet assistance for his floundering Nationalist Party.[12] On Saturday, they published a joint statement outlining a new framework for Sino-Russian relations.[13] Secretly, the Soviets agreed to help reorganize the party along Leninist lines, found a military academy, and back a military expedition to conquer the country.[14] Through this pact, Sun brought the Nationalists into an alliance with the small Chinese Communist Party, a decision that would have enormous ramifications in the history of East Asia.[15] Two currents of the postwar moment had collided in the city of Shanghai, culminating over the course of a single week. New techniques of politics and communications set China's technopolitical process on a new path. Unfortunately for Sun and his parties, however, in January 1923 Shanghai was still a semicolonial city within a complicated country, filled with powers who would try to neutralize him and the radio station with whom he had collaborated. Like Sun, the station must also therefore be understood within wider political contexts.

Broadcasting and the Technopolitics of Semicolonialism

"There is always something new, and nothing is impossible," announced the *China Press* on December 19, 1922. "These two adages are the keynotes of modern life, and they apply with great emphasis to the small army of men who are now teaching the world the wonders of wireless telephony." Ernest George Haywood Osborn—a New Zealand–born journalist, entrepreneur, bigamist, and conman—certainly placed himself chief among them.[16] He had arrived in Shanghai from Tokyo three days before and was already busy promoting his new company, the Radio Corporation of China, and arranging that country's first broadcasts. He would transmit from a casually appointed room on the roof of a nine-story commercial office tower. "Contrary to the general idea that it is full of whirring machinery and dynamos," the office "has the appearance of a bungalow drawing-room" reported the *Press*. He placed his equipment inconspicuously in the corners and hung the walls with heavy broadcloth to dampen echoes. A piano, a phonograph, and stylish wicker furniture completed the setup.[17]

Osborn had come to Shanghai precisely because of its semicoloniality—a kind of shared political, economic, and cultural sovereignty that hid immense imperialist power behind the facade of local control. The system encompassed national bureaucratic institutions and small, mostly urban, extraterritorial enclaves where Chinese authority did not apply at all. Thus the concession areas of cities like Tianjin, Wuhan, and Shanghai (awarded first to Britain after the mid-nineteenth-century Opium Wars, then to a succession of other foreign nations) functioned under internationalized institutions with limited interference from the central government in Beijing.[18] Here, the question of who could control communications technology was frequently obscure and confused. In cities like Shanghai, neighborhoods were divided among different foreign powers, as well as a local Chinese administration.

Different telephone exchanges, post offices, and police systems operated, sometimes along different sides of the same street.[19] The fact that no single imperial or national body had absolute authority meant that, unlike Tokyo (where Osborn had most recently operated), a kind of bureaucratic stasis prevailed, with each institution preventing the others from taking concrete action. These particular political conditions created a fertile patch of soil for a small, independent station like his.

Though secured by the politics of the city's distinctive semicoloniality, Osborn placed the development of Shanghai broadcasting squarely in the context of radio's global moment. He predicted that Shanghai would be able to broadcast to Seattle and San Francisco, eventually even New York and London. Best of all, he told readers, China was well-positioned to leapfrog ahead of its competitors, avoiding obsolete "stage[s] of her development."[20] Shanghailanders would soon embrace his optimism. "Bringing Shanghai into line with the world's modern cities," Osborn's station, operating under the call sign XRO, sent out its first transmission on Tuesday, January 23, 1923, at 8:00 in the evening. Broadcasting at the two-hundred-meter wavelength through a fifty-watt American-made transmitter, the initial program consisted of a violin solo by the famous Czech musician Jaroslav Kocian (1883–1950), a singing quartet, and news bulletins from the city and the world.[21] Two fashionable hotels, the Astor and the Carlton, gave dinners in honor of the occasion, while wealthy expatriates hosted private parties "in anticipation of the unique event in Shanghai's history." The number of receiving sets in the city was estimated at five hundred, though the number of listeners was reckoned to be several thousand, both foreigners and Chinese.[22] Ships at sea, as well as wireless stations in Beijing, Suzhou, Nanjing, Hong Kong, and even distant Fengtian, some seven hundred miles away, reported hearing the music and speech quite clearly.[23] Over the following days and weeks, the station built a regular program of classical music, news, editorials, and storytelling, but it was the American bands and their jazz that really set the city ablaze. Shanghai's "jazz-hounds" rejoiced as dances with radio music came into vogue.[24]

The audience on those opening nights included both foreign and Chinese enthusiasts, reflecting the mixed international city. Some of the latter gathered at the Shanghai YMCA, where Professor C. H. Robertson, a well-known radio expert, held a demonstration—perhaps using his own mobile transmitter, an American military model that had been marched in New York's 1919 victory parade and later sold as war-surplus.[25] Prominent members of Shanghai's Chinese community, including the former president of the Chinese Chamber of Commerce, as well as successful merchants and businessmen, stood in the crowd. The *China Press* reported, with a tinge of orientalist condescension, that some of the listeners could "hardly believe their ears and at the close of the concert kept Prof. Robertson for an hour having him explain certain points concerning the invention. The lights had to be turned out before some of the audience would leave."[26] To some, radio broadcasting did seem like a kind a miracle, but it was not without its teething problems. Some of the early nights' programs were marred by interference from ship- and land-based wireless stations. The crackling and interruptions from wireless operators, the sounds of unregulated overlapping radio frequencies, necessarily made radio listening a politically informed experience, a sonic demonstration of the lack of a singular governing authority in Shanghai.[27]

Echoing the clashing of electromagnetic waves, Osborn's radio station touched off a multifarious struggle among people and parties to control, promote, and utilize the new technology. The rump government in Beijing, the imperialist powers, and (as we have seen) anticolonial revolutionary forces in the form of Sun Yat-sen's Nationalists vied to alternately suppress or manipulate wireless communication. At this fluid moment, when many new forms of technological and political organization seemed possible, each strove for advantage. Though these technopolitical struggles also occurred elsewhere in the world, the political situation in Shanghai, with its foreign concessions and complicated administrative structure, made the question of control more existential. As a rule, early radio stations in North America, Western Europe, the USSR, or colonial territories did not threaten to ally with foreign powers to topple the internationally recognized government of their territory.

Sun's broadcast provocation on January 25 could not go unanswered for long. But the first reaction from his rivals came from someone with little power—the president of China. In early February, the station received word that the titular president of China, Li Yuanhong, had requested information about how he could tune in to the broadcasts from Shanghai. Osborn quickly went to work assembling a specially constructed set for the president.[28] Unluckily for the Radio Corporation of China, President Li, though an early radio enthusiast, was a powerless figurehead.[29] True power resided in the warlord faction known as the Zhili Clique, and they did not like radio. Though the area under direct administration of this government was rather small, centered on a few small provinces in the north of the country, they held a trump card: the old imperial capital of Beijing. They would have been of little significance for broadcasting—there was little national administration to speak of—if not for that fact and the China Maritime Customs Service.

Despite its unassuming name, the Customs Service was the most powerful bureaucratic institution in China from the late Qing period until the Second World War, playing a significant role in the regulation of both politics and technology. Nominally a bureau of the Chinese government, in reality imperialist interests staffed and largely controlled the Service. It was of use to both sides, however. The funds it remitted to the Chinese government represented Beijing's largest and only truly reliable source of revenue. For the Western powers and Japan, the Service worked to siphon off capital used to repay various loans and settle the massive reparations "owed" by China according to the treaties that ended its nineteenth-century wars and the 1900 Boxer Rebellion. Essentially, the Service was a customs bureau, a finance ministry, and a colonial office for the international imperialist regime, all rolled into one.[30] Though it could influence the various warlords by controlling the financial tap, it remained technically beholden to the Chinese government. There were at least a half-dozen "governments" claiming sovereignty over the entire nation at this stage. So which government to follow? The foreigners, by general consensus, agreed that the only regime they would recognize would be the one exercising control over Beijing at any given moment, no matter the larger national picture. From 1922 to 1925, the Zhili Clique controlled Beijing and so won formal power over the Maritime Customs Service. It used this authority to order a ban on the importation of radios from abroad.

Indeed, radio made military men the world over nervous. For two decades, radio transmissions had been used in warfare to coordinate between forces in different parts of a territory. As the technology improved, so did the possibilities for weaponizing the new medium. Intelligence and timely coordination won wars, and instantaneous, invisible communication across often-hostile terrain provided both. Moreover, as the popularity of ham radio demonstrated, it was relatively simple to use the parts of a standard radio receiving set to build a transmitter of one's own, giving high-powered communication technology to just about any person or military force that desired it. The fractured political situation in China increased the Beijing government's paranoia about such uses for radio equipment. A government that controlled all, or even most, of the country would perhaps have been more relaxed, but the Zhili Clique ruled very little territory. Much of the region surrounding Shanghai was controlled by its enemies, and the south of the country, with its many international trading ports, was unfriendly to the regime. Moreover, none other than Sun Yat-sen, the North's sworn enemy, had already used broadcasting technology for propagandistic ends, criticizing their government and demanding their disarmament. For these reasons, the government in Beijing dusted off decade-old regulations and insisted that the importation of radio equipment was illegal. It gave the order to the Maritime Customs Service, through its Anglo-Irish Inspector General Frederick Maze, that all such materials should be seized in the ports.

Unsurprisingly, this ban frustrated radio-crazed Westerners and more liberally minded Chinese, and they soon sought ways to work around the regulations. Given the weakness of the Beijing government and the unsympathetic attitude of the foreign officers staffing the customs houses, this was not a terribly hard task. But the consortium of foreign powers that regulated the largest Chinese cities had to deal with the fact of radio's essential illegality. Their disdain for the Chinese government's position on the matter can be seen in the meetings that foreign consuls held in Shanghai. In these informal conclaves, representatives of foreign powers gathered to form policies that became de facto law in the country, without input from any Chinese person or body. The minutes of these meetings, therefore, contain snapshots of the people and practices that constituted the semicolonial technopolitical regime of China.

One such meeting was held on September 29, 1925, and chaired by an Englishman, Leonard Arthur Lyall, Maritime Customs Service Commissioner for Shanghai. Having joined the Service some forty years previously, Lyall was as experienced and knowledgeable a China-hand as one could find. Born in London, he followed a rich family tradition to the East. His father had founded a successful trading company in Hong Kong in the mid-nineteenth century and profited handsomely from the trade in opium and coolies. The senior Lyall had also been one of the new colony's first Legislative Council members, and his first three sons were born there. By 1925, L. A. Lyall had surpassed his father in length of service and power accrued; as commissioner for Shanghai, he was the second-ranking man in Maritime Customs Service, the officer responsible for the taxes and trade in the busiest port in China.

It was not particularly momentous for Lyall to have representatives of nine nations assembled at his offices, as they were that day.[31] The consuls had gathered to discuss orders from the Ministry of War in Beijing regarding the importation of arms into China: guns and explosive materials, but also telescopes and binoculars, water bottles and radios. Anything that could conceivably be used by an army was banned. To a government fighting constant wars, and often besieged, this made eminent sense. To the imperialist consortium who controlled the borders and the largest cities, it was a tiresome bother, an indication of the backwardness of the Chinese state. The foreign consuls had petitioned Beijing for a change, but the government had not budged. "Wireless apparatus generally is not Munitions of War, and the Chinese Government has no right to treat it as such. If it is desired to frame rules concerning the use of wireless, these should be negotiation between China and Foreign Powers," said the British consul to those assembled. Mr. Lyall testified that radios were "the item that is giving me personally more trouble than all of the rest of the Regulations put together."[32]

The American representative suggested that there should be a distinction between military portable telephones and radios, but the French delegate protested that Customs was already ignoring the law anyway. "Am I to understand then that you can import radio sets, and that the Customs will pass them?" he asked. "I had one or [more] cases some months ago," Mr. Lyall responded; "I was very tired of this subject, and I passed several and reported it to Peking, and suggested that I might continue to do so pending the final decision of Peking. Peking turned me down and said I was not to pass them." In the end, the council voted to ask the Beijing government for further clarification, despite its obvious intent that all radios be banned.[33] As early as the spring of 1923, the Beijing Ministry of Communications had repeatedly stated its position that private radio receivers were illegal in Chinese territory.[34] As the Ministry of Foreign Affairs put it, foreign radios and broadcasting "harm our national sovereignty, and impede the work of administering the airwaves."[35] This opposition did not translate into reality, however. As one of the consuls at Lyall's meeting pointed out, radios were very common in Shanghai regardless of the regulations. Through its control of the apparatus of state in the guise of the Customs Service, the imperial powers simply ignored the will of the internationally recognized Chinese government for years, and radio broadcasting and receiving were tacitly accepted.

This does not mean that things went smoothly for E. G. Osborn and the Radio Corporation of China in the first months of broadcasting. In fact, things went spectacularly badly for the founder of radio in Shanghai. Like many other early broadcasters, the business model of Osborn's station was based not on the sale of advertisements but on the sale of receiving sets. Sets were not moving fast enough, and though the company was still quite young, Osborn had promised quick returns for his numerous creditors. Francis Xavier Lopes was the least patient. Hearing that Osborn was to leave Shanghai on his honeymoon, he went to the British Supreme Court for China to demand his former business partner's arrest under the charge of embezzlement in the sum of US$9,062. The arrest was executed immediately after his wedding.[36] Though Osborn later countersued for false arrest, the incident seems to have put him in poor stead with the rest of his business partners—not to mention his new in-laws. On March 22, *China Press* announced that Osborn had left the Radio Corporation to be replaced by two engineers, one American and one Chinese.[37] The Radio Corporation of China continued broadcasting for a few more weeks but ceased operations in early April, having collapsed financially. All the radio equipment reverted to the Chinese investor who had backed the venture, a certain Mr. Zeng.[38] The station had lasted a little over two months.

Osborn was not put off. He announced the formation of a new radio company, later named National Radio Administration Ltd., incorporated under the laws of British Shanghai.[39] The company built an entirely new broadcasting station on top of the Wing-On Department Store building on Nanking Road.[40] This station encountered government resistance: the Beijing government ordered Shanghai's international authorities to shut it down.[41] Osborn and partners attempted to use colonial legalism to evade this, reregistering in Hong Kong—a full colony under solely British jurisdiction—rather than in the Shanghai Concession. He was undone by the Chinese manager of the new station, a certain Mr. Guo, who appealed to the American and British consulates to secure the company a license to operate. Since he had made a formal request, the matter could not be overlooked, and when the Beijing Ministry of Communications expressed its displeasure, the Municipal Council ordered the equipment dismantled. All that was left was the iron tower looming over the Wing-On building.[42] Despite this setback, the National Administration company and a growing host of others continued supplying radio amateurs with the illegally imported parts to build their own sets. Throughout the summer and fall of 1923, aerials sprouted up on the rooftops of Shanghai, both inside the concessions and out—this in spite of the lack of commercial broadcasting and the fact of radio's illegality.[43] To some, therefore, Beijing's regulations seemed "but a scrap of paper."[44]

Scientists, Amateurs, and the Frustrations of Radio

We have seen how anticolonial revolutionaries, militarist governments, and semicolonial imperialist systems responded to the technopolitical shocks of broadcasting in Shanghai. But what did that current feel like to men and women who did not lead revolutions or armies? To those just trying to promote the science of radio? Or to those simply trying to listen in? Answers begin with China's first enduring radio station. The Kellogg Switchboard & Supply Company, an international corporation based in Chicago, rented all of Mr. Zeng's broadcasting equipment and erected two steel transmission towers in an empty lot in the French Concession.[45] It began broadcasting on May 15, 1924. The Kellogg Corporation, whose main line of business was the sale of radio equipment, leased airtime to four other companies—including *Shenbao*, the most influential Chinese language newspaper.

Two days after the launch, the radio engineer, lawyer, and later Guomindang official Fang Ziwei (alias George T. W. Fong) arrived in Shanghai with an ambitious agenda for his country. Fang saw radio as an instrument that could resolve China's crisis—that of disunity, poverty, and general chaos. Like Sun, he attributed China's lack of prosperity to its poor transportation and communication. Radio, much more economical than the telegraph or telephone without the need for poles and wire, could help China advance.[46] He anticipated the development of a private wireless industry to transmit music and important speeches across the country. Equally, though, Fang believed that the government needed to provide laws to enable and encourage the growth of wireless broadcasting—the technology could not advance without the politics. If this unitary process of technopolitical development were to occur, he had "boundless optimism for the Chinese wireless industry."[47] Fang was in a good position to judge. He was probably the most accomplished Chinese radio engineer in the world. Born in Shanghai in 1900, he was initially educated in the Chinese classics. Later, Fang's father, a Shanghai financier, sent him to receive a Westernized, scientific education. He graduated from the Ministry of Communication's Nanyang University (predecessor of today's Shanghai Jiaotong University) in 1919, after which he and his young wife, Helen, went to the United States. Fang took degrees in electrical engineering from Harvard and Michigan, where he also studied law. After graduation, he began work at Westinghouse, where his innovations won several patents. In 1923, he was part of the team that conducted the first transatlantic wireless communication between London and Pittsburgh.[48]

Fang was cutting edge, not just for China but for the world, and he brought a clear message to his countrymen: only close cooperation between government, technology, and business can bring success. In the United States, the encouraging attitude of the government aided scientists while interfering little in research. Regulations there aimed at ensuring public safety and improving the quality of broadcasts. According to Fang's utopian framework, one person could report on an issue of public health, and then the whole country would immediately comprehend its significance. "This kind of marvelous speed is far from even the dreams of the ancients," the engineer observed.[49] He urged the government to issue regulations sanctioning radio while also avoiding the cession of sovereignty to any foreign body. Rather than restricting imports, the government should reward people who install receivers. Likewise, important organizations in every large town should be encouraged to install a radio for the mutual exchange of news.

Soon Fang began carrying out a public relations campaign on behalf of Westinghouse and his own business interests to bring the vision of a radio-networked China to life. The corporation had recently run into trouble getting its shipments through customs. Fang, therefore, brought a Westinghouse radio with him to Shanghai and gave it as a gift to the Chinese Science Society (*Zhongguo Kexue She*) to stimulate interest in their products and pressure decision-makers. Fang made sure to publicize the gift through a letter to the Chinese Chamber of Commerce and reports in various newspapers. He gave interviews in *Shenbao* and published articles on recent radio developments in the society's magazine *Science* (*Kexue*).[50] The largest Chinese scientific journal (with a monthly circulation of approximately one thousand copies, though we can assume the readership for each issue was several times this), it played a key role in publicizing advances in radio technology, publishing fifty-six articles about the wireless between 1920 and 1925.[51] These often included translations of new experimental research and novel methodologies into Chinese.[52] In 1925, a new journal, *Telegraphist's Companion* (*Dian You*), specifically dedicated to wireless professionals and enthusiasts, launched, marking the start of a boom in radio publications. The first edition of *Telegraphist's Companion* carried news concerning radio research, electronic science, and international amateur societies.[53] The first three editions of the magazine all included articles detailing an easy method for constructing a homemade radio receiver, featuring woodcut diagrams of the vacuum tubes, crystal diodes, battery apparatus, and aerials required.[54] Chinese science, the author of these articles believed, remained "in its infancy." Building one's own radio was the patriotic thing to do in this context, spreading knowledge while contributing to the economy.[55]

This concern about China's domestic technological industry, broadcast infrastructure ownership, and general level of scientific development was common in radio circles. One defender of the government in Beijing stated that these patriotic worries were the true reason behind the continuing refusal to promulgate concrete radio regulations. Because China lacked a sufficient number of qualified wireless experts, he argued, "if we were to first issue the regulations, who would administer it? Would it be Chinese people? Or foreigners? Even if we call it autonomously run, the true power would still be in the hands of the foreigners, which is worse than not having a wireless industry at all."⁵⁶ Similarly, the journalist and radio enthusiast Cao Zhongyuan feared that what an American businessman had told him would come to pass, that the Chinese wireless industry would become an American monopoly.⁵⁷ "Pondering these words makes one tremble with fear," Cao said. "The domestic wireless community does not lack intelligent or capable men, and time will not wait for us." He urged cooperation to prevent "another new industry fall into the hands of foreigners."⁵⁸ To men concerned about foreign domination, radio held deep significance for China's wider technological economy, its science, and its national sovereignty. Developing domestic radio manufacturing capabilities was therefore of the utmost importance for the nation's political development.⁵⁹

Fang Ziwei, for his part, advocated the use of domestically manufactured crystal receivers as one solution.⁶⁰ One troop of Boy Scouts in Tianjin took up this call and organized a radio electronics study group. The group lay China's supposed technological backwardness at the feet of its youth, who allegedly did not enjoy scientific exploration. Western boys, by contrast, "all have a natural disposition towards research," they claimed. Working against these stereotypes, two of the Boy Scouts had recently constructed homemade radios, however, and were listening in. Hoping to increase the number of participants in these activities, the group addressed the newspaper-reading public: "Do you want such knowledge? Do you want such happiness? Do you want to increase the honor of Chinese young people? If so, please join us immediately."⁶¹

Yet none of this mattered if the government would not set down laws legalizing and encouraging the spread of radio. Beijing remained on the offensive. When the advertising of radio schedules in *Shenbao* caught the eye of the Ministry of Communications, they forced the newspaper off the air. The Kellogg station itself kept broadcasting nonetheless.⁶² It stopped advertising openly, but in defiance of Beijing's orders, and with the tacit consent of concession authorities, programming continued. This double standard infuriated radio-minded Chinese. "The Westerners who like to play with radio are completely safe within their concessions," wrote Cao Zhongyuan.⁶³ Though he despised the Beijing government's myopia when it came to the wireless, it frustrated him that the foreigners could act with such impunity. "China's bureaucracies cannot penetrate the concessions. Consequently, more and more aerials are hung, and there are few limits to installation. Our Military Commissioner and Police Bureau cannot do a thing.... In our sacred and inviolable country the consent of our Ministry of War is a useless formality."[64] Despite efforts at a crackdown, "installers still install as before. Businesses still do business as before, and broadcasters still broadcast."[65] The only ones harmed, in Cao's estimation, were the Chinese public, who bore the brunt of this policing and who suffered most when *Shenbao*'s program was forced off the air.[66]

Despite these frustrations, the radio market continued expanding. Though the customs blockade did cause problems—notably forcing the collapse of the Electronics Supply Company and the near bankruptcy of Kellogg—enough components slipped through for trade to continue. Smugglers trafficked parts from offshore boats, supplied false cargo manifests to inspectors, or trusted a friendly guard. Radio dealers like Kellogg relied on this gray market, keeping only a thin "facade of legality."[67] Even China's Foreign Ministry became involved, sending letters to the British consulate complaining that its rights under the 1922 Washington Treaty (which guaranteed China's sovereign monopoly over its airwaves) were being violated.[68] Still, nothing changed. A consequence of this ambiguous situation (wide acceptance and *de jure* illegality) was the rise of counterfeit manufacturing. Small radio parts, mostly American, were relatively easy to obtain. But because of the import ban, full radio sets were rare. Wireless businesses, therefore, commissioned Chinese artisans to turn the parts into replicas of the more expensive Western-brand radio sets.[69] Though the price and legality of receivers limited their number, interest in the new medium was immense.[70] Audiences for Chinese-language radio demonstrations put on by the Young Men's Christian Association (YMCA) scientist-in-residence Professor C. H. Robertson and his Chinese assistant C. H. Han routinely packed auditoriums. They traveled the length and breadth of China, spreading the Gospel and enthusiasm for radio; perhaps two million people attended over the course of a decade.[71] In one visit to Xiamen, seventy-eight hundred people turned up the first day. Within ten days, almost thirty thousand enraptured residents had experienced the demonstrations.[72]

This intense interest in radio was a latent resource, untapped by either governments or the people themselves. But changes were on the horizon.

For China, the chaotic, vibrant period from 1922 to 1925 marked a turning point in the technopolitical process from which new systems of politics and communication would emerge. In those years, just as Sun Yat-sen began refounding the Guomindang along Leninist lines, entrepreneurs, scientists, and amateur enthusiasts began to reorganize the country's communications infrastructure around radio. As we will see, the consequences of both reorganizations played out over decades, paralleling and influencing one another. Sun dreamed that these new technologies would assist in the reunification of China, knitting politically and geographically disparate cities and regions together into a new national fabric. This farsighted man was not wrong. The new wireless listening systems were as yet embryonic, to be sure, but the seeds of a mass media and mass society had been sown. Only the lack of supportive government regulations and public finance held radio broadcasting back. A Chinese regime would soon begin to change that—and from an unlikely place.

THREE

The Manchurian State Constructs a Newsscape, 1922–1931

The great Manchurian warlord Nurhaci had been dead for three hundred years by the time Captain Ronald Spear, a British spy, entered his throne room. Little had changed. Twin golden dragons still spiraled around the gateposts of the hall. Ten rectangular pavilions maintained sentry in the forecourt, five to a side, around a cobbled courtyard overgrown with grass. The pavilions had once been the offices of Nurhaci's military bureaucracy, the Eight Banners and Two Wings, ancient institutions disbanded two decades before Spear's arrival in May 1925. Yet the palace was far from deserted. Footsteps and foreign tongues still resounded off the walls, just as they had done in centuries past. The courtyard's chambers were no longer the ceremonial center of a seventeenth-century horde; instead, they now served as something much more modern: a radio station. The central throne room, an octagonal pavilion of red, yellow, and gold—long ago christened the Hall of Grand Governance—housed a humming radio receiver. Under the name World Radio Receiving Station, Mukden, it gathered missives from Paris and Berlin, San Francisco and Washington, all under the watch of a German overseer. Two Russian operators took the night shift. In one of the old Banner pavilions, a smaller receiver and transmission set communicated with Beijing, Shanghai, and Saigon.[1] Together, these facilities helped the Manchurian state project military influence, build alliances, and signal political power and technological modernity to subjects and strangers. The old palace complex was still at the heart of a northeastern kingdom, one nearly contiguous with its antecedent, but it served a different dynasty—that of the warlord modernizer Zhang Zuolin and his son.

Fengtian, the capital city then still known to foreigners by the old Manchu name of Mukden, lay at the center of a rapidly changing, state-sponsored reconstruction of communications infrastructure. Faced with the challenge of fast-developing mass political states like the Soviet Union and Imperial Japan, as well as a newly Leninist Nationalist Party under Sun Yat-sen (and after 1925, Chiang Kai-shek), a more consciously organized newsscape was key to the survival of the people and institutions that administered the Chinese Northeast provinces. Indeed, the imperatives of war and political competition informed all aspects of this Manchurian state's perspective on the technologies and practices of news.[2] In this, it was hardly unique; the orientations of the global newsscape provoked concerns in many regions. As Heidi Tworek demonstrates, for example, the early twentieth-century German state used investment in information infrastructures to exert power on the world stage and within China itself.[3] German wire news services like Transocean competed with the likes of Reuters and United Press to supply news items to Chinese papers, including through the Manchurian wireless network. (In fact, the German telecommunications giant Telefunken plays a role in our story as the builder of Fengtian's most powerful transmitter.) The Germans were not alone in this competition, however. Daqing Yang has shown a parallel collection of concerns within the Japanese Empire, particularly when it came to Northeast China. Beginning with their demands for formal colonial rights in 1915 (the infamous Twenty-One Demands) and continuing with the Ministry of Communications agreement that helped spark the May Fourth Movement, the Japanese sought influence over the Chinese newsscape, both its infrastructure and information.[4] The Japanese had a direct competitor in this realm: the American corporations described by Michael Krysko.[5] With the Soviets, French, and British also in the mix, the battle for influence within the Manchurian newsscape was more of a mêlée.

This book has already demonstrated how politics and technology are aspects of a single process that expresses itself through the infrastructures and practices of news. The case of Manchuria shows how competition with outside entities is one driver of this phenomenon. Competition with mass-oriented states in advanced stages of the technopolitical process (USSR, Japan, Britain) and inchoate mass parties (the Nationalists) compelled the northeastern government to begin consciously developing and manipulating the newsscape with an eye toward education and propaganda. Indeed, like all subsequent Chinese regimes, it placed the construction of mass communication infrastructure at the heart of its state-building program, pioneering plans for a national broadcasting and radio-listening system, promulgating the first broadcasting and radio ownership regulations in China, and enforcing a government monopoly on radio transmission. The Manchurian government built two of the largest and most powerful broadcasting stations in Asia and used them to prosecute one of history's first radio propaganda wars. The state—that is, the elite, decision-making bodies within society—desired to create an information-saturated newsscape in order to protect itself. It correctly trusted that people would eagerly take up new news practices in response, turning themselves into participants in the mass society. As we will see, however, the government lacked a program to make use of this engagement. It built the technology and practice of a mass society—and the *possibility* of mass politics—without actually forging a mass-participatory state. That potential was clearly seen when the region finally fell to foreign invasion, and Zhang's stations play a key role in mobilizing *Japanese* popular opinion in favor of the occupation.

Given its importance to the history of the period, the story of the early Manchurian broadcast system helps decenter the narrative of China's technological and media development away from Shanghai and national governments. In recounting that story, this chapter joins a flood of scholarly investigations on China's northeast that have appeared in recent decades. These works have investigated the region from a multiplicity of perspectives: Japanese imperialism, Chinese nationalism, identity construction, industrial development, and energy exploitation, to name but the most prominent.[6] All of these have identified the region as a crucible for multiple forms of East Asian modernity, drawing connections across the many regimes that governed the region in the twentieth century (Qing, Russian, Japanese, Warlord, Manchurian, Nationalist, Soviet, Communist). Koji Hirata and Victor Seow, for instance, have demonstrated continuities between Japanese industrial, technological, and political modernism in the region and subsequent Chinese communist institutions.[7] In this chapter, I push the story of the Northeast's precociousness in another direction, demonstrating that an advanced technological modernity also developed outside of Japanese influence—and, in fact, in opposition to Japanese power—within the regional Chinese regime. Indeed, for ten years a modernizing warlord state in the Northeast led China into a new age of wireless technology and radio statecraft.[8]

The Contexts of a New Radio Newsscape

Marshal Zhang Zuolin and his son Zhang Xueliang ruled a fiefdom dubbed "the promised land of Asia." The size of Britain and France combined, with a population approaching thirty million, Manchuria was a military, political, and economic power unto itself. Driven by agricultural expansion, vast mining operations, timber and coal reserves, and convenient rail connections with the USSR, central China, and the Japanese Empire, this northeastern region underwent rapid industrialization. As in the United States, railroad companies encouraged the settlement of rural lands along their lines, offering discount or free passage to immigrants. Millions poured in from regions of China south of the Shanhai Pass, settling the wilderness and pressing into the cities; figures ranged from 300,000 to 1.5 million a year. Russians, Japanese, Manchus, and Koreans mixed in with the overwhelmingly Han Chinese population.[9] Manchuria became a coveted prize—the Russo-Japanese War of 1904–5 was largely fought on its soil and for its control. After their victory, the Japanese held the most important railroads and economic concessions. The company that ran them, the South Manchuria Railway Company (*Mantetsu* or SMRC), became known as the East India Company of Manchuria for the degree of semiofficial colonial authority it exercised. Legally part of the Republic of China, the region retained de facto independence under the Zhang regime, though the Japanese Kwantung Army—theoretically tasked with guarding the railways—held military superiority in most places.

Geographically isolated from the rest of China and politically unified, the Northeast avoided the direct effects of the wars that plagued the country during this period. For Manchuria, the 1920s were a boom time.

This ethnic, economic, and military firmament incubated China's earliest government radio system. The life of Zhang Zuolin, the eventual patron of China's first government broadcasting, exemplifies the martial context of the new newsscape. Zhang began as a poor bandit leading a group of toughs in the lawless late-Qing countryside. During the Boxer Rebellion of 1900–1901, government authorities recognized his gang as an official army unit and pressed its members into service. Over the course of successive rebellions and wars, Zhang continued to improve his standing. By 1920, he was the sole power in Manchuria, with the title of Governor-General and Marshal of the Fengtian Army. It was not until 1922, however, that this warlord's thoughts turned to the airwaves. That year, Zhang attempted to extend his dominion by pushing south of the Great Wall. He was soundly repulsed. Limping back to Manchuria, Zhang nursed a fierce determination not to be outshone again. He set about modernizing the administration of Manchuria, resolving to raise its industry, agriculture, transportation, and wireless communications up to international standards. He brought home with him the means to accomplish this—the captured equipment and hired expertise that would form the basis of the wireless industry in the Northeast.

One of the experts, a radio man named Liu Han, is representative of the background, education, and career trajectory of the engineers who built the first radio systems. Growing up in a time of great social and technological change, and fascinated by communications, he found an interesting and profitable career managing state wireless networks. Various Chinese governments (Qing, Republican, Warlord) sponsored his career at every stage, from education to maturity. They demanded and cultivated radio talent. Liu's father, employed by a railway line and thus part of a Westernizing middle class, had ensured that his son received a classical education—studying ancient Chinese philosophy, history, and poetry. At the same time, Liu *pere* hired a neighbor, Mr. Song, who worked for a telegraph company and spoke good English, to tutor his children in the language. Mr. Song eventually brought the teenage Liu to his telegraph office and explained the mysteries of telegraphy, Morse code, and character encoding.[10] The latter especially piqued the student's interest since they required feats of memorization, not unlike the process of memorizing Chinese characters themselves. Studying in his spare time, the young Liu quickly became an expert in Chinese telegraphy. In 1907, he graduated secondary school and became an apprentice at the telegraph office where Mr. Song worked. Soon he was running the office by himself. Then, in 1909, the telegraph administration transferred him to Harbin, a city that would be the focus of his later career and where he would raise his family.[11]

Liu was still in Harbin when, in July 1913, public notices advertised an entrance examination for *Beijing Jiaotong Daxue*—Beijing Communications University. The centuries-old imperial dynasty had fallen only eighteen months previously, and a new republican government was eagerly pushing reforms. The promotion of science, engineering, and communications ranked high on the list of goals. That year, the Ministry of Communications founded a wireless department at *Beijing Jiaotong* and began seeking students to fill its first class. Already twenty-four years old, Liu had a wife and children to look after. But being perfectly qualified for the opportunity, he gambled on a return to school and matriculated in September.

The students received a thorough and advanced education at *Beijing Jiaotong*. The professors, mostly foreign-educated "returned students" with a deep understanding of Western learning and fluent in multiple languages, taught specialized classes on the "art of electrical communication."[12] Wireless made up only part of the curriculum, though. Students took required courses in mathematics, physics, chemistry, geography, English, Chinese, cartography, electrical theory, circuitry, and wired telegraphy. Higher-level students also attended further technical courses, as well as courses in economics, German, and French. The first class, thirty-two men, including Liu Han, graduated in April 1914.[13] If one includes schools in Tianjin and Shanghai, Chinese telecommunications schools graduated around 530 students over the first ten years.[14] This cohort supplied the knowledge, labor, and social-scientific networking necessary to found a world-class wireless system in northeastern China, including Manchuria, Beijing, and Tianjin over the coming decades.[15]

After graduating, Liu Han took a post in Shanghai for two years before returning to Beijing, where he began teaching popular classes on wireless topics at his alma mater in 1916. He was a comfortable and respected professor when, in July 1922, he received a letter from Fengtian. Zhang Zuolin's forces had just retreated back to their northeastern base, and the marshal sought to rebuild his military capabilities. He needed the latest technology. The Army Organization Department (*Lujun Zhenglichu*)—with the marshal's eldest son, Zhang Xueliang, as chief of staff—would oversee a new wireless bureau. Who better to turn to than Beijing's foremost expert to help build it? Liu could not resist the invitation. Radio was his career and his passion.

Zhang Zuolin, for his part, valued radio as a military tool. Even before the 1922 war, Zhang Zuolin had encountered incidents of unreliable or leaky intelligence transmission by telegraph and post, underlining for him the importance of telecommunications. Encoded wireless was a much more secure and efficient way to communicate with his armies and officials. The marshal soon installed wireless stations in every strategic locale.[16] This investment in radio technology also aided the rise of Zhang's influence on the international stage. In 1925, Captain C. Ronald Spear, the British spy, came to observe Manchuria's rapid military and technological development and divine its implications for the future of China. After inspecting the artillery, small arms, and poison-gas arsenals, but before reviewing the air force, Spear took a day to examine the state of wireless in Manchuria. He went to Nurhaci's old palace to see the receiving station—the spot likely chosen as much for its location at the center of the Chinese city, away from Japanese authorities, as for its symbolic significance. The transmitter—with a radius of three thousand miles it was in constant contact with all of China and much of the world—was similarly defensively situated, some distance outside the city, in the vicinity of a military camp.

The armaments purchases and radio facilities not only offered defense and communication but, in fact, constituted a form of alliance with foreign powers. Like most governments, Zhang's purchased largely on credit. With each purchase of a transmitter or tank, foreign powers became more invested in his continued success and his continued ability to repay. Additionally, the purchases contributed a self-reinforcing narrative about development and industrialization in Manchuria. They demonstrated Zhang's power, his state's modernization, and the likelihood of his perseverance in a deeply unstable political environment. Spear's report, therefore, fed into an ongoing debate within the British government about the arms embargo to China, which had been agreed on at the 1922 Washington Conference. Zhang's success in all elements of modern warfare—artillery, chemical weapons, airplanes, wireless communication, and propaganda—inclined the British government to support lifting the embargo in favor of Zhang, even if he ceased to control Beijing. Demonstrating modernity through wireless, as well as through arms, brought its foreign policy advantages. Britain recognized Manchuria as the most advanced power in China and contemplated acknowledging it as the nation's legitimate government.

Though military needs drove the sense of urgency surrounding his ambitious wireless network, radio also seems to have genuinely interested Zhang. His obsession with wireless continued, it was rumored, even into the afterlife. When a bomb shattered his body in the summer of 1928, the ghost of Zhang Zuolin struck back at his Japanese assassins, haunting the airwaves in spite. His antagonists, after all, had wanted to seize complete control of his fief, including the radio. Dalian "grows birch, poplar, and the pretty pseudoacacia," wrote one journalist, "all whispering to each other, each in its own language, in the summer breeze. The radio station on [Dalian's] observation peak whispers too, telling of [Fengtian's] affairs, Harbin's scandals, Tsitsihar's markets, and the latest armed rising in Mongolia. The late . . . Generalissimo [Zhang Zuolin] took much interest in wireless, and they say his uneasy ghost has been tampering with the radio waves on observation hill."[17]

This interest evinced itself in life as well. In 1923, Professor C. H. Robertson came through Manchuria as part of his radio tour. He lectured daily at the officer training school, demonstrating a two-hundred-mile radio transmission and conducting laboratory work with the radio corps.[18] Yet to Robertson, nothing seemed "more significant than the quiet evenings with General [Zhang] and his keen, intelligent group of officers, holding informal discussions on science, on religion, and on the moral issues of life." Still, given the unstable political environment of the period, he could not help pondering "how much all this will count in the potential struggle between Russia, Japan, and China of which Manchuria is the future stage."[19] In the end, Marshal Zhang's interest in radio technology *would* play a key role in the fate of Manchuria and China as a whole, though he would not live to see it.

Shifting the Newsscape, 1922–1925

On December 31, 1924, at quarter past four in the afternoon, a train pulled into Beijing railway station bearing a dying man. More than a thousand banner-waving students met it. They had waited for the better part of a day. The police had allowed the students to throng the platform, so when the train steamed into the station, chaos ensued, and the leader of the Guomindang, the nation's foremost revolutionary, had to wait a quarter of an hour before he could alight. Students scuffled with the police, braving accidental cuts from bayonets in an effort to glimpse this man, Dr. Sun Yat-sen.[20] But they were disappointed. Too ill to make a speech, Sun's Soviet minders whisked him immediately off to *Hotel de Pekin*, where doctors and nurses waited. It was his first time in the capital since he unceremoniously fled in 1912. Despite their idol's silence, the students seemed satisfied in the end. Sun Yat-sen had come to make a final effort at achieving his life's work: a free, united Chinese Republic.

Sun had spent 1923 and 1924 in Canton, reforming his Nationalist Party along Bolshevik lines and preparing for a Northern Expedition to challenge the authority of the Zhili Clique's chief warlord in Beijing. That military expedition collapsed before it had even begun, when Zhang and his allies pulled off a coup and took Beijing. With the aura of Sun's blessing, they attempted to establish a national government. Through the winter and spring of 1924–25, as Sun lay dying and Zhang tried to consolidate power, the new administration would carry out Beijing's first broadcasts, attempt to liberalize wireless laws, and plant the seed of a national system of broadcasting and radio listening.

Zhang, for his part, was well-prepared. He had spent two years constructing a radio newsscape with his engineers. As far back as January 1923, the military authorities in Fengtian had announced that various wireless stations in the Northeast would be unified "to quickly spread news and to merge military needs into a single system."[21] The system established a newsscape encompassing the local, regional, national, and international while conflating technological, martial, and propagandistic goals within a single process. The radio network connected all the major towns of the Northeast by wireless; the two largest metropolises, Harbin and Fengtian, communicated directly with Shanghai and other Chinese cities.[22] The numerous town-level substations utilized twenty-five portable radio sets—surplus British materiel from the Western front—seized as war booty from the Beijing government.[23] The centerpiece was a multifunctional wireless station in Harbin, whose transmitter (an old Russian machine retroceded along with the China Eastern Railway) Liu Han had repaired and modified.[24] Despite their ability as technicians, Liu and his colleagues still relied on outside sources for much of the equipment. Though at first they approached Japanese wireless interests for assistance, in the end, the radio regime came to rely on technology from German, American, and British corporations, much to the chagrin of the Japanese.[25]

Besides transmitting propaganda and coordinating military affairs, the Harbin Station established an office that received news from around the world. Information from Germany, France, England, and Italy arrived in Harbin on the same day or no later than the next day. A multilingual staff managed the international news flow.[26] The station hired at least two foreign-born employees, one Russian and one Japanese, both women and both of whom later became radio announcers. The station monetized these foreign and domestic news reports by selling them to newspapers and the public. In addition to sending and receiving commercial telegrams, the new station installed radio (*wuxiandianhua*) equipment in important institutions in Harbin, like the Chinese and Russian theaters.[27] Manchurian authorities hoped to "utilize the Harbin wireless station to spread Manchuria's politics (*chuanbu Dongsheng zhengzhi*)."[28] Whether, in the first two years of its existence, the northeastern radio spread political ideology through spoken (as opposed to Morse code) broadcasting is uncertain and controversial.[29] But the process of constructing a radio network certainly increased the prestige of Zhang Zuolin's state. The stations strengthened his ties with German corporations, French officials (who sought to combat German influence), and foreign scientists, as well as Chinese engineers and intellectuals.[30] It signaled Zhang and Manchuria's place at the forefront of scientific, political, and industrial development in Asia. The investment in wireless also shifted the newsscape, placing Zhang's state and military at the nexus of a substantial network reaching from world capitals to Chinese newspaper readers and army brigades. When war broke out again, Marshal Zhang's radio system featured prominently in his campaign.[31]

It also took center stage in his efforts to win the peace. In November, just two weeks after Sun departed Shanghai on his journey north, Zhang Zuolin and his army entered Tianjin.[32] Within an hour of his arrival, he installed modern radio apparatuses in the old telegraph office by the main rail station, establishing a connection with his capital, Fengtian, as well as with his fixed and field wireless stations.[33] Throughout the recent war in North China, Zhang had never been far from his communication network, having fitted his private train with a mobile radio center. Indeed, the chief of the signal corps had brought with him to Tianjin a large number of operators, fitters, and mast men to convert existing facilities from purely military to private and commercial applications, including some broadcasting. The former chief of the Fengtian wireless station took up the post of Tianjin telephone bureau head, a post with a far larger purview than the title suggests; from this position, he would oversee the development of radio in Beijing and Tianjin.[34] The new stations south of the Great Wall would be part and parcel of the Manchuria radio network; Beijing and Tianjin would have to abide by the wavelengths, regulations, procedures, and personnel decisions set by Fengtian.[35]

Under the influence of the Zhang regime, the national Communications Ministry in Beijing began to display a much more open attitude toward radio, encouraging its public use and renaming the Telegraph Bureau the Radio Transmission & Reception Office in recognition of its new purpose.[36] At the same time that the Communications Ministry mooted expansion and liberalization, it also cracked down on the illegally installed radios and private broadcasting emanating from Shanghai.[37] The Beijing authorities, despite noises about future changes, still wanted to enforce the strict radio laws. Doing so emphasized that China was not simply bowing to the foreign pressures from the foreign consuls and Customs Service. Beijing would liberalize on its own terms.

It is unclear whether Sun approved of this diffidence. In the years since he had inaugurated the age of political radio in Shanghai, Sun had never lost his enthusiasm for the technology, making sure to install the latest model receivers in his headquarters. One correspondent in southern China reported that "Dr. Sun, with his usual clear vision, realizes the part radio will play in the future development of China. . . . [He] is very enthusiastic about radio and in spite of his constant 'preparations to flee' Canton and the time spent in 'taking refuge on gunboats' and in 'railway cars' he has given considerable time to the inspection of his radio system."[38] Nor was this enthusiasm limited to the doctor. His wife, the inimitable Song Qingling, became the first Chinese woman to go on the air when she broadcast some quotes of Confucius from Canton.[39] It was fitting, then, that the establishment of radio in China's capital intertwined with his final months. In February, the Beijing-based *Telegraphists' Companion* reported that the Communications Ministry authorities planned to experiment with new broadcasting machines in Central Park, within the walls of the Imperial City.[40] The plan called for the installation of loudspeakers for music, news, and speeches—the majority in Chinese but with some English to appeal to foreign residents.[41] The authorities refitted the erstwhile telephone station for broadcasting, even installing a "special microphone for sending Chinese music."[42] On Sunday, March 1, 1925, just twelve days before Sun Yat-sen's death, the Ministry of Communications carried out Beijing's first experimental broadcast. One American and one Chinese engineer supervised, assisted by corps of experts from the Ministry.[43] Chinese and Western-style music played on a phonograph in the radio station, emanating from a loudspeaker set up in front of the park's main restaurant. No previous announcement was made, and, apparently, passersby paid little attention to the contraptions,

> until suddenly a Chinese opera number floated through the park, seemingly coming from nowhere. The voices and the tones were clear and faithfully reproduced. Everyone in the park at the time was mystified, while many rushed to the plaza. In response to the many enquiries *What is it?*, the operator in charge explained in detail. One man insisted it was a phonograph, but admitted it was the largest he had ever seen. Some were skeptical, some curious, and many interested in the music regardless of its source. All, however, remained until the program was completed.[44]

The Beijing broadcasting experiments were much more than a demonstration of technical prowess. They were a blueprint for future policy. Radio's potential as a force to restructure personal behavior and wider politics became evident immediately. "It will become the custom of the president, at fixed hours, to broadcast his own mandates," wrote a longtime China hand. "With the installation of receiving sets in the cities and towns, together with loudspeakers, it will be possible for people far distant from Peking not only to hear the voice of the president, but also to learn at first hand the doings of the government.... If the people can be accustomed to listen to Peking daily they may in time be brought to the point where they will obey its edicts."[45]

Sun Yat-sen had foreseen this two years before, when radio first came to China. He had said then that radio would "greatly assist in knitting the various cities and provinces of the country much more closely together."[46] Although he did not live to see that fulfilled, he did live to see its premise proven. After several months of declining health, Sun died of liver cancer on March 13, 1925, two weeks after the experimental broadcasts began. The Ministry of Communications sent out the news by wireless telegram.[47] His funeral rituals featured sophisticated communication innovations. Loudspeakers carried the ceremonies to the large crowds outside the hall where it took place. While Dr. Sun's body was lying in state, a phonograph record of his last public address was transmitted over the Central Park broadcast system.[48] These technologies multiplied the amount of information passing by soundwave through the newsscape, vastly increasing the number of participants in the events.[49] Gone were the days of 1919, just six years before, with their shouting and cardboard megaphones.

Sun's radio vision would carry on, even after his death. The use of loudspeakers for amplification, the broadcasting of political propaganda, and the emphasis on unification through sonic indoctrination lay the groundwork for a national program of broadcasting and radio listening that, sustained and expanded by subsequent regimes, would shape twentieth-century China. But as 1925 dragged on, Zhang Zuolin still did not have a radio station exclusively devoted to broadcasting. He could understand its potential well enough. He need only have listened in to the Shanghai Kellogg station, which regularly broadcast talks on scientific expeditions and discoveries, musical programs, and lectures by notables passing through the city. "Thus are interior places," wrote the *China Press*, "inaccessible to the railroad or even the telephone and telegraph brought into direct communication with the outside world and many a dull evening enlivened by news of the world and the best in art and thought." But, continued the *Press*, "our only worry is that some ambitious *tuchun* [warlord] may hear of this and take it into his head to put in a broadcasting plant in order to tell the listening world what he thinks."[50] Heeding this advice, Zhang Zuolin soon constructed two of the most powerful broadcast stations in the world.

The Most Powerful Radio Stations in China: 1926–1928

After Sun's passing, China endured another major civil war, from October 1925 to April 1926, which saw Zhang's Manchurian regime consolidate control over North China.[51] Only then was Zhang able to turn his attention to the long-planned expansion of radio broadcasting. The wireless administration, led by Liu Han, began to draft regulations putting all radio—message transmissions, broadcasting, and personal listening—on a firm legal footing. His rules, based on the 1924 British bylaws that helped establish the BBC's system, gave the Manchurian government a monopoly on the right to broadcast, collect listening fees, and levy taxes on the import and sale of receivers.[52] In theory, the import tax and listener's registration fee would support the development of broadcasting. In reality, a considerable portion of these funds was diverted to other uses; indeed, an American official judged that spending on radio by the Beijing government was insufficient to support broadcasting.[53] Perhaps the expenditures were insufficient compared to American standards, but the Manchurian government was in the process of building four modern radio stations, easily in the same class as those in Japan, French Indochina, or British India. It was certainly not falling behind. More important, collecting the import tax asserted independence from the international consortium that managed Chinese tariffs. By setting its own rate, the Manchurian state struck a blow for Chinese sovereignty from the semicolonial regime overseeing China's financial affairs, angering both the British and Japanese, and establishing its own place in global affairs.[54]

Finally, in 1927 and 1928, Zhang opened his new broadcasting stations. Tianjin began experimental, technically challenging broadcasts in late March 1927.[55] Chinese opera, performed in Beijing at the Grand Theater, traversed the same telephone wires that had carried the news of the student revolt eight years before.[56] But the shape of the newsscape had shifted since then. Sound and information still passed through newspaper offices and canal boats, but it now traveled instantaneously through the air as well. The Tianjin station picked up the signal from the telephone wire and transmitted it to the estimated four hundred listening households of the city. By December, the number of radio sets registered in Tianjin would shoot to more than two thousand.[57] Audiences as far as Fengtian, Harbin, and Dalian listened in, bringing to the northern Chinese public the promise of widespread radio.[58] The station transmitted "commercial quotations in Peking, concerts, vocal songs, Chinese theatrical programmes, speeches [and] lectures, Chinese and foreign press news, weather and time reports, besides all world radio news picked up from different stations in Europe and America the same day."[59] Twice as powerful as the Shanghai Kellogg station, Tianjin's facilities contained the finest equipment and furnishings money could buy. Green felt covered the broadcast booth, enclosing a piano and the musicians' seats in a muffled quiet. The station retained the services of a Beijing woman (her accent was thought to be very "clear") to recite the broadcast programming and announce the domestic news.[60] The prominence of and preference for female announcers in Chinese broadcasting would continue for decades, in marked contrast to the male-dominated airwaves of other nations.

The Fengtian broadcasting station, complete with its own self-playing piano, opened on August 7, 1927.[61] The Beijing Station opened on September 1, 1927. Harbin opened officially in January 1928, though it had begun trial broadcasts earlier. Under call sign COHB, it broadcast three times a day from "the most up-to-date [station] in the Far East."[62] Indeed, the stations in Harbin and Fengtian, at one thousand and two thousand watts respectively, were two of the most powerful in East Asia. Only one station in Japan, and one in the Japanese-leased territory around Dalian, could compete. The next largest station in China, in Nanjing, had but a quarter of Fengtian's power. Listeners across all China, Japan, eastern Siberia, Korea, and the Philippines could tune in.[63] And tune in they did. Radio listeners in Shanghai quickly began listening to Russian Opera from Harbin, broadcast twice a week. Radio fever spread across the Northeast itself as well.[64] The *China Press* summarized the wireless florescence: "The wide wastes of Manchuria, which once resounded to the drumming patter of the hoofs of Tartar hordes, are now equipped with radio stations which link up with all Asia and most of Europe. Millions of dollars are being spent in radio equipment in North and

South Manchuria ... [and] wireless schools are springing up everywhere.⁶⁵

A Chinese engineer who had been sent to America to purchase radio equipment for the Manchurian government outlined the vision of this fresh, dynamic newsscape best. It would, he promised, center on social improvement, providing talks on hygiene and science, issuing weather reports and "news of world happenings" side by side in the native language.⁶⁶ Content would flow, not like the drips and drabs of print mail up the Yangtze or encoded electricity through a wire, but fluently, every day. Nor would this new technological, political, and informational process touch only the rich, the literate, and the leisured. A new government program, the engineer noted, would help "poorer villages" install community radio receivers and loudspeakers so that everyone could hear.⁶⁷ In fact, by the end of 1928, dozens of counties in the three northeastern provinces had installed a public radio, either outside the county office or inside a school, to facilitate the political education (*chuan zheng jiao*) of the people, as they claimed happened in Europe, Japan, and America.⁶⁸ Elements in the government hoped such education, a vertical integration and reorientation of the community newsscape, could also strengthen the regime in its struggle with Japan.⁶⁹ This dream of a newsscape refashioned with government at the center, acting as benevolent pillar and facilitator of information throughout society, presaged similar programs sponsored by later Communist and Nationalist regimes.

In the summer of 1928, however, at the time of Zhang Zuolin's assassination—five years after the country's first broadcasts—radio ownership rates remained stubbornly low. According to official statistics, there were approximately three thousand radios in Beijing and Tianjin that year. Numbers like these obscure as much as they reveal, though. The figure does not include the sets that remained unlicensed and therefore untaxed, so the actual number was likely much higher.⁷⁰ Furthermore, interest in radio listening and ownership was expanding rapidly. The value of American radio exports to China had increased from $15,257 to $34,279 between 1926 and 1928.⁷¹ In the spring of 1928, a Chinese company in Tianjin claimed that it was "selling upwards of $10,000 worth of American radio equipment each month to the Chinese public in that city." This did not reflect the full scale of the industry, as "the bulk of radio material imported into North China [was] of Japanese manufacture." German imports were also substantial.⁷² Given the economic and legal strictures it had to overcome, the Chinese radio industry was booming.

Listeners could hear broadcasting from Tianjin, Beijing, Fengtian, Shanghai, and Harbin, along with a great number of foreign stations. Programming produced by the Zhang regime's stations included content for both entertainment and edification. They heard Chinese theater, including opera and prose performances, Western opera, and classical music performed by Russian exiles in Harbin, as well as news and propaganda from across the country and around the globe. Lecturers delivered talks on education, self-improvement, "the elementary principles of sanitation and health . . . business methods, farming, and similar subjects."[73] Variety kept the audience tuned in. An exaggerated, if instructive, headline five months after Marshal Zhang's death read, in part, "The Fengtian Station Is Number One in the World" (*Fengtian diantai wei shijie di-yi*). Though the claim is debatable, it is not far off. There were only two transmitters of the latest type from Telefunken—one in Nauen, Germany, and one in Fengtian, China. The Chinese papers reported that their machine was even more advanced than the one in Nauen since it was manufactured more recently and improved on the initial model. Indeed, "the Fengtian shortwave station leaves nothing to be desired, and is so good that it is seizing back overseas communication rights" from foreigners.[74] Moreover, the technicians were Chinese-born and -trained; the whole Manchurian wireless industry claimed to use only one German technical expert. Every other employee was Chinese, many coming from Fengtian's own wireless school.[75] It was a feat of organization and engineering unrivaled in Asia, excepting Japan. Zhang Zuolin and his team of radio experts had built one of the world's most advanced wireless systems in only six years. It fell to his son, however, to make use of this inheritance.

Radio Wars: 1928–1931

On January 10, 1929, Zhang Xueliang, the eldest son of Zhang Zuolin, invited several of his father's top generals, including the elder Zhang's longtime lieutenant Yang Yuting, to a game of mahjong at his home. By three in the morning, the generals all lay riddled with bullets, executed on the son's orders.[76] The arrests and nighttime trials allowed Zhang Xueliang to consolidate power, to replace the old guard with dependable men in all the most vital offices of his father's state—the army, the railroads, the factories, and arsenals—and, of course, the radio administration. The night of the fateful mahjong game, he made sure to arrest Zhang Xuan, its director.[77] The young marshal needed the broadcast network, and the leverage it provided in the domestic and global newsscape, to save Manchuria and himself.

Despite the official acknowledgment of suzerainty, the Manchurian radio stations never assented to anything more than nominal supervision by Nanjing. In only the most obvious indication of this, from 1928 to 1931 the stations continued to use their own call signs. Under international treaty, every country had received a set of call signs under which their radio stations should operate. China was assigned X. Those transmitters under Nanjing's control followed this rule: Radio Nanjing became XGZ, Hangzhou XGY, etc. But the radio stations in the Northeast continued to use call signs beginning with C: Fengtian COMK, Harbin COHB, Beijing COPK, and Tianjin COTN. The radio station in Canton, run by a Nationalist faction opposed to Chiang Kai-shek, also followed the C pattern. Similarly, foreign-owned stations followed foreign call-sign patterns—for instance, Kellogg's KRC in Shanghai.[78] The authorities in Nanjing were reduced to pleading for Fengtian's cooperation in unifying the country's radio system. In a letter pointedly addressed to the staff of the northeastern telecommunication offices (not their superiors, who might be less receptive to persuasion), Nanjing wrote:

> Our national industry has already unified, how can telecommunications send a message of discord? Even if in the past, the Three Provinces Telecommunications did not integrate, in the last year telecommunication authorities have proposed a plan of unification, with the pooling of profit and risk, with an administration that would consider each side to the mutual benefit. It would take telegraph, radio, and even long distance telephone, and combine them so that the two sides balanced out one another, and were connected in a way such that development would advance exponentially. Such an administration would be a brilliant model of national achievement. Actualizing such methods would be sufficient to make Three Provinces Telecommunications grow more prosperous by the day, to realize its dreams, with equally favorably rights.[79]

Though the letter closed by asking the employees to kowtow and ask forgiveness of the central government, the tone is more pleading than imperious. Nanjing needed the support of the Manchurian telecommunications system more than the other way around. Even in December 1930, Nanjing still relied on Manchurian radio equipment and expertise to maintain outbound radio-telegraphy traffic.[80]

Zhang Xueliang maintained Manchuria's radio edge not only through technology but also through propaganda. His government led the way in information warfare, prosecuting one of history's first "radio wars"—a struggle between two powers to shape the newsscape, to influence its flow, patterns, and content. In 1929, the Soviets, trying to gain the upper hand in the renegotiation of a railroad treaty, deployed their newly reorganized army against Zhang's regiments, leading to a brief but fierce conflict, the largest ever fought between the Chinese and a Western power. The battle carried over onto the airwaves. Zhang Xueliang "expressed indignation at the malicious propaganda broadcast by the Soviets, remarking that the TASS Agency had tried to convince the world that Russian Whites, aided by Chinese troops, have invaded Soviet territory."[81] His radio stations replied in kind. The Russo-Chinese War in Manchuria settled down into a protracted propaganda battle. "Although accounts of border clashes are too frequent to permit doubt that the situation approaches hostilities, with considerable loss of life on both sides, the main concern of each party appears to be to fix the blame for these incidents on the other." [82] In an effort to engender international sympathy and righteous anger domestically, the northeastern authorities tried through broadcasts and newspaper stories planted by radio to give widespread circulation to charges the Soviets had invaded Chinese territory, looting, burning, and sabotaging along the way. Neither side felt contained by the truth in the race to influence world opinion. "The fireworks of war preparation along the Manchurian border have their counterpart on the ether," wrote one correspondent on the scene: "Since beginning of the dispute between Russia and China, the Russian radio station at Khabarovsk has daily broadcast news and propaganda about the Chinese Eastern Railway Situation. Mukden [Fengtian] has taken the hint and is using its radio station for similar purposes. Adherence to the facts of the dispute does not appear to be the guiding principle of these rival stations."[83]

Manchuria directed radio propaganda not only abroad but also to the rest of China and its own residents. For instance, the Fengtian station broadcast speeches praising the development of the region, and Zhang Xueliang in particular, while also appealing to their fellow Chinese for assistance. For instance, in late August 1929, as the Soviet border war raged, a prominent Hong Kong businessman, Sir Robert Ho Tung, gave a talk that was heard across the country on the importance of the work being done in Manchuria.[84] As part of a larger project to develop modern social and industrial infrastructure, the Northeastern Education Commission broadcast self-improvement programs and moralistic Chinese opera.[85] Reformers also used broadcasting to amplify the message of an opium suppression campaign, an ironic development given the young marshal's own addiction to opiates.[86] In the end, though, none of these grand plans or social campaigns had time to yield much fruit.

The Manchurian Incident

In the third week of September 1931, dark rumors began to float through Manchuria. Japanese troops were quietly entering the cities. Somehow, someone had ordered the reserves mobilized. Artillery appeared on the outskirts of the capital, where, by agreement, it was forbidden. A Japanese friend came to Marshal Zhang Xueliang and whispered a warning: Do not go to Fengtian, General. It is not safe.

The morning of Saturday, September 19, Liu Han, semiretired in Beijing, awoke and tried to tune in to the Fengtian station, as was his habit. Like many Chinese that day, he found the airwaves empty. He was confused, then panicked. Something was terribly wrong.[87] At 10 p.m. the previous evening, a Chinese nightmare had suddenly manifested. Japanese troops began an assault on Chinese soldiers and civilians in Manchuria. When firing began in the North Camp, which Captain Spear had inspected six years before, the commander ordered his troops to pile their arms and not to resist. Fearing a provocation, and not expecting the coming invasion, Zhang had previously commanded that his troops not fire on the Japanese if an incident arose. Now the rifle fire was continual; shells fell every ten minutes. Still, Chinese forces made no reply. By midnight, seventy or eighty Manchuria soldiers had been killed at that camp alone.⁸⁸ At the same time, shellfire began raining down on the small suburban town of Beidaying, too. Over the course of the night, Japanese artillery destroyed the town, killing many civilians.

Within the city of Fengtian, Japanese railway guards and police cordoned off the Japanese residential district for protection, while Kwantung Army troops assaulted the walled Chinese district. Machine-gun bullets interrupted a Friday night dance at the Mukden Club in the International Settlement. A Chinese general, driving past in a car, was shot just outside. His chauffeur was killed instantly, but passersby carried the general to the house of a German doctor, where he died in the morning.⁸⁹ Elsewhere in the city, Japanese forces swarmed the airfield, the arsenal, the Bank of China, the post office, and the radio station. The guards at the mortar arsenal put up a fight, but, unsupported, they were wiped out by the Japanese.⁹⁰ By 6:30 a.m. on September 19, Fengtian had fallen.

Meanwhile, Zhang Xueliang sat in a Beijing hospital far away from his capital. At 1 a.m., aides had awakened him, handing him "a radiogram stating that Japanese troops were attacking. . . . Sleep, of course, was no longer possible, and the Young Marshal, pale and anxious, waited restively for further news. A few minutes later, another message was brought to him and a number followed in succession till, shortly after three o'clock, there were no more. This absence of news told its own tale. The Japanese had taken control of the telegraph and telephone offices."⁹¹ In reply to these messages, the marshal sent an urgent telegram ordering all the Chinese soldiers in Fengtian to cease resistance. He would not fight. And just like that, Zhang Xueliang, the virtual ruler of Manchuria, found himself without a domain.

Before the Japanese military occupation, three Chinese radio transmitters operated in the Fengtian station: a long-wave German-built plant that facilitated communication within Manchuria, the RCA station that maintained direct communication with San Francisco, and a German Telefunken transmitter that kept direct communication with Berlin. Among these, the long-wave station was "completely wrecked on the night of the Japanese occupation."⁹² The Chinese employees and guards had apparently given such substantial resistance at the station that the resulting damage rendered the facilities inoperable for three weeks. The two other stations, unharmed on the night of the invasion, were immediately shut down to prevent them from transmitting news. Furthermore, the Japanese cut "all telegraph wires leading into China proper south of the Great Wall . . . forcing all external communication over the Japanese lines running from [Fengtian] to [Dalian] and thence by cable to Japan, from where messages are relayed abroad by cable or Japanese radio."[93] In the immediate aftermath of September 19, nothing but Japanese censored news and information could get out. They had quarantined Manchuria, however briefly, from the global information network in which it had become enmeshed. The ensuing confusion gave the Kwantung Army time to consolidate control.

During this crucial period, the stations, still the most powerful in China, proved invaluable for the Japanese.[94] On October 6, the repaired Fengtian broadcasting plant began transmitting Kwantung Army propaganda back to Japan, aiming to swing public opinion behind a military action that had, after all, occurred without government order or popular mandate.[95] The information campaign succeeded spectacularly. The Japanese government refused to censure the Kwantung conspirators. Rather, Tokyo swung to backing the action of its rogue army. Radio Tokyo began regularly broadcasting live transmissions from Fengtian.[96] Five years later, far to the west in Xi'an, Zhang Xueliang would mirror this propaganda coup, storming a radio station and turning it against its master as he fomented an anti-Japanese resistance—and precipitated the Second World War in Asia.

For ten years, from 1922 to 1931, the Manchuria-based regime of Zhang Zuolin and Zhang Xueliang sought to shape the regional, national, and international newsscape to their advantage. Tapping into the deep wells of professional capacity produced by the wireless telegraphy schools, their modernizing state, the most rapidly developing region in Asia, established a network of wireless transmitters and radio broadcast stations to rival others anywhere in the world. After founding the first broadcast station in China's capital, Beijing, the radio regime also promulgated China's first official regulations for broadcasting and radio listening. In so doing, the Manchuria state made a claim for China's radio and tariff autonomy, contrary to the wishes of the colonial powers. Furthermore, by cultivating ties with American and German companies, the Manchurian radio system implicitly rejected the growing influence of the Japanese. Zhang Zuolin (at least in his later years) and Zhang Xueliang were hardly the Japanese puppets portrayed in their rival's propaganda.

Perhaps most important, the Zhang regime pioneered plans for a national broadcasting and radio-listening system supported by the pillars of a central government broadcasting monopoly and a network of subsidized or free public listening stations in villages, towns, and cities. The system, taken up by Nationalists and Communists in succession, would prove immensely powerful, becoming the facilitator of the mass political mobilization that would come to define China. The Manchurian state, therefore, presents a story of Chinese technopolitical modernization outside of Shanghai and the Leninist parties. It did not, however, carry out any enduring effort at mass politics—either in terms of party organization, mass conscription, or widespread indoctrination—trying, in a way, to avoid the unavoidable corollary of the technological process it sought to initiate.[97] Though it pointed the way toward a state-led mass media in China, the Manchurian government ultimately failed under Japanese assault. The Nationalist regime in Nanjing that replaced Manchuria as China's most important technological and political power would, over the coming years, try to avoid the same fate. To do so, it would allow and even encourage a widespread restructuring of the ways in which people received news. Taking up the perspective of individual experience, then, we will hereafter largely leave the state behind. Having examined the motivations, resources, and methods of the elite, decision-making groups within society, this book will, from this point forward, endeavor to look at how the changes they facilitated were experienced by everyday people, often far away from the centers of power.

FOUR

Reading the Radio, Listening in the Streets, 1927–1937

On July 17, 1937, Generalissimo Chiang Kai-shek went to the Central Broadcasting Station in Nanjing, the seat of the Nationalist government. Amid the flags and heavy drapery of the broadcast studio, he declared China's intention to resist the Japanese invasion. Throughout the previous week, throngs of people had crowded the doorways of every shop and square with a radio. The loudspeakers poured forth a constant stream of news updates, stirring songs, and exhortations into the streets. They listened in silence, full of emotion and indignation at the repeated provocations of the Japanese army. No work was done, no business conducted. People did nothing but listen to the news of the Marco Polo Bridge Incident—a series of skirmishes with the Japanese army—and the beginning of what would become the Second World War.[1]

In large cities, the news was immediate and everywhere, heard directly through public loudspeakers in teahouses, theaters, and stores. But it also arrived in some of the most remote villages, even reaching households still without a radio. Years later, a young schoolboy from Anhui named Meng Yiqi remembered the patriotic excitement and violent hatred of Japan aroused in him that summer: "Though our family was in a district cut off

Reading the Radio, Listening in the Streets, 1927–1937   85

from the outside world, and we could not get newspapers to read . . . in the market town there was a family surnamed Xiong (who were landlords and businessmen) who had a crystal radio set, so we could hear of [the events] by word of mouth."[2] Although he could only hear it secondhand, the news of war arrived to Meng Yiqi as it was happening throughout that summer. Radio had become the means by which information arrived, directly or indirectly, to even the most distant places.

In many other rural locations, the news arrived through a more formal channel: the nationwide radio information network established over the preceding decade by the Nationalist government. In Sichuan's Qu County, an isolated district one hundred miles from the nearest city and nine hundred miles from Nanjing, Magistrate Zuo Xunyou paid close attention as tensions worsened. He and the trained radio monitor listened in patiently every night, transcribing the most important news and speeches, which they published in a small mimeographed paper called "Radio News." The post office delivered it daily to every town, district, and village under his jurisdic-

FIGURE 4.1 A man in rural Hebei reading a blackboard of transcribed news during the War of Resistance against Japan, 1938. Source: Haldore Hanson's China Collection (1937–38), Gould Library Special Collections, Carleton College. Reprinted with permission.

tion. In the county seat, paperboys took them to teahouses and taverns. One night, in the middle of July, Zuo heard Chiang Kai-shek's speech declaring war. The magistrate filled with emotion, an excitement beyond sleep; before daybreak, the night's big radio news was already plastered on the walls of the main thoroughfares. When the first rays of morning broke through, the city was seething. The people read: "The war has begun. Regardless of whether North or South, young or old, all have a responsibility to protect our soil in this War of Resistance. All must be resolute in their spirit of sacrifice."[3] Hearing the news, the old scholar Zhu Bowen, head of a genteel and ancient family on the north side of town, wrote an ode to the young soldiers departing for war, urging courage in the face of national crisis. Other citizens pasted patriotic couplets on the doorposts of their homes, boasting of their commitment to China's cause. The magistrate himself walked around town, observing the spirit of the people and reading the outpouring of patriotic poetry. He felt moved to write a poem of his own, praising the radio:

> How highly useful! Supplementing my own eyes and ears
> Its merits so outstanding! Calling out to me knowledge for
>   the people.[4]

When radio broadcasting first appeared in China in 1923, it immediately became popular among foreigners and wealthy Chinese in port cities such as Shanghai, Tianjin, Harbin, and Hong Kong. But the penetration of radio technology into the interior of the country and into the working classes was slow. Neither factory workers, day laborers, shop attendants, nor even the lower rungs of office workers could afford pricey radio receivers. Reception outside the ports was poor. Radios were simply too expensive, and broadcasting facilities too weak, to justify the expense. It was not until the Nanjing decade (1927–37) that China started experiencing "radio fever" as prices dropped and broadcasting technology improved. Per capita radio ownership remained low, however; by the start of the Second Sino-Japanese war in 1937, there were fewer than one million radios in a country of four hundred million people. This begs the question: did radio influence the experience of the newsscape outside the urban bourgeois classes? Was "radio fever" a nationwide phenomenon?

The answer, in short, is that radio did indeed foster change on a national scale but not in the ways one might expect. We must put aside images of families gathered around a living room radio for a fireside chat. The Chinese experience was different. First, China's radio-listening culture was predominantly communal. Teahouses and theaters installed radios. Shops of every description hung loudspeakers from their storefronts to attract customers. Streets became impassable because of the number of people who huddled together on a sidewalk to listen; even illiterate urban laborers could now become consumers of mass-produced news. Communal radio listening also spread to smaller cities and villages, where groups of families jointly raised funds to purchase a radio set and install it in temples or schools.

Second, the experience of radio lay at the fluid intersection of written information and mechanically reproduced sound. It was not a purely auditory phenomenon but one that bent the entire intermedial and multisensory newsscape in new directions. A government program to place listening stations (*shouyinzhan*) in rural counties played a central role in this restructuring. Radio monitors (*shouyinyuan*) at these stations were responsible for listening to news and propaganda broadcasts twice a day, transcribing and printing their content in new local papers, posting news items to public walls (*bibao*) and blackboards. The radio-listening stations brought news and information on a near-daily basis to a vast area of the country, which previously had waited weeks for the arrival of print news.

This radio-facilitated newsscape revolution profoundly changed citizens' experience of news and politics outside of the major port cities. Whether the receiver was located in a school, government office, education center, or private home, news and information arrived more quickly and more frequently than before. In that sense, the presence of rural radios illustrates Marshall McLuhan's famous and controversial dictum that "the medium is the message."[5] While the Chinese made conscious choices about how to deploy and develop wireless communications techniques, making them agents in radio's effects, McLuhan's underlying theory that the presence of a technology can be as transformative as the content of its message finds substantiation at this moment in the technopolitical process. The presence of a radio could change the way whole districts (especially in more remote areas) received and processed information and, by extension, responded to national and international events, regardless of the specific messages that radio brought or even whether individuals heard the sound directly. Indeed, for the interior, the nation and the world went from being at a week or more's remove to being of immediate and urgent import. Similarly, in large cities, formal news (its content edited and its delivery structured, in contrast with rumor) was no longer restricted to the elite but was supplied directly to the often illiterate population. Whether rural or urban, the change in technology induced a change in behavior. The possibility of more frequent and more voluminous news incentivized attention, organization, and a reorientation toward wider things: the sprouts of what would later become a mass society.

Street Listening and the Urban Working Class

At the start of the Nanjing decade, only the wealthiest citizens—the same well-informed elites who already read newspapers, who had access to news, and had fully formed political opinions—could afford to install a home receiver. A radio could cost an exorbitant sum. The vacuum tubes used to convert radio waves to electric signal, highly advanced technology for the time, had to be imported from the United States, Europe, or Japan.[6] So, while a larger number of vacuum tubes improved the strength and sensitivity of the radio's reception, they also substantially raised its price.[7] A vacuum tube set made in Europe or the United States cost upwards of Chinese $100—representing far more than a year's salary for a laborer.[8] Imported Japanese sets dominated the lower end of the market and cost about one month's unskilled labor.[9] Clearly, such an investment would not be possible for most people. Crystal receivers, which used a crystal diode (hence the name) instead of a vacuum tube to detect radio waves, offered a less expensive alternative. Easy to construct at home from commonly available parts, it cost only about two weeks' wages for an unskilled laborer.[10] The sound was weak, especially during the day, and could not be amplified—one used headphones to listen—but because they did not require batteries or another source of electricity, they became popular in rural areas where access to dry cell batteries or a dependable electric grid lagged.

In cities, however, the middle and working classes had access to radio through other means; individuals could visit a teahouse or theater, or they could listen for free from storefront loudspeakers. The proprietors of teahouses and theaters used broadcast performances of traditional art forms, like storytelling and opera, to draw crowds to sit and spend money in their establishments.[11] But not everyone could afford to lounge in a teahouse on a regular basis; for the indigent or stingy, there was streetside listening. Shopkeepers hoped that public loudspeakers in front of their stores would drive foot traffic and so improve business. This practice, based on the assumption that people would seek out opportunities to hear information, created a positive feedback loop. Competition drove more businesses to install wireless sets, while the public came to expect radio access. According to contemporary accounts, receivers and loudspeakers became common to the point of ubiquity. A writer named Ming Xin observed in 1936 that "the number of listening households increase daily, and businesses are scrambling to get broadcasting. Every day from morning until night, it doesn't stop for a moment. You can hear the sound of radio no matter whether you are on a large avenue or in a small alley. Radio-waves are ever increasing, like a cobweb in the air."[12] An account from Tianjin describes the resulting conditions on Chinese city streets:

> When you walk by a store that has installed a radio you are at risk of being squeezed. It's more like walking down a *hutong* [alleyway] of people! After trying hard to escape, your eyes will settle on the contemptible object: the crowd is only staring at an ugly wooden box, the radio. Its delicate singing dazes them. You see this phenomenon all the time. Near the corner of Hedong district's Dongfu Bridge, there is a store with a very humorous note plastered to the radio box: *Those standing in the audience, Beware of the Trolleys!*

Listeners crowded the storefront to such a degree that they spilled off the sidewalk and into the path of oncoming trams. The only solution for this congestion was the regulation of public radio loudspeakers, the observer went on to argue. Indeed, authorities banned street-side radios in Shanghai's French concession, allegedly resulting in a much more orderly flow of pedestrians.[13]

In other areas of Shanghai, as in other major Chinese cities like Beijing, Wuhan, and Canton, radio continued to proliferate to the point where residents were often not required to step outside to listen. One simply had to open the window to hear broadcasts playing from the shops below.[14] Such noise and congestion inevitably piqued law-and-order cranks. But a ban on street-side radio would have profound repercussions, cutting off access to radio—and news—for large numbers of people. Anticipating this criticism, one writer retorted, "Some people say, *Broadcasting is a tool for inculcating knowledge in the people; it is not something just for entertaining the bourgeoisie.* This is very correct, but we have to put aside high-sounding words and speak some truth." His snobbish truth turned out to be that poor people are inconstant, without judgment or taste in their listening, and satiated with popular music. "However, when the radio station broadcasts lectures or cultural programming, the temporary audience is disgusted and begins to melt away. Even good topics and moving speeches do not attract the people's interest."[15]

There is often a stark difference between what is deemed "educational" or "good for you," according to social elites, and what most people consider entertaining. The fear that the masses were obsessed with "mere entertainment" became a common trope among intellectuals in the 1930s. The distinction, however, between news, propaganda, and entertainment is in the eye (or ear) of the beholder. Entertainment (even privately produced and broadcast) and propaganda were sometimes indistinguishable, as music and theater promoted political themes.[16] Song is one of the most ancient and universal forms of information sharing. Regardless, the casual listener could not avoid news since bulletins commingled with "entertainment" segments. Police reports, war news, and political propaganda were intentionally interspersed with the programming so that the audience was forced to hear them if they wished to continue listening.[17]

This fact did not stop complaints about the numbers of illiterate people listening in, especially when they intruded on bourgeois public spaces: "Even the silent libraries have installed radios. Thus you can see radio's influence is too great," recorded Ming Xin, the Tianjin-based journalist. Recently, a library on the city's north side had installed a radio and played it at full volume every night in the reading room. "Is this supposed to be a blessing or a curse? Or a bid to solicit more customers like a shop? I couldn't even guess the purpose of it." The library was rather small, with just a newspaper room, courtyard, and book stacks. Both spaces were extremely crowded, especially in the evening. Even before the radio was installed, there often was no place to sit and read a newspaper, and the presence of the radio exacerbated the situation. "Now, there is nowhere to even stand in the newspaper room. The house is crowded with people, and the air smells awful. But have these people come to read newspapers? No, they have not! More than half come to listen to the radio, and read in name only"—for they are unlettered. "Those who want to read the paper can't, and illiterate people have the paper in their hands [pretending to read]. What is the meaning of this?! Is sitting down to listen that much more comfortable than standing in the street to listen?"[18]

Illiterate people, by definition, leave few traces in the archive. But we can see them in the observations left behind by others. The writer's criticism of common radio listeners as undiscerning and shallow is a frequent one, but it inadvertently illustrates the popularity of free public radio broadcasting among the working classes and the illiterate. And as we will see, the "uncouth masses" paid close attention to important affairs when they *did* concern them, in times of national crisis and war. Not paying attention to dry, "educational" programming was a result of boring content, not ignorance.

What did listeners tune in to? It depended on where they were located (distance from the broadcasting station, weather, and geographical surroundings all affect reception), the time of day (radio waves propagate more efficiently at night), and the sensitivity of their radio. Radio stations were most numerous in Shanghai, which by 1937 had almost twice as many stations as New York—more, in fact, than any other city in the world.[19] Shanghai housed the great majority of China's private radio stations, which tended to cluster within the foreign concessions, where they were relatively immune to government interference. These broadcasters were privately held companies, usually undercapitalized and dependent on advertising.[20] Some existed solely to promote their owner's other business interests, for instance, the amusingly named jazz station RUOK that served to promote American-run nightclubs and cabarets.[21]

Most listeners outside Shanghai, however, could only reliably receive the more powerful government-administered stations; by 1937, there were more than a dozen across the country. The Nationalist Party (*Guomindang* or GMD) had comprehended the power of radio from its earliest days. We have already discussed Sun Yat-sen's affection for radio; his immediate subordinates differed little. Chen Guofu, a powerful GMD figure and later director of China Central Broadcasting, recalled in his memoirs: "One day in 1924, I was in Shanghai listening to a broadcast station report market prices, when suddenly I thought of propaganda. If our party could have this kind of tool, how could it not be far more efficient than running a newspaper?" In the summer of 1925, Chen wrote a letter to Canton, then the headquarters of the Nationalists, asking if they desired individuals skilled in radio work. After a short while, he received a response from Whampoa Military Academy headmaster Chiang Kai-shek, saying: "These are very much needed. We hope to enlist [radio men]."[22] It was not until the autumn of 1932, however, that the Guomindang opened a powerful station in its new capital, Nanjing. Designed and built by Telefunken Company of Germany, it was for a time the third most powerful broadcasting station in the world, after Nauen, Germany, and Daventry, England.[23] When it opened, on November 12, 1932, Chen Guofu spoke at the ceremony, praising the possibilities of the new station. From now on, China's broadcast news would not only reach its own borders but also distant parts of the world, transmitting government directives, promoting Nationalist ideology, encouraging literacy through education, and advancing culture. "Reporting on the current affairs, it will send the world an understanding of our country's true situation, extending righteousness around the globe, more so than any comparable propaganda work," bragged Chen.[24]

The new station and broadcasting administration lay near the heart of Nationalist Party politics, dominated by men like Chen, who were personally loyal to Chiang Kai-shek.[25] These powerful oligarchs had a vision for radio in China. First, public loudspeakers would relay educational, cultural, and propaganda programming in cities. Starting in 1933 in Nanjing, the Guomindang government began to place loudspeakers at all important intersections and in all government buildings and parks using wired broadcasting networks—a telephone-wire-like system that carried the broadcast signal from a central receiver overland instead of through the air.²⁶ Every Monday morning, a unified broadcast would lead citizens in rituals commemorating Sun Yat-sen and reaffirming his Three Principles of the People. At other times, the loudspeakers would transmit news or political speeches. The Guomindang intended this program of joint public listening as a model for the rest of the country, though such a project would not be fully enacted until the 1950s, when it was taken up by their Communist successors. The second half of their program, the plan to bring news to the countryside, had a more immediate effect. Still, this reconfiguration of the rural newsscape was not entirely a government-directed affair. The widespread demand for information meant that individuals and communities also set about reorganizing themselves when the technological means became available.

Private Radios in the Countryside

In the 1930s, radio technology penetrated the small towns and villages of the Chinese interior. Private enterprises played a role in bringing radio listening to the population there, just as they had in coastal metropolises. Businessmen in smaller cities invested in radios for their storefronts or teahouses, betting that the radio fever infecting Shanghai and Canton would spread to their hometowns. For example, in September 1936, a businessman named Ren Ziwen from the remote city of Fenyang in Shanxi Province purchased a receiver from Shanghai "for the sake of removing obstacles to popular knowledge, and promoting local culture." He installed it in a theater, intending to sell tickets to local citizens. The people of Fenyang would thus be able "to hear world news, the speeches of famous people, and well-known actors from Tianjin and Shanghai."²⁷ Similar storefront and teahouse radio apparatuses likely appeared in small cities and large towns, anywhere large enough to support shopping streets. But this commercial information practice is poorly attested. The communal purchase of radios stands out more in surviving accounts of rural communication in the period. Influential families would pool their money to buy a receiver from faraway Shanghai, Wuhan, or Canton. Installing it in a local institution like a temple or school, everyone could gather to hear warnings of unrest or market prices or music, generally alleviating the psychic and material burdens of being isolated. Writing to the provincial government in 1930, the people of Nanling County in Anhui Province explained:

> Consider our county's position in south-central Anhui, only a little more than 100 *li* from the commercial port of Wuhu. Yet our locality's facilities and the level of our society and culture is still underdeveloped. There is not a single cause behind this, yet the difficulty in communications is truthfully one of the main reasons. Look at the countryside on our eastern, southern and western borders—touching on Jingfan and Qingtong counties. Mountain ridges make it difficult for travelers in either direction. It is especially hard to be well-informed of news. There is only a mountain stream connecting the county seat to Wuhu. At the beginning of summer, the mountain torrents rise suddenly and small steamboats can travel back and forth. In the fall, when the waters recede and the river becomes shallow, the small steamboats stop coming. Transportation is thus obstructed.... Additionally, in the cold of winter traders from our county's rice producing areas transport rice out to be sold in foreign parts. However, market prices fluctuate, which they cannot anticipate beforehand. Therefore they frequently take a loss.[28]

More worrying in the eyes of the locals was the increasingly unsettled political situation and "stealthily multiplying bandits." The telegraph was too slow and unreliable—wires could be easily cut, and telegraphists at key junctures were undependable—to hear news or send for help when outlaws threatened. Discussing the problem in a joint meeting, the assembled representative groups of the county decided a shortwave radio, "whose costs are easy to bear" was the only economically viable option for their community.[29] The fundraising and installation results for this particular community are not recorded.

Villages in places as diverse as Guangdong and Shandong also participated in the communal radio phenomenon. In 1930, the Guangdong village of Huangbang Ridge decided to install a wireless receiver. "Since the invention of radio by Marconi, all parts [of the country] are installing a set, one after the other," observed a report on the village. "We look forward to the masses understanding the omnipotence of modern science, to breaking down their complacency and conservatism." The citizens of Huangbang Ridge helped in this modernizing effort by advocating for the installation of radio, enabling the people to receive "scientific entertainment." The residents sent around requests for contributions, and donors responded enthusiastically.[30] Similarly, in 1932, in rural Yishui County, Shandong, all parts of the community helped purchase and set up a radio in the local education center (*renmin jiaoyuguan*):

> Given the poor state of local transportation and communication, important news of national affairs is extremely delayed. It was felt necessary then to set up a radio receiver. Having consulted the Education Department, the matter was referred to the County Council. The whole county raised funds to purchase a wireless set. There was deep support from all walks of life. The head of the education center personally went to Tianjin to purchase a ten-vacuum tube radio, which has already been properly installed. They are now performing tests, and within a few days it will be playing publicly, disseminating information. The influence on the knowledge of the people of the whole county will be significant.[31]

In Yishui, as in many small places across China, local people gathered together to raise funds (*choukuan*) to install a radio in their town. They sent a community member a long distance to purchase the equipment. They set up the radio in a common institution so that everyone could listen to propaganda-information (*xuanchuan*) and important national affairs (*guoshi yaowen*). The residents of Yishui, Nanling, and Huangbang Ridge did this of their own accord, with their own money, without central government intervention. These villages, situated in very different areas of the country (the far south, the north coast, and the central interior), show that the communal installation of radio receivers was not limited to one region or type of settlement; it was not just the suburbs of major international ports taking up the practice. Indeed, they point toward the conclusion that radio and the access to information it provided were of paramount importance to the people of rural China.

Compelled by the desire, or even need, for information, people purchased whatever kind of receiver they could afford. Businesses and communal institutions invariably bought the more powerful and expensive vacuum tube receivers. These were capable of amplification and, therefore, could be played via loudspeakers to large public audiences. (Additionally, they had more reliable reception.) Rural individuals and families preferred crystal receivers because they were inexpensive and did not require electricity or batteries. Given these advantages, the practice of building crystal radios quickly spread through experience and word of mouth. A 1934 report recounts the story of a country teacher residing in the Yangtze River Delta, not far from Shanghai. Frustrated that he had no way of receiving domestic or international news except for papers that were three or more days old, he began to teach the students how to build crystal radios in his hands-on engineering class. Within a short time, they built more than seventy sets. "The sound was quite good and the listeners all very happy," says the report. People from rural Anhui Province heard this and imitated the schoolteacher's example. Soon more than ten crystal receivers had been constructed in the Anhui district, and the school there began "gathering together the farmers in the area to listen during every day off and after class, promoting science and instilling knowledge."[32] With simple designs and cheap materials, crystal sets spread to more locations and to more segments of society than tube radios, bringing even farmers into the age of mass communication. By the outbreak of the war, more than half of the one million radio sets in China were crystal sets.[33]

Available evidence does not allow us to statistically analyze the extent of private radio ownership in rural areas. Certainly, radio penetration varied from location to location, depending on factors like the wealth of local families and the town's proximity to international market centers like Tianjin or Wuhan. But anecdotal evidence does suggest certain regions saw a boom in radio ownership in the latter half of the 1930s. An American businessman who frequently traveled to Shandong observed aerial wires, supported by bamboo poles, coming from the roofs of many of the more modest dwellings. He found them attached to homemade crystal radios, with headphones rigged up from telephone receivers, which he thought had probably been picked up from secondhand dealers. The sets had surprisingly good reception. Another American observer noted that "formerly when traveling through the rural areas, practically no antennas were to be seen. Today a network of bamboo poles and wires on the humblest of buildings in small villages attests [to] the extent to which radio had penetrated the country."[34] At the beginning of 1936, a Chinese reporter traveling in the countryside of Ba County, Hebei, found that "one of the most encouraging things is that the radio sets are very common (*hen pubian*) in the villages of the County now. Everywhere, on any day, you can hear drama and speeches from Tianjin, [Beijing], Nanjing, and even Japan. Chiang Kai-shek's New Year's speech, broadcast from [Nanjing] Central Station, was heard clearly everywhere. Many people could still describe his main points and intonation, which shows the average person's attention to current affairs. This is a good phenomenon." On the subject of the radio's potential, the reporter opined:

> If the central government wants to popularize education and publicize national policies, it is necessary to use the wireless. However, the central government also needs to have a comprehensive plan for the establishment of rural radio, or set up a general office [to manage] public funds in every place, or set up a factory to manufacture a lot of radio sets to sell to the public. Then there is a possibility of universalizing radio. Otherwise, if you allow people [to] buy [radios] by themselves at will, there may eventually come a day when they are universal, but the speed with which that day arrives will be much slower, and the effectiveness [of the policy] much worse.[35]

In fact, the central Guomindang government had already started carrying out a plan for rural broadcasting. Indeed, by 1936 this plan was already meeting with substantial success.

Radio Monitors and the Transformation of the Rural Newsscape

Reminiscing from exile in Taipei in 1968, Wu Daoyi, erstwhile director of China Central Broadcasting, recalled the inception of the radio-listening station (*shouyinzhan*) initiative, a program that would eventually bring many rural people into direct contact with fast-paced, ever-changing technopolitical modernity for the first time. The number of radio sets in China in the summer of 1928, he wrote, did not exceed ten thousand altogether. So although the Central Broadcasting Station had been established to transmit party ideology, news, education, and music, the infrastructures and behaviors needed to receive them did not exist. To begin remedying this, in July 1928, they drafted the first plans to train "radio monitors" (*shouyinyuan*), people whose job it would be to listen to broadcasts and disseminate the information received. At this early stage, their primary duty was to promote the party by manipulating newspaper coverage in major cities. As few Chinese newspapers maintained reporters in other areas of the country, the reports these individuals received via party radio became highly influential. "At night the monitors listened to the Central Station programs, especially to the news and the main ideas of the speeches. These were separately transcribed, copied, and delivered to the local newspaper office, to be published the next day," Wu Daoyi recalls of their practice. "At the same time, public newspapers were hung up for the neighborhood people to peruse."[36]

Initially the program was extremely small, with monitors and their American-made six-tube radios sent to just ten major cities. The November 1932 opening of the new Central Station in Nanjing enabled a rapid expansion and shift in direction, however. Previously, the Central Broadcasting Station's transmission power had been a paltry five hundred watts (one-quarter the power of Zhang Xueliang's station in Fengtian). Even though the announcers tried to clearly annunciate and repeated their broadcast scripts, radio monitors in distant places still had difficulties writing it all down, leading to numerous mistakes and omissions. With the new, more powerful station, reception would be much improved. Now "radio monitors could be used across the nation's 2,000 counties—the goal being to install a radio-listening center (*shouyin zhongxin*) in each county."[37] The monitor program requested county-level governments from every area of China except the Southeast (which had its own separately run radio system and monitor program) select high school graduates to send to the capital for five months' training in science, technology, and radio repair. Afterward, the monitors would return to serve their hometown governments with a six-tube radio and other equipment necessary for maintenance of the receiver and dissemination of the news, sometimes including a mimeograph machine. This iteration of the monitor program graduated three cohorts of trainees, 439 *shouyinyuan* altogether.[38] At least twenty-five of these were women, but the great majority were men.[39] Ending in about 1935, the program theoretically provided a trained radio monitor for about one quarter of Chinese counties. If we exclude occupied Manchuria, the Southeast (Guangdong and Guangxi), and the border provinces of Xinjiang and Tibet, the percentage would be substantially higher, though wealthier provinces were overrepresented in the number of *shouyinyuan* they dispatched for instruction.

In addition to this national program, some provincial governments also introduced their own radio-listening-post systems. In 1930, the Nationalist stronghold of Zhejiang organized a *shouyinyuan* training program so that every locale could be "well informed of military and government news."⁴⁰ By the following year, there was a trained individual in each of the province's seventy-four counties, listening (according to regulations) to the provincial and national stations three times a day, for a total of nine hours.⁴¹ The Sichuan provincial government selected trainee monitors for its extensive program through an exam.⁴² Guangxi Province, governed by a Nationalist Party faction independent of Nanjing, built its own electronics factory, training center, and radio station so that "government orders and current news items could be heard throughout the province."⁴³ Thus, even in remote mountain districts like Baise, on the Vietnamese border, public loudspeakers broadcast transmissions from Manila, Nanjing, and Japan, in addition to the local provincial station. "It is interesting," said one visiting correspondent, "to see a group of ignorant peasants in one of the more out of the way cities, far back in the mountains, many of whom do not even know what the world on the other side of the hills looks like, and have never seen a river launch, motor car or aeroplane, standing amazed in the presence of one of these loud speakers, listening to some political speech delivered at Foochow or Nanking, or some concert given in Hongkong!"⁴⁴

The New Life Movement (1934–36), a movement for national rejuvenation directed by Chiang Kai-shek, invigorated these programs. In June 1935, the Ministry of Education ordered the installation of a radio in every existing school and in a new type of institution called a *minzhong jiaoyuguan* or "People's Education Center."⁴⁵ Designed for all ages and types of education, from literacy campaigns to instruction in new agricultural methods, they became key nodes in the new rural newsscape. A newspaper usually hung on the wall, its content transcribed from nightly broadcasts. Communities gathered there to listen to music or important speeches from the radio, drawing in residents who might otherwise have avoided the facility. If the town or village lacked a dedicated school, temples were often rechristened for this civic duty, while retaining their religious function. According to a survey by the Ministry of Education, before its program began, 975 schools or educational centers had radios installed. By the end of the first year, an additional 810 schools had installed sets, nearly doubling the number of official public radio receivers.⁴⁶ By January 1937, 3,285 sets had been installed in rural *jiaoyuguan* across the country.⁴⁷ One cannot assume equal geographical distribution, however. Sichuan, for instance, had installed radio-listening stations in 108 counties out of some 146 total, while Gansu, Ningxia, Qinghai, and Guizhou had installed none.⁴⁸ But most inner provinces made substantial progress. War-torn Jiangxi, recently the scene of vicious fighting between Communist and Nationalist armies, succeeded in installing radio-listening stations in half of its eighty-three counties.⁴⁹ Given its temporal and financial constraints, the program placing radios in rural schools was a success, especially when one considers that "schools" usually referred to People's Education Centers, which served the entire community, not just the youth. In Jiangxi, Song Meiling (Madame Chiang Kai-shek) remarked on the operations of one education center:

> In order to incorporate the village school into the very heart of community life, the children are taught during the day, and at night the adults go to classes. While farmers also gather to listen to current news, which is patiently and painstakingly explained to them by the teacher. The people consequently not only know what is happening in China itself, but also what is happening in other parts of the world, and visitors are surprised to find how well informed the average farmer is.⁵⁰

In Hubei, most counties installed the required radios in *minjiaoguan* rather than in middle or high schools, indicating the radio's wider purpose.⁵¹ The program did not merely aim at having a class or high school assembly listen to government broadcasting. Instead, it attempted to create clearinghouses for information in local communities and thus transform the rural newsscape.

In 1936, a teenager named Hu Baoren living in the remote Qinling Mountains of southern Shaanxi witnessed such a transformation. The installation that fall of a radio in the rear-courtyard assembly hall of his school allowed students and teachers to listen, for the first time, to programming from the outside. They heard recitals of Sun Yat-sen's Last Testament and Chiang Kai-shek's admonitions, as well as the latest news and entertainment from Beijing, Tianjin, Shanghai, and Wuhan. When the Xi'an incident occurred, vast numbers of teachers, students, local militia, and community members came to hear the news as it happened. The government may have placed the radio there, but it was the drive for information that brought local people swarming around it. To better serve the immense demand, students placed blackboards and wall newspapers with transcribed radio stories inside and outside the school.[52]

Laiyuan County, in mountainous western Hebei bordering directly on Shanxi Province, experienced a similarly enthusiastic reaction when the radio-monitor program began there. The local histories record that, in 1936:

> The government ordered Laiyuan to send a college graduate to Nanjing for training for one year to learn about the maintenance of radio technology. At the time this was a very unusual assignment, and many people competed to go. After much competition, it came down to Ma Daqi of North Stone Buddha and Wang Xuehe of South Slope, who were of the same ability. The magistrate had no way of choosing between the two of them, and because Laiyuan was a mountainous, remote place, to be considerate he asked for instructions from [the prefectural seat of this area]. In the end they both went.
>
> After the two studied and returned they brought with them a radio which they installed on top of the south city gate. It was called the "Laiyuan County Wireless Listening Station." The two listened each night, and transcribed important news, which they posted to the city gates. The people fell over each other in eagerness to view these reports. Afterward the listening station moved to the Temple of the God of Wealth within the city. Every evening the people went directly to the temple courtyard to listen. Around the time of the 7/7 [Marco Polo Bridge] incident, the political situation was precarious, and everyone paid careful attention to national affairs. The listeners often filled the Temple's courtyards.
>
> At that time looking after radio was a cushy job. Ma and Wang every day received a salary of 1 silver dollar each, two times higher than the salary of an elementary school teacher, five times higher than the salary of a policeman. If we use the equivalent grain prices, their salary was 600 or more yuan per month in today's money. We can see how short supply these radio overseers were and how backward science and technology was![53]

Not all places had as smooth an experience as Laiyuan, however. Sichuan's Tianquan County, for instance, struggled to afford a radio. In 1935, the provincial government ordered every county or city to select a scholar through examinations for *shouyinyuan* training in Chengdu. It also requested that every county contribute the training fee and radio purchase price, which amounted to 320 silver yuan. This was a comparatively small figure for most county governments, but because Tianquan was so poor and mountainous, and had suffered from famine for a number of years, "the treasury was empty, and finding 300 yuan was very difficult."[54] The magistrate, therefore, requested the provincial government subsidize the cost, but they refused, sending back a letter insisting that each county must purchase the radio themselves. In 1936, a new county magistrate went in person to the provincial capital, Chengdu, to beg for the aid in person. Only then did the provincial government provide funds to purchase a six-tube superheterodyne radio, along with dry-cell batteries and some parts. Magistrate Zhang Zhaodi personally brought the equipment, altogether worth 160.9 yuan, back to the county seat. After the radio arrived, however, there was no money to retain a *shouyinyuan*, so ordinary office workers performed the maintenance and propaganda duties. The following year, the radio was moved to the People's Education Center, located inside a teahouse. There "they listened every day to music and speeches of famous people."[55]

The scale of audiences like that in Tianquan varied according to the nature of the radio, the quality of the loudspeakers, the location of the listening post in town, and the content of the programming. The crowds could grow quite large, however, such that foreigners found it remarkable. One observer noted that while the percentage of listeners was not as great as in Western Europe, the Chinese listened in the hundreds or even thousands at a time. He described loudspeakers set up in public spaces that "when heard 20 yards away, [gave] such a terrific roar that a programme announcement sound[ed] more like a bombardment than the voice of the announcer." Indeed, he said, "when radio plays are broadcast, the crowds of listeners become so boisterous that a European onlooker feels as if he were at an international football match just after a goal has been shot."[56] Similarly, the Chinese consul general in Sydney related the story of a man who had gone fifteen hundred miles to the interior of China. The man arrived at a public square at ten o'clock and saw about five hundred people standing around. "Not knowing the cause he asked questions and was informed that the Government in [Nanjing] was about to broadcast the news of the day. So here," the consul concluded, "1500 miles from the coast were people who once waited five to seven days to obtain news, which was now obtained through the Government Broadcasting Stations, within twenty-four hours."[57]

Having just one radio at the listening center in a county-town did not necessarily confine its influence to that settlement. Besides word of mouth, news was sometimes distributed to surrounding towns and villages by means of mimeographed newspapers published at the listening station or education center. These were usually single sheets with a summary of news announcements. This chapter opened with an example of the practice in Qu County at the outbreak of war in 1937. Fliers were not only posted all over town but sent to the outlying villages and distributed at teahouses and theaters. In 1936, in Enshi, Hubei Province, an official proudly reported that "we listen twice every day on the seven-tube radio to the broadcast transmitted from the Central Station. The wall newspapers at every *minjiaoguan* in the county, as well as the news in the local daily newspaper *Qingjiang Ribao*, are all provided by this office."[58] When even wall newspapers and printed handbills were absent, a wealthier family in a neighboring village might have a crystal radio set and quickly share the news with friends and neighbors; access to such a device and such information also reinforced one's status. We saw an example of this in the experience of Meng Yiqi in rural Anhui at the beginning of the chapter.

The creation of programs to place radio and trained radio operators in rural areas ranks among the most significant legacies of the Nanjing decade, alongside its nominal political reunification of the country, the bureaucratization of the government, and its programs to prepare China for conflict with Japan.[59] When a county installed a radio, students, teachers, magistrates, or radio monitors could use it to transform the speed, volume, and breadth of information arriving in an area. Even if most people did not hear or read it directly, it would inevitably reshape the oral network of a society that lacked other reliable sources of fresh information. News attracted people, changing their behaviors and introducing the ritual connections of a rudimentary mass society. Obviously, such energy was not universal, but a single reception point still had the *potential* to revolutionize the local newsscape, connecting an entire village, town, or county much more closely to national and international events. In other words, radio's influence should not simply be measured by the relatively small number of receivers but rather by the way it changed the geographies and practices of news.

Political Crises in a Changing Newsscape

What effects did these changes have? By 1937, a piece of news or propaganda could reach many residents of China within twenty-four hours, and that news continued flowing day after day. In large cities, news was no longer restricted to the small fraction of the residents who constituted the reading public but was supplied directly to the masses, even if they were illiterate, through street-side loudspeakers. The nation and the world became more proximate, news more urgent. Events seemed to move at far greater speed. A school in Datong, Shanxi Province, wrote that its newly installed radio would provide access to important news from Nanjing, Shanghai, Beijing, Tianjin, and Wuhan "three days earlier than print media, which was a big help to those concerned with the current political situation."[60] Datong was the second largest city in Shanxi Province, immediately west of Beijing. Even there, among presumably privileged high school students, print news from the capital had arrived several days out of date. Radio accelerated the news and shrank space, transforming the newsscape just as the political disasters of the 1930s erupted. Comparing the popular experience of two particular crises—the Shanghai Incident of 1932 and the Xi'an Incident of 1936—helps reveal this acceleration and tightening, along with radio's political repercussions.

In early 1932, soon after their takeover of Manchuria, Japanese forces invaded Shanghai. That Shanghai was the site of battle was significant; it held more radio stations than any city in the country. As violent fighting progressed street by street, radio stations carried live accounts of the combat to all corners of the nation and the world. In Geneva, China's representative to the League of Nations entered negotiations to relay "the roar of the guns broadcast by Chinese stations" on a Swiss transmitter so that the sound could "ring into the ears of the world's representatives."[61] More important, the sounds of the fighting resounded domestically. A reporter in Hangzhou wrote of the impact: "News of the Japanese bombardment of Shanghai fell upon Hangchow like a thunderclap. An ominous silence prevailed and crowds of people flocked round the shops where special loud speakers were in operation for the public service. It was interesting to witness this new phase of modern Chinese life, for now, the man in the street receives the latest items of news briefly served up over the wireless, and he listens in amazement and lingers in expectation."[62] Radio also supplied the only up-to-date news reports for major papers outside Shanghai; it was much cheaper than telegrams and more reliable given the chaos in the city.[63] The reports and sounds of the fighting immediately became the most pressing topic for people across the nation. "There is a subdued joy rioting in the eyes of the people," observed another journalist, "as the wireless bears the news of the continued successes of the armies of the Middle Kingdom. Loud-speakers are continually pouring forth tales of valour from the front, and there is a palpitating silence between the items of news."[64]

In rural areas, the radio reports not only encouraged fierce nationalistic feelings but also stimulated the demand for more radio technology. Radio salesmen reported that interest in their wares reached new peaks with the invasion of Shanghai. They indicated that people in all parts of the country were listening with unprecedented avidity to reports of the fighting. "Radio brought immediate pictures of the combat before the eyes of the people" and "unite[d] the nation in common sentiment," wrote one.[65] Salesmen traveling the interior were pressed to set up all available receivers, and purchase orders exceeded the number of available sets in the country. During the broadcasts, townsmen who gathered around the sets received the reports with an enthusiasm that drew comparison to Americans listening to radio reports of football games.[66] Radio's impact manifested through the visceral reaction of Chinese people but also through foreign fear. A report in a British-run newspaper viewed the nationwide mobilization and radicalization caused by radio in a negative light:

> The advent of broadcast has brought places, days away in terms of actual travel, within a few seconds of Shanghai and the result has not been altogether good. In the provinces people, who by no probability at all could

have been involved in the Sino-Japanese dispute, considered the transfer of their capital in the event of Japanese invasion, military inadvisable and politically undreamt of. Throughout the length and breadth of the county the population was elevated to the heights of patriotic jubilation one moment, and dashed into the profoundest depression the next. . . . Unfortunate movements in the past have spread with comparative slowness: there has been, generally, sufficient time to make [plans to protect foreign nationals]. . . . Wireless broadcasting improperly controlled is a danger, the potentialities of which cannot be overestimated. Inflammatory reports can be spread throughout the country on a second's notice, and the feelings of the people affected, not by a wave of public opinion, but by a sudden and overwhelming tumescence.[67]

The Shanghai Incident and its accompanying radio reaction occurred before the greater part of the radio boom of the mid-to-late 1930s. By 1936, radio and its accompanying newsscape revolution had penetrated even more intensely into Chinese society, especially in rural areas. Zhang Xueliang, the former marshal of Manchuria, understood through experience the power of this technology. Like his father before him, he had been greatly interested in promoting it within his fiefdom. When the Japanese moved to take Manchuria from him, they stormed the Fengtian broadcasting station, immediately cutting the city and region off from the outside world. Broadcasting their own propaganda to the home islands, to China, and to the Western powers, the Japanese sowed enough confusion to delay response for days or, possibly, weeks. What action could have been taken had those reports been received sooner? We cannot know. But when it was time for Zhang Xueliang to execute his own coup, he used the very same tactic as his Japanese enemies.

On the night of December 12, 1936, Zhang's forces kidnapped Chiang Kai-shek, who was staying at a hot-spring resort outside Xi'an, in western China. Zhang held Chiang hostage, hoping to force a war against Japan. The same evening, in the center of the city, Zhang's soldiers stormed the radio station and began broadcasting to the nation and the world. The Xi'an radio station had been established in Beijing in the spring of 1936 by the Central Broadcasting Authority, but the continued encroachment of Japanese forces encouraged a relocation farther inland. Although not among the most powerful stations in the country, its broadcasts could still reach the majority of

Chinese territory.[68] The Xi'an station was, therefore, a vital tool. Luckily, the sympathetic station chief did not resist.[69] Zhang Xueliang could thus make speeches explaining the causes of his mutiny and his goal of uniting the Chinese parties to oppose the Japanese invasion.[70]

The propaganda from Xi'an also targeted an international audience to convince foreign powers of the righteousness of the plotters' cause. Agnes Smedley, an American communist sympathizer and journalist, happened to be in Xi'an. The night after the incident, she was awakened by soldiers bursting into her room. After going through her papers to make sure she was not a spy, they recruited her to help with propaganda. John Service, an American foreign service officer in Beijing, remembered hearing her voice. "I was fiddling with [the radio] during the Sian [Xi'an] affair, and suddenly I realized I was hearing Sian, Sian calling, an English voice. It was Agnes Smedley who happened to be in Sian. She was broadcasting the news. So, I went down to the embassy the next day and told them about news being broadcast from Sian, and they were absolutely staggered."[71] According to Service, Smedley reassured the foreign powers that Chiang was safe, that it was not a true coup but an attempt to unite the country.

The broadcasts were successful enough to cause near panic in Nanjing over their effects. According to the memoirs of Wu Daoyi, the Xi'an station "incessantly broadcast ridiculous and unfounded opinions, sowing confusion in people's hearts across the country." During the crisis, Central Party Office Secretary Ye Chucang met with Wu to plead that he quickly suppress the Xi'an broadcasts, "to make it so that in the area around Nanjing and Shanghai people could not listen to their voices." Wu explained the principles of radio broadcasting to the secretary and described a method of jamming that could be used to make their content inaudible in a small area. But to eliminate Xi'an's voice entirely, they could only have the air force bomb the station. "I don't care to understand any of these principles [of radio], just go block it immediately!" screamed Secretary Ye. Wu, therefore, disassembled an unused two-hundred-watt transmitter in Nanjing and transported it to Luoyang on a truck. By December 23, the central government had a temporary station in Luoyang jamming the Xi'an station's transmissions.[72]

Over the course of the crisis, people across the nation had been glued to the radio. In Zhengzhou, the capital of Henan, residents gathered in front of every store with a radio each night. They stood listening attentively, not daring to move despite a steady rain, lest they miss something.[73] In rural areas, radio monitors copied down important speeches denouncing Zhang from the Central Broadcasting Station, and almost every listening post was packed with people. Monitors mimeographed the broadcast transcripts, distributing copies to associations and schools, posting notices in streets and alleys.[74] One such place was Qu County, Sichuan. One freezing December night, the magistrate sat listening intently with his radio monitor. Sounds rumbled forth from his radio again and again, sometimes clear, sometimes muddled. Although the magistrate sat next to the brazier, he could not help shivering. There was nothing to do but wrap himself in a quilt and persevere. The news was of a crisis in Xi'an convulsing the whole nation. Early the next morning, the magistrate fanned out large sheets of paper and dashed down the news of the mutiny and Zhang's manifesto "Urging of the United Front," to be pasted around the town. Seeing there was leftover paper and recalling the circumstances of the previous evening, he carefully composed the following verse:

> Busy all day,
> Never a moment of leisure.
> A severe winter, a cold night,
> My whole body wrapped up,
> I hear the singing of the rooster!
> Sometimes clear, often muddied,
> When the air becomes quiet next to my ear, that is the most
>   bitter time.
> And tonight,
> Gloomy and deformed,
> How dare I take the headphones off?[75]

When Communist and Nationalist parties concluded an anti-Japanese armistice, Chiang flew to Luoyang and safety. That evening the Nanjing central station broadcast the news. When the glad tidings of Chiang's release arrived through the speakers in storefronts and public listening venues, there was jubilation in the streets.[76] Firecrackers broke the gloom that had hung over the country for the past fortnight. Huge crowds jammed the streets, listening to radio broadcasts of the generalissimo's release, beating gongs and blowing bugles, in what contemporaries called an unprecedented celebration. Festivities lasted well into the night.[77] People across China united in a communal celebration, indulging in a moment of national relief. Through these acts, and the pursuit of information, they delivered themselves unto the mass information order. It may still be true that the state forms the citizenry according to its own designs, but only their behavior makes that possible. Studying the newsscape, then, we see a history of information and politics not merely focused on the state's collection or perception of data but one centered on participation, a story of engagement in a mass society from the bottom up.

The shift in experience occasioned by broadcasting was profound. For the first time in Chinese history, news came regularly, at times within twenty-four hours, to many formerly isolated corners of the country. Wu Daoyi, for his part, concluded in 1968 that the Nanjing government's radio initiatives, from the perspective of their subsequent exile, may seem insignificant. "But thirty years ago, it truly gave a great deal of benefit to places across the country.... It contributed to the people's understanding of national policy and gradually deepened their unified conviction."[78] Radio—even if just the source of oral rumor—became the means by which information arrived at some of the most remote places. Although he could only hear secondhand, Meng Yiqi, the student from Anhui, heard war news as it was happening throughout the summer. Inspired, he became part of a student propaganda team, using crystal sets and mimeographs to transmit information. News had produced feelings of indignation, patriotism, and unity—as well as action. It had induced individuals from diverse classes and locales to repeatedly participate in a mass social behavior. And in the coming war, news would help keep the country in the fight. In mountain hideaways and city basements, in hurriedly whispered conversations and organized groups, people kept listening for reassurance that the resistance continued, that the Chinese people struggled on.

FIVE

The Occupation of the Mind, 1937–1945

After a half-century of tension, ever-increasing in intensity and pace, the battle for China finally commenced in the summer of 1937. The Empire of Japan, for decades an aggressively expanding power, now pushed for hegemony over the most populous nation on Earth. Nationalist China, tied in an uneasy alliance with the Chinese Communist Party, threw itself into the existential struggle. All sides recognized radio as a crucial tool in the contest, but for the defenders, it was more immediately important. Broadcast songs, speeches, and ceremonies rallied the populace to resistance. Wireless news connected the country and brought hope and encouragement to its listeners. Indeed, the drive for news and information lay near the heart of the wartime experience in China. For civilians, quality information was a valued commodity superseded in importance only by physical safety, food, and shelter. Displaced persons and those in occupied zones desperately searched for something to grasp beyond rumor; radio or radio newspapers were often the primary sources of outside news, even if the content was sometimes misleading.

To the Nationalist government, radio became the means through which it kept the idea and hope of a free, united China alive, for as the Japanese army had occupied the land, Japanese-friendly attitudes had occupied the minds of the Chinese people. News of allied fighting brought reassurance that what felt like an impossible situation was, in fact, transient and surmountable. It undermined the relentless feeling, promoted through Japanese propaganda, that China would inevitably lose the eight-year-long war, that it was far better to get along with the occupiers than resist, that the Japanese had defeated Western imperialists on China's behalf, or that the collaborationist regimes were preferable to radical Communists or corrupt Nationalists. Subverting the occupation of the mind, radio helped destabilize the psychological foundations of Japanese and collaborationist regimes. Broadcasting, therefore, remained a top priority for the Nationalist government. They would risk many lives and much treasure to maintain this vital link, as we will see.

Yet, at the same time, Japanese-controlled regimes dominated the airwaves, attempting to exert control over the geographical and social spaces of China through wireless technology, as Daqing Yang has argued.[1] Radio, in other words, was essential to the Japanese re-formation of the newsscape. Imperial institutions and collaborationist governments built powerful new broadcasting networks and promoted radio listening and ownership, creating the informational infrastructure and mass social behavior exploited by subsequent Chinese governments. In 1949, fully half of radios in China were still standard-issue Japanese receivers, and most broadcast stations were of Japanese design. The People's Republic built its political communication machine on the scaffolding of Japan's wartime system.

This chapter provides new insight into the technological infrastructure of the wartime Japanese Empire and demonstrates that, with respect to communications, the occupation was constructive rather than destructive. The word *development*, especially in China, has an overwhelmingly positive connotation, and attaching any positive subtext to the Japanese-controlled collaboration governments of China is difficult. The influence of the Japanese occupation is, therefore, popularly seen purely in the negative; it caused the militarization of society, the destruction of political institutions, infrastructure, and the economy. Above all, what is remembered is the residual trauma of a brutal military regime. These things are all true. But in the eight years it administered the most populated, industrialized regions of China,

Japanese and collaboration governments also undertook social, political, and infrastructural development. A growing literature points to the rapid industrialization of Manchuria and North China under Japanese rule.[2] I argue in a similar vein that one cannot describe the rise of mass society in China without addressing the development of radio listening and other forms of mass social participation under the collaborationist occupation regimes. In some cities where statistics are available, the number of registered radios jumped tenfold over the course of the war. The broadcasting network was strengthened and unified. Concerted campaigns copied from those in Japan and Manchuria spread the idea that radio was essential to the body politic. Unlike the Nationalist government, the Japanese had the means—radio manufacturing and transmission equipment industries—to make this a reality.

Just as important, collaborationist governments promoted the idea that radio was essential to the construction of a mass-mobilized state, to the unity of government and people. "Without radio there is no nation," ran one wartime motto. Concerted campaigns, based off those in Japan a decade earlier, spread this philosophy along with the radios sets themselves. By spreading these ideas and building physical infrastructure, the war *did* contribute to the formation of a Chinese state but not in the ways that Rana Mitter and others have outlined in studies of the wartime Nationalist government.[3] This construction occurred through the efforts of Japanese-collaborationist governments, which used their infrastructures and production capabilities to advance the technopolitical process. In identifying this phenomenon, I highlight a historical thread that scholars like Mitter and Timothy Brook have identified when they point out that historians can learn much from looking beyond the "superimposed moral map" of *collaboration*.[4] Even while retaining the ability to judge a violent occupation regime, we can appreciate the ambiguities of experience and mixed legacies that any power dynamic leaves. In this instance, an examination of the collaborationist administration reveals a technopolitical process that sits outside the narrative of domestically led nation-building. Instead, an occupier drove the technopolitical process in China, surmounting the country's vastness, its condition of communication and transportation underdevelopment, and its internal divisions to unite the nation into a single newsscape. Radio technology realized its utility—for good or ill—in the trials of total war.

Conquest

On August 8, 1937, several thousand Japanese soldiers marched down Chang'an Avenue, past the Forbidden City, for five hundred years the palace of China's emperors. Japanese army planes escorted the force overhead, dropping leaflets on the impassive crowds that turned out to watch. The raining paper promised that the Japanese army would drive out wicked rulers and wicked armies. Tanks and cavalry followed in procession. Earlier in the day, Japanese soldiers had raided the offices of the Chinese government radio administration near the embassy quarter and confiscated their equipment, cutting off the city's last Chinese-controlled link with the outside world. The land wires had gone out of commission weeks before.[5] "Beijing is deathly quiet," despaired a Chinese writer.[6]

Yet the airwaves were *not* quiet; radio ceaselessly disseminated resistance propaganda to besieged populations like that in Beijing. Though the Nationalist communications system struggled for survival under the conditions of war and conquest, through determined effort broadcasting officials maintained a Free Chinese connection with the outside world and with its own people during the chaotic retreat west. In areas left behind, radio became even more important as Japanese forces began to exert control over the urban newsscape.

Surrounded on all sides by enemy troops, Beijing had surrendered less than a month after the outbreak of hostilities, sparing it the destructive urban warfare later visited on Tianjin and Shanghai. On entering the ancient capital, the Japanese Kwantung army quickly moved to assert control. They seized police stations, secured intersections, and occupied offices. They also muzzled the Chinese press, though most publications had "voluntarily" ceased printing as employees made good their escape. The Central News Agency's office continued to publish copy until stopped by Japanese police on the evening of August 6. Within a few weeks, every newspaper and magazine still in print carried Japanese content.[7] By September 2, the local radio station carried simple Japanese language lessons to reinforce the continuous stream of propaganda issuing from the presses. Propaganda teams set up radio loudspeakers in some villages outside the city walls. A foreign journalist touring these settlements observed "groups of Chinese civilians cluster[ed] around them to listen to assurance of a new and beneficent rule."[8] The pull of news was stronger than the repulsion of conquest.

Tianjin fell soon after, and as in Beijing, the Japanese moved swiftly to control the newsscape. They arrested reporters and editors and began putting out their own official paper, which contained only Japanese propaganda. Frightened Tianjin residents sought out newspapers like *Shibao* or the English-language *North China Herald* if they could. Yet these had to travel all the way from Shanghai, and fresh news turned stale by the time of arrival. Thus, Chinese citizens turned instead to the many "special edition" news-sheets, which sprang up like mushrooms in the darkness of the beleaguered port. Usually single-paged or double-sided fliers, the news-sheets carried information from the Central Broadcasting Station in Nanjing, sometimes mixed with a little gossip. As soon as an edition was printed, it flew off the shelves, a best seller. "The editors of these papers are 'shady characters' (*touxi fenzi*), but all are also patriots," wrote a witness. The illegal news-sheets constituted one of only two ways to obtain information about the war. "The only [other] hope is to listen a bit every night to news broadcast by the Central Broadcasting Station. The radio industry has done an incredible business lately."[9]

By late September, a news drought prevailed over northern China. The demand for radio receivers grew exponentially alongside. A witness reported that inland, in Hebei Province, close to the fighting, no papers could be delivered because of ruptures in the transportation system. Strikes and official suppression had halted Beijing and Tianjin's presses. Shanghai's papers had not been sent out since fighting broke out there on August 13. Some publications made it up from Wuhan, but the supply did not meet demand. Starved for information, people fell into a depressed, worried state and became, according to the observer, more susceptible to rumors. Though many small places were still without radio, existing receivers were crowded until late at night without, it was said, exception. Audiences listened to every word broadcast and were moved by every syllable, the observer breathlessly reported.¹⁰ Broadcast news was at a premium. According to one source, in Tianjin's foreign concessions, each and every family tried to buy a set to listen in. "In the Chinese city," by contrast, "residents only listened with the utmost care because of the enemy's restrictions on tuning-in."¹¹

The Japanese authorities were well aware of this as they attempted to enact their occupation of China's newsscape. In Beijing and Tianjin, the regulation of radios became quite strict. Police confiscated any sets with four or more tubes—that is, those capable of receiving signals from areas still under Chinese control. The *Kempeitai* (military police or gendarmerie) could inspect receivers at any time. Some districts carried out regulations more strictly than others. For a period in 1938 in Tangshan, no radios could be used at all, and people were forced to tear down all antennas. Despite these regulations, some radio owners still listened to China's Central Broadcasting late in the evening or at four-thirty in the morning, a convenient hour for both reception and secrecy, when residents could quietly get up to hear the first reports of the day.¹²

Fearing the consequences of this practice, Japanese engineers established a transmitter in Tianjin soon after the occupation, tuning it to the same wavelength as Nanjing and transmitting false reports or static noise to scramble sound waves from the Central Station.¹³ The effort came to naught, since listeners could "tune into Jiangsu, Shanghai, Changsha, and Wuhan stations and hear the same news broadcast from Central Station," according to one defiant report. "Every night in streets and alleyways of the city, people are crowded around the radios of households and businesses, holding their breath and quietly listening, hoping for the beautiful sounds of the receiver."¹⁴

Hundreds of miles south, in Shanghai, fighting between Imperial troops and Nationalist forces began on August 13. If the Japanese expected another quick victory, they were mistaken. A bitter resistance pinned them down for weeks, then months—Chinese armies held out until the end of November. During this period, the government and citizens' groups used radio to organize and rally the population. Local Chinese radio stations transmitted news and patriotic songs. Broadcast fundraising drives drummed up charitable contributions.¹⁵

The powerful transmitter at the Shanghai Municipal Station led the resistance. On the sixth anniversary of the fall of Fengtian, September 18, 1937, it organized a mass ceremony in which citizens throughout Shanghai simultaneously swore an oath to support the government and obey the leadership, to sacrifice everything and prosecute the war to the end. The local *dangbu*, or party office, ordered all flags lowered to half-mast. At 11:30 a.m., the city's Chinese radio stations began relaying anthems, instructions, and exhortations through storefront loudspeakers. Shops had already posted copies of the oath along both sides of the street for passersby to read. At noon, troops of Boy Scouts deployed throughout the city ensured that all pedestrians and vehicles had the propriety to stop. Store managers and factory directors, headmasters and teachers instructed their wards to cease work. After three minutes of silence, at the radio's signal, the city raised its right hand and recited the vow.[16]

On the day of the oath, the Municipal Police of the International Concession worked feverishly to circumscribe the protests. Several Chinese radio stations had removed to the settlement for safety. The most important among them, the municipal station, sat proudly at the heart of the International Concession, at the Peace Hotel, on the corner of Nanjing Road and the Bund. There, the Nationalist's greatest propaganda asset was safe from Japanese bombs—for the time being. But it was not safe from attempts by concession officials, mostly British, to bowdlerize the station's songs and plays in a vain attempt to deflect Japanese anger. In their capacity as censor, the concession police ruled a song entitled "The Chinese Race Will Not Be Ruined" was "not objectionable," but "Defend National Territory" offended with the words:

> Brethren, Rise up! Defend the National Territory.
> Launch a counter-attack upon the enemy.
> See the Japanese have commenced operations.
> Listen to the guns of the aggressors; they are roaring
> To retreat means death. The only way to live is resistance.[17]

British authorities continued to warn off the station throughout the months-long battle. On October 10, the police interviewed Tao Baichuan, the member of the city's standing committee in charge of propaganda, at a

Chinese restaurant on Fuzhou Road. "It was explained to him the embarrassment he was causing the Settlement authorities through demonstrations and poster distribution and the injudicious way he was using the Settlement for means of propaganda," recorded a British official. The policeman also impressed on Tao his belief that such activities might lead to incidents and mob violence that could destroy worldwide sympathy for China.

Thus, Shanghai presented an entirely strange and anachronistic situation. In the midst of a war between two world powers, the concessions stood as islands of peace, believing themselves untouchable, their haughty officers harboring few doubts about their ability to, in the end, avoid the fray. It is extraordinary that Tao took the time (or had to take the time) to engage with such legalistic complaints in the heat of battle. He did not really bother to deny the propaganda offenses perpetrated by the municipal radio, replying sardonically that "he would do everything possible in an endeavor to persuade those responsible for the demonstration and poster distribution to desist, but he was not responsible for that part of the programme, only the slogans to be broadcasted and an appeal to the people through the press to be patriotic."[18]

Smaller Chinese-owned or -operated stations in the concessions were not so immune. After the outbreak of hostilities, the radio station at the Grand Hotel—despite its English name, a Chinese theater—ceased broadcasting and instead lent the facilities for use in "national salvation" propaganda. Perhaps because they were an obvious target, the authorities in the International Settlement pressured the station not to broadcast anti-Japanese programming. On the anniversary of the Mukden Incident, while other stations relayed the oath and accompanying speeches, concession police officers listened carefully to their wavelength and confirmed that no agitating speeches were broadcast between 11 a.m. and noon.[19] The intimidation campaign had succeeded in this instance. Police harassment of stations broadcasting anti-Japanese plays and songs continued over the following months.[20]

As the battle ground to its conclusion, members of the municipal party leadership continued to offer comfort and encouragement to Shanghai's citizens, which became especially necessary as the situation grew increasingly bleak. Tao Baichuan personally led the efforts to fortify the spirits of the local Chinese, to lead them through temporary defeats and reassure them of inevitable victory. In November, he broadcast a speech from the municipal station insisting that Shanghai would not be abandoned, denouncing rumors to the contrary. "Our brave soldiers . . . continue to fight for the defense of Shanghai," he insisted.[21] The propaganda chief then urged civilians to organize into volunteer brigades to assist in the defense of the city. By the time of this broadcast, however, defeat in Shanghai had already been assured. Nanjing ordered employees of the Municipal Broadcasting station to evacuate the city. Their skills were too valuable to leave behind.[22] The front had moved west, to a line near the capital, 125 miles upstream along the Yangtze.

Nanjing had been under air attack for nearly as long as Shanghai. Air defense drills, announced to the citizenry via radio, began in early August. Bombing commenced soon after. The central broadcaster had moved to a war footing immediately after the Marco Polo Bridge Incident. Regular broadcasting programs were suspended in favor of war coverage. Pleasant songs were replaced by martial tunes and emergency instructions. As the most important Chinese radio station, Nanjing played a key role in the transmitting of this patriotic propaganda, which other stations then relayed.

The importance of China Central Broadcasting in Nanjing made it an obvious target. Wu Daoyi, the official in charge of the station, took immediate precautions to protect the precious broadcasting facilities. The station's walls were repainted with camouflage, an ugly grey color. Workers piled earthen ramparts against the skirt of the buildings to protect against bombs. Within the structure, they erected sturdy frameworks to protect the most important transmitter parts from direct and indirect hits.[23] But the situation in Nanjing only deteriorated. On November 20, the central government ordered a retreat to Wuhan. Though the radio station was still broadcasting, sources of programming gradually ran dry. On the night of November 23, the station's duties were transferred to Changsha. Only ten employees remained in the capital. "We had operated this station—first in its class in East Asia, a powerful propaganda tool—for five years," Wu Daoyi recalled in his memoirs. "We sank into indignation and despair as we got to work demolishing it." Anything of value that could not be moved, like a large generator, was smashed. Movable equipment was placed in wooden boxes and brought to the docks for shipment, but in the confusion much had to be left behind.

Building a New Radio China

In China's southwestern provinces, the Nationalist government finally found a secure bastion to hold out for the rest of the war. Distance and natural geography hampered Japanese efforts to advance. The invaders were already spread thin in the vast swathes of territory they had occupied; conquering cities is one thing, but holding a subcontinent is another matter entirely. Though a massive Chinese counteroffensive would never come, the capital at Chongqing became a base from which to launch propaganda attacks and to organize and sustain the will of the resistance in occupied territories. All of this required a radio station. Though the Telefunken-manufactured broadcasting equipment taken from Nanjing was reerected, the resulting transmitter decreased in power from 75 kw to 10 kw, sharply circumscribing the penetration of Nationalist broadcasting.[24] Luckily, however, a powerful shortwave station manufactured by the British Marconi company had been in planning for several years. In 1936, the construction committee had asked permission from the central government to move the station, originally planned for Nanjing, to Chongqing. Even at this early date, Sichuan was already being viewed as a safe haven, far from possible sites of conflict. It began broadcasting with the call sign XGOY to North America, Europe, and all of East Asia in February 1939. "Every day in the small hours of morning," Wu Daoyi recalled, "from 2 a.m. until 4:20 a.m. they provided the 'recorded news' program"—that is, slow dictation-speed news to be written down and distributed by "our countrymen listening in occupied areas."[25]

As a powerful force undermining their domination of the newsscape, the Japanese frequently targeted the station in air raids.[26] But Chongqing is a mountainous city, full of slopes and ravines, spiraling upward from the confluence of the Yangtze and Jialing rivers. This topography offered a solution. The government set about digging (by means of backbreaking manual labor) an extensive series of caves to serve as air-raid shelters. They placed all sorts of critical infrastructure underground, including factories, government offices, and, eventually, the radio station. The damp environment was not conducive to maintaining the state-of-the-art Marconi electronics, but at least they could not be destroyed wholesale.

Having secured broadcasting capabilities, the Chongqing radio administration needed to address the reception of its messages. Throughout the retreat west, the government had sought to continue some of the radio initiatives of the Nanjing decade. In a document dated June 6, 1938, the Ministry of Education encouraged the installation of a public radio-listening network in the rear. It propagated many (unrealistic) guidelines. Every county, school, and educational institution was required to install a radio. More reasonably, each district's *shouyinyuan* had to continue listening to the news in order to publish reports in the wall newspaper and put out mimeographed news fliers every day or, at the least, every three days. The *shouyinyuan* were also instructed to gather audiences to listen to educational speeches. Every shop and private residence that had a radio was encouraged to make it available for public listening. Finally, radio manufacturing was to be moved to the southwestern redoubts, and governments should solely patronize domestic radio parts to stimulate the industry.[27]

Of course, much in the plan was unfeasible. There was almost no radio manufacturing occurring in China, so there were no factories to move to the southwest. Every school and institution would probably have installed a radio if they could, but sets were rare and expensive. The wealthy owners of private radios were unlikely to welcome many strangers into their homes to listen. But the radio-monitor program did strengthen substantially in the province of Sichuan in the early years of the war. From 1937 to 1938, the province installed 537 radios in schools and public education centers (*minjiaoguan*).[28]

According to reports from *shouyinyuan* overseeing these radio installations, attendance varied. While a neighborhood *minjiaoguan* in Chengdu might attract as many as five hundred people for the daily broadcasts, the average audience for this urban listening post was around three hundred. When rural Jiangyou County, in Sichuan's far northeastern reaches, first received broadcasts in its *minjiaoguan* in February 1939, fifty-one people attended. Over the next two months, clarity varied there. Sometimes, the six o'clock news suffered from interference. Sometimes, a downpour intervened, and people did not tune in. No one would venture out, and besides, the storm guaranteed spotty reception—the form of the newsscape once again depending on the weather. But even good reception did not necessarily guarantee clear understanding, since the Chongqing announcers spoke the dialects of

Beijing or Shanghai, not Sichuanese. "The audience was agitated [*xingfenxing*] by the lectures on the war," reported the monitor in Jiangyou County, "but many speakers used accents from other provinces, and so there were spots they did not understand. Sometimes the *shouyinyuan* would have to explain, but since the *guan* was crowded and the broadcast was only an hour long, time was short." This problem was resolved, however, whenever the audience listened to the Chengdu municipal station, which the local people understood and listened to with deep interest. When the radio broke down in March, however, the monitor sent it to the broadcast education office in Chengdu for service. "Upon its return we will continue our broadcast work," said his report.[29] It is not recorded when or whether the *minjiaoguan* got its radio back.

Jiangyou County's issues with its equipment exemplify larger problems that plagued the Chongqing government's communication technology throughout the war. The climate of Sichuan and Yunnan, the heart of wartime Nationalist China, was not conducive to a radio newsscape, degrading equipment with its damp, and muddling electromagnetic waves with its rain. Electronics deteriorated with startling rapidity—a situation worsened by the severe lack of replacement parts or even repair tools. Moreover, no manufacturing base existed that could remedy the deficiencies. Government agencies repeatedly issued grand plans for a broadcasting network that would encompass every organization, county, and school to increase morale and strengthen the war effort.[30] For instance, in 1941, the Central Broadcasting Industry Association in Chongqing issued a plan to install more than a quarter-million radios in the areas under its control.[31] But in 1942, it was laying plans for a manufacturing plant that would eventually be able to produce a paltry two thousand sets a year.[32] It is doubtful that the plant even approached that number in the end.

The problem of supply was apparent across Nationalist territory. A series of letters between the Shaanxi provincial government and Chongqing show some of the squeeze. The first, dated December 1939, urged the national government to send batteries and replacement parts for broken radios. Because of wartime shortages, it said, both were hard to obtain. By June 1941, the provincial government wrote that "formerly in the city of Xi'an you could buy many types of equipment. Now you cannot get them for any price."[33]

Beyond batteries and vacuum tubes, basic tools like soldering irons and voltage meters were impossible to find, making repairs difficult. So desperate were things in Hubei (that is, the areas not occupied by the Japanese) that the government resorted to appropriation. In a decree of May 1940, every private person, organization, or group possessing a radio had to provide it for public use. The name of the former owner, the make and model, and price information were all registered with the Central Broadcasting Administration so that the set could be returned or the owner reimbursed at an unspecified later date.[34]

Crowded around the decreasing number of working receivers, private citizens acutely felt the lack of radios. They missed the comfort and reassurance of the news and the feeling of connection to which they had become accustomed in the years before the war. Some recalled the pedagogical opportunities that radio purported to offer. In a letter to the Ministry of Education, one wartime refugee expressed the frustration of many in Nationalist China when he remembered how women and children in his old apartment building in Nanjing had learned to read with the assistance of broadcast courses. Now he had repaired to his ancestral town in Sichuan, where there was no radio. The town sat on the route between Changsha and Chongqing, and the stream of refugees constantly asked after war news: "Because there is no same-day news, and there are no radio transmissions, the only information is contained in newspapers over seven or eight days old.... The need for radio in this place has become most urgent ... so that the people can receive accurate war news, and can understand the government's ideas and decrees, and can listen to Resistance propaganda."[35]

This refugee's radio would never come. Whereas before the war, cheap radio parts had been imported from Japan or from the United States through eastern Chinese ports, now little could be sourced. Airlifts transported some American-made equipment, but this amounted to little more than the bare necessities to keep transmitters, military receivers, and newspaper office sets operating. The lack of replacement parts led to a steady decline in the number of working machines. The American Office of War Information, in charge of propaganda and communication in the China theater, estimated toward the end of the war that there were fewer than ten thousand radios in all of unoccupied China.[36] Even this estimate seems generous, however.

In 1944, the mayor's office in Hengyang estimated that only about twenty radios remained in that city. Changsha and Qujiang mayors estimated ten each.[37] Another American report estimated that only eight hundred usable sets remained in Chongqing at the end of the war.[38] Adding to the frustration, an unreliable electric grid and expensive batteries made operating the remaining radios more than a few hours a day a daunting task.

The dire communications situation in Nationalist China stood in stark contrast to the situation in Japan and, indeed, to that in occupied China, as we will see. Certainly, individuals in Chongqing were aware of this. A newspaper article in September 1942 highlighted the fact that in 1940, there were officially more than five million sets in Japan, owned by urbanites and farmers, rich and poor. The difference between southwestern China and Japan could not have been more stark. There was opportunity in this asymmetry, though. As the writer of the article rhetorically inquired, "Can the Japanese thought police patrol more than five million radio sets without missing a bit of propaganda rapping on the eardrums of the Japanese people?"[39] With this in mind, the Chongqing station broadcast a great deal of Japanese-language propaganda aimed, in part, at audiences in Japan.

The primary targets, however, were enemy soldiers fighting in China and elsewhere. Japanese prisoner interrogation reports indicate a mix of mistrust and curiosity about these efforts. The soldiers could pick up Chongqing broadcasts in places as distant as Wuhan, Hangzhou, and Guangdong, even on a standard-issue radio. Sometimes they simply listened to streetside loudspeakers. Reports show that they regarded such broadcasts as "fun to listen to" though "sheer nonsense." "It didn't matter whether it was a broadcast, or pamphlet, the willingness to believe Japan among Japanese soldiers was dominant," one said.[40] But the Japanese soldiers also describe how news from home became precious as censors held back more letters; only postcards made it through to these young soldiers, scared and alone in a faraway country. They never heard of Japanese defeats or home-front deprivation through approved channels and thus sensed they were missing something. When this information did come, it came as a shock. One prisoner recalled a friend telling him that a Japanese submarine had been sunk near Taiwan, news the friend had heard from Chinese radio. "At that moment it was like I had been struck by someone. From then on, although I could not listen to Chongqing directly, I heard it from my friend."[41] As deserters and prisoners chatting with enemy interrogators, this group cannot be considered a representative sample of Japanese soldiery. But it is equally reasonable to assume from their testimony that there was some level of curiosity about Chinese and Allied news in the Imperial ranks. The thing they most desired to know—the state of things at home—was also the most sensitive. They sought this information out, regardless of the perception that listening to the enemy was a treasonous act. The pull of news and the ubiquity of radios in the Japanese army and in Japanese-held cities meant that many soldiers could and would tune in.

Beyond broadcasting propaganda to the enemy, the Nationalist radio network fulfilled two additional roles. The first task was transmitting to the world an important message: we're still in this. Certainly, some listeners around the world tuned in, but most listeners were just curious radio hobbyists. Outside East Asia, there was no significant population in any country tuning in to Chongqing. The radio broadcast itself, however, was a message of continuity and perseverance to Allies. Chongqing's programs signaled the same message to its own population under Japanese rule. We're still here, fighting. In fact, the ways this population listened to radio and interacted with the newsscape were central to their experience of the war.

Radio Newspapers in the Rural Occupation

Jie Ru, a student from Canton, escaped the fall of his hometown by following the Sui River west to the medieval walled-village of Guangning, seventy miles upstream.[42] The guesthouses filled with exiled city folk, and the timeworn walls became plastered with slogans, posters, and notices advertising recently arrived organizations. The locals saw unfamiliar sights in the streets: government functionaries in uniform, women in qipao, girls with short hair; disheveled activist youths; refugees looking for a place to stay or simply moving through. There were few goods to be had at the market and even less food. Japanese airplanes occasionally bombed the town, collapsing homes and businesses, strafing the ground with bullets. Hard currency and banknotes were scarce, as were quinine and other basic medicines. "In a time and place cut off from news and communication"—in other words, a place like Guangning—"you long for a daily newspaper," Jie observed. Thus, armed with a radio and a pistol, he began printing a single sheet 16 *kai* (A5) news flier, which "quickly became the nerve center of the whole mountain town and the surrounding districts."[43]

Jie provided an indispensable service: connecting the people of Guangning to the broader national struggle. So did numerous comrades in similar situations across the country. By 1943, an official from the Chongqing broadcasting station reported that "radio is now the only source from which over 500 newspapers and thousands of mobile newspapers receive their news and make it available to the people."[44] While it was not quite true that it was the *only* source of news—information from the local and neighboring districts could be gathered in traditional ways—for thousands of places and people like Jie Ru, it was true enough.

By the end of 1938, Japanese forces had occupied most of China's important central and eastern regions. But, in truth, the occupation was far more limited, felt most acutely in urban cores, along the railways and roads, and more in the daylight than the night. Though Japanese forces could strike practically anywhere in China, exercising effective control was another matter. Much of the country, therefore, fell into a painful, confusing "in-between" space. In Guangning, Jie Ru lived under conditions of war but not under daily occupation. He distributed his paper across three counties but still felt the need to hide his writing space and radio high in the mountain forest, his printing operation in a valley hamlet. He had the trust of local farmers but still felt the need to carry a gun. Jie's experience was, in some ways, typical. Like the majority of the Chinese population, he lived in a liminal, gray area of occupation—without a daily experience of Japanese administration but also outside of Chongqing's sphere of political or military power projection. In the midst of this disorientating situation, the radio—and the newspapers that derived from it—anchored him and the newsscape of his rural county to a wider resistance. Radio blurred the occupation. You could live in a city under Japanese rule and still get instantaneous news of enemy defeats a thousand miles away. You could be cut off from any substantial force of your troops and still feel the daily pulse of your cause. You could live in the margins of occupation, or at its very heart, and the voice of free comrades would still bleed through.

Surprisingly, underground newspapers were not limited to outlying districts like the area Jie Ru occupied. They were also an urban phenomenon. Even in places like Beijing, the first major Chinese city to fall in 1937, illicit media operations dared to spread radio news, working to counter a dense Japanese-collaborationist propaganda network. The Japanese authorities had moved quickly to snuff out obvious channels through which uncensored information might flow. They shuttered newspapers and magazines, then reopened them under compliant ownership. Japanese censors occupied the broadcast stations and telegraph exchanges. Backed by arms, they oversaw a largely Chinese workforce. "In every county on every front, not a few Japanese or traitor propagandists (*xuanfu yuan*) have forfeited their lives," complained one underground Chinese activist in a 1939 report on the propaganda war. Radio listening was harder to control and, therefore, became invaluable to the resistance. "Central Broadcasting sending out true information is our most useful tool," wrote the partisan. "The enemy have forbidden listening. They have jammed broadcasts. It hasn't worked a bit. Our secret newspapers in Beijing and Tianjin have developed quickly, and never fail to shatter the false propaganda and ridiculous theories of the enemy newspapers. We use a Japanese-manufactured mimeograph machine and a German rolling printer worth several tens of thousands of Japanese yen to engage in the struggle."[45] Indeed, some underground presses in Chinese cities claimed to have been capable of near-instantaneous news releases. For instance, as our Beijing partisan proudly recounts, "When our [Nationalist] air-force made an expedition to Taiwan to engage the Japanese, the news was transmitted by our secret newspapers to uncountable masses within an hour of Central Broadcasting [in Chongqing] announcing it."[46]

Broadcast news punched through the barriers of occupation, where through the efforts of conscientious volunteers, it had the habit of multiplying. This could happen even without the printing operations of people like Jie Ru, the mimeographs used elsewhere, or the expensive German presses of Beijing. One account from Fenyang County, Shanxi, illustrates the compounding power of a single receiver. A teenager from that rural county, Sun Puyi, built radios as a hobby. Returning to his hometown for summer vacation in 1937, Sun found people there anxious to get news of the situation. The town had no newspaper, and it took two or three days for newspapers from

Beijing or the provincial capital to arrive. So Sun immediately set up his radio in his family's living room.

> The news got around Fenyang very quickly that Sun Puyi had a radio. Day after day the numbers grew of those who came to our house to hear news of the fighting, and soon there was no way for everyone to sit down in the room. The only thing to do was to move the radio out in the yard, and we helped him by carrying a table out to serve as a platform for the radio. Every evening huge crowds of people gathered around the radio. When the news broadcast began, there wasn't any need to try to create order. Everyone became very quiet, and the only the announcer's voice was heard reverberating through the evening quiet. The listeners' reactions rose or fell according to the news of the action.[47]

The availability of news had a powerful effect. It caused the townsfolk to discipline themselves into participants in the inchoate mass society in many ways. The radio listeners gathered and quieted themselves. Some copied down the broadcast reports and posted them to the gates of the school. Crowds huddled to read them, and someone always transcribed the poster to spread the news further. On the occasion of a Chinese victory, Sun's brother saw a distinguished elderly man perform the extraordinary gesture of kneeling on the ground to copy down the news, appearing to bow before the information. "That evening in Fenyang was just like having a festival, with people flying flags and firecrackers popping everywhere," he remembered.

Such optimism would not last. The Japanese army, eager to secure valuable coal fields, invaded Shanxi. "Fenyang fell under a 'dark cloud.' Those who came to hear the news every evening became more and more concerned. Should we stay in Fenyang, or should we leave now?" The radio warned of approaching danger, of the desperation of the struggle. Some, like Sun himself, took the voice drifting through the courtyard over those dark months as a call to action. The young student packed up his radio, got on a bicycle, and rode more than three hundred miles to Xi'an, in neighboring Shaanxi Province. He eventually made his way to the city of Chengdu in Nationalist China.[48]

The deep desire for broadcast news evinced by Sun Puyi's neighbors was evident across the country. Groups of all stripes, including partisans, youth propaganda teams, local and provincial governments, harnessed this interest (and need) to win trust and affection. Whereas slogans and soldiers were not always welcome, news always was. Meng Yiqi, the high schooler from Anhui who had heard the news of the Marco Polo Bridge Incident from a family with a crystal radio in his village in 1937, thoroughly appreciated the importance of information in the crisis. Like Jie and Sun, he left home to join up with the resistance. He joined a youth propaganda squadron (*tuidong-dui*) organized by a Nationalist mass mobilization program. The provincial government planned to send a squadron, each comprising five to eight students, into every county. It dispatched Meng's to Shou County in central Anhui, the hometown of the squad leader, a Fudan University student. The remainder of the group, mere high schoolers, frenetically prepared themselves, taking up pistols and donning gray uniforms. Despite the veneer of organization, much was chaos. "No one knew who was leading us above the squad leader," Meng remembered, "or even if there was a supervisory structure. Whoever showed up could lead."[49] The war had injected a great deal of fluidity into society; a college student could declare himself a soldier-activist and make it so. The squad leader announced that they would have two primary goals: first, propaganda work for the resistance; second, the organization of self-defense forces among the villages and towns.

Traveling to Shou County, the group anticipated many ways to be useful. But an ominous sign greeted them on their arrival in the county seat. Naturally suspicious in the wartime atmosphere, the townspeople locked the gates in front of the students and demanded an introduction letter from the government as the price of admittance. Having finally talked their way in and settled in town, the team began going to outlying villages to do labor or propaganda work, which they called "going down to the country" (*xia xiang*). They painted resistance slogans all over while facing scowling disapproval from the town fathers and bureaucrats, who wanted nothing to do with these troublesome young men (who were rudderless *and* political). How could they win the trust, gratitude, and allegiance of the townspeople? Meng and his comrades found the solution in news. The radio expansion of the 1930s had reached even this out-of-the-way place, and now it came to play a role in the war effort. A shop in the little town had installed a crystal radio receiver, complete with a large antenna on the roof. Using headphones, one could listen to broadcasts from Nationalist radio stations. After some discussion, the shop owners agreed to let the General Mobilization Squadron, as the students now called themselves, send team members to listen to the news that night. "Every evening two of our [classmates] went to listen to the news and important information about the resistance. They barely missed a single word transcribing the broadcasts. The next day at 8 a.m., the news was plastered all over town. Each poster was surrounded by many people reading. They were very well received, so the students worked even harder."[50] Radio reports had succeeded even where laboring in strangers' fields had failed. News was a most welcome commodity in uncertain times.

In addition to the small handwritten productions of Meng and Sun, larger-scale professional operations also survived in the margins. A Chongqing news and propaganda worker, Liu Canqi, reported a conversation with a newspaper editor from the Guangxi borderlands:

> His paper had a circulation of 7,000 copies of the same size as any big paper, of which 6,000 copies were daily "smuggled" to Kwangchowan [Guangzhouwan], Hongkong, and Canton by organized Chinese underground workers. It was sold in these enemy-occupied places to Chinese people at a higher cost than fixed prices, showing the hunger for Free China news there. He asked me to send him more material and if possible telegrams, as his paper could only copy CNA [China News Agency] broadcasts, and, being 12 days' distance away from either Kukong and Kweilin, it has no [other] opportunity to get any fresh news from Chungking and the outside world.[51]

Both this editor and the people of Guangdong wanted more and better news. Because of this desire, the news posters put up by Meng Yiqi, the handwritten bills copied by Sun Puyi's town, and small broadcast report-based papers like Jie Ru's had thousands of analogues throughout China.[52] One only needed a radio—even a crystal set costing just a few yuan would do—some paper, and ink. The scale and geographical dispersion of these operations made them broadcasting's most pervasive manifestation in the war years and the most important factor in the maintenance of Nationalist influence over the newsscape.

Underground Listening

On July 9, 1943, American and British forces from North Africa landed near Licata, Sicily, the first Allied breach of Nazi-occupied Europe. The news, received instantaneously from Allied broadcasts, well before any announcement by local authorities, sent the price of gold skyrocketing to more than four times its prior value on Shanghai exchanges. Chinese stock, real estate, and commodity values also soared as people tried to dispose of their Central Reserve Bank notes—the currency of the Japan-backed puppet government. In Beijing, a doctor named Wang was heartened by the news of the landings, as he later was by Churchill's speech in August from Quebec, during which the prime minister urged the allies to turn up the heat on Japan.[53]

The gyrations in the speculative markets and the report of the Beijing doctor indicate just some of the ways that secret radio listening, which persisted in occupied Chinese cities throughout the course of the war, impacted Chinese listeners and the larger public. In the available Chinese archive, listeners are often silent. But the period of the Pacific War offers a counterpoint. From 1941 on, the American Office of War Information (OWI) branch in Chongqing collected attestations to radio-listening practices in occupied areas. Seeking to increase the efficacy of its own propaganda programs (it broadcast in Chinese from both San Francisco and Chongqing), OWI compiled interviews, letters, and commentary that illuminate listening practices in this period—preserving images of the wartime newsscape in action.[54] A number of letters from listeners in Shanghai, sent by a circuitous route, reached American broadcasters and propaganda workers. In fact, both people and letters made it through the lines of control with surprising regularity. As might be expected, some of those arriving in Nationalist China were refugees. But some, incredibly, seem to have been traveling for business.[55] Both types of travelers provided accounts of economic and political conditions, as well as the news and propaganda environment in major Chinese cities.

A series of letters exchanged with a high schooler from Shanghai named Ye Xianglong constitutes the most extensive surviving correspondence. It begins with a letter of April 1943, written by Ye in response to a broadcast call for feedback from listeners. Ye's family originated in the Western part of

Zhejiang Province; his father and older brothers were merchants. He was in the spring of his senior year and dreamed of going to "Free China" after graduation to pursue a college education.[56] In the event, Ye did not make it to Free China because of financial constraints, but he did send at least six letters over the course of two years describing conditions in occupied Shanghai. He most frequently touched on the economy, broadcasting, propaganda, and morale. Education was also of particular interest, as he eventually matriculated at St. John's University, where he majored in chemistry. Ye's family must have been relatively comfortable. Nonetheless, he asked for broadcasts that "report the work and life of our students in America, with special emphasis on the ways and means of self-support for the poor students in the colleges. I shall need this guidance for my advanced studies in three years."[57]

After the outbreak of the Pacific War, economic problems were common across Shanghai and occupied China. Inflation caused the standard of living to drop precipitously. Average prices multiplied more than a hundred times. Ye astutely observed that the causes behind this lay in the unlimited issuance of fiat currency (Central Reserve Bank notes) by the government, as well as the scarcity of goods and the monopoly the army enjoyed on their consumption.[58] Morally flexible businessmen, black market profiteers, collaboration officials, and corrupt Japanese army officers continued to lead the good life, patronizing riotous bars and restaurants where their every whim could be indulged. Meanwhile, the homeless and destitute populations seemed to increase tenfold. What had passed for the middle class all but disappeared. Japanese civil and military administrators seemed unable or unwilling to cope with the mounting supply problems. For much of the population, even access to food became tenuous. The "co-prosperity sphere" became "a sarcastic phrase which the Chinese use to annoy local Japanese officials."[59] After 1943, electricity was available only two hours per day, usually from 8 p.m. to 10 p.m., conveniently also the hours of the main evening broadcast of the Japanese radio network. Newsprint was still available, but paper and ink had deteriorated greatly in both quality and quantity. All metal from statues, railings, fences, and other public adornments had disappeared into the maw of periodic scrap drives.[60] No wonder Ye earnestly desired that the American broadcasts "tell us about the home front situation in the United States and England."[61] Were things as bad on the other side of the world?

The constricted and restrictive conditions that existed in the rest of the consumer market also applied to radio receivers. The Japanese banned and confiscated shortwaves and radios with more than five tubes—that is, those which could receive broadcasts from Allied stations. The more powerful sets were available on the black market, however, where wealthy Chinese and Japanese businessmen were frequent customers. This profiteering class benefited most directly from radio news, as they could take advantage of developments in the war to play the market. The exchanges thus became an epicenter of illicit information.[62] Such was the concentration of rumor flowing from the markets that the Japanese began "paying much attention to the brokers in order to curb such news."[63]

The high prices paid by such people on the black market fueled demand for stolen radios, both legal and illegal models. Dozens of case files from the Beijing courts testify to the frequency of radio theft during the Japanese occupation. The radio set was often the most expensive item in a middle-class home, excepting cash and jewelry. They proved a tempting target for thieves. One instance occurred in late July 1939 when a poverty-stricken, thirty-eight-year-old man from Hebei named Liu Yu climbed over a wall and into a courtyard home. Seeing that no one was there, he entered the rooms. A radio lay on the table. He grabbed it and took it to a pawn shop on the western side of the city, receiving only eight yuan for the stolen merchandise.[64] This was a fraction of the price the pawnbroker could have sold it for. In other cases where pawnbrokers or radio dealers were swindled (for instance, through the passing of a bad check), they variously quote the value of the sets at 130 to 180 yuan.[65] Thieves held no leverage in price negotiations. Obviously, the sale of illegal radios does not appear in court records; no one would have gone to the police to complain of such a theft. Curiously, though, Beijing holds no record of prosecution for illegal radio ownership, though whether this is an omission of the archive or evidence of absence is indeterminable.

Of course, the outrageous prices on the black market were out of reach for most Chinese. Where, then, did one obtain a radio set, especially an illegal one that could receive shortwave broadcasts from Chongqing, Delhi, or San Francisco? Under such tough economic and legal conditions, many Chinese built their own radios from commonly available parts, or else they converted old equipment to their purpose. For instance, a Shanghai resident wrote in February 1944 that "a year and a half ago, I paid $70.00 to buy a broken receiver set from a junk shop. It was later repaired and equipped with a short-wave apparatus. So after my class hours, I often listen to [the American broadcasts]."[66] Our friend Ye wrote that he built his own radio (with four tubes) after school in his spare time. He could thus listen every evening between 8 and 10 except when the weather was bad. "To listen to International shortwave is against the 'LAW.' But I don't care about the danger of being caught, because I want to receive correct news reports. Moreover, I do not only listen to it myself, I also report its contents to my friends, schoolmates and relatives," he claimed.[67]

These homemade shortwave sets could not be registered and were, therefore, quite dangerous to own. If the illegal set was not homemade, the owners could occasionally rely on a friendly (or bribed) repairman to service it and not report them. Still, others solved the problem of secrecy with converters, which could be hooked up to legal medium-wave sets and then safely stowed away when not in use.[68] Officials in the puppet government could legally retain shortwave radios and, like the Japanese soldiers, often turned to foreign news after hours. The doctor in Beijing knew of one police officer who kept such a set in his office.[69] Yet everyone, regardless of their station in life, had to listen in secret. Ye writes that "to listen to shortwave radio in Shanghai one has to risk military persecution and court-martial. When I listen to your broadcast, I always hide myself in a remote studio during the depth of night."[70] An account from Harbin, in the puppet state of Manchukuo, describes "ten people in a room, door and windows tightly shut, taking turns listening. It is the only way to receive news from the motherland."[71]

The amount of illicit information one received (directly or indirectly) through radio listening varied according to class. The average person could not listen to Nationalist broadcasting in the streets through public loudspeakers as they had before the occupation. Compounding the situation, evidence of the circulation of underground newspapers in urban areas declined over the course of the war.[72] Most of the informants who testified to listening directly to Chongqing and Allied news came from an intellectual or business background. A Peking University student who arrived in Chongqing in 1944 claimed that the most common source of news "was not from newspapers or Japanese-controlled broadcasts. Many people still possessed short-wave radio sets, and they listened to stations outside enemy control."[73] One can presume, however, that he is referring to students and intellectuals, not necessarily society at large. "Most people" did not have access to shortwave radio sets. In one of the few glimpses of the working-class newsscape, a restaurant worker reported that "the world news is obscure to the people of Shanghai." For instance, he had heard that there were naval battles going on around the Philippines but did not hear that there were actual landings on Luzon until entering Nationalist-held territory.[74]

Which Allied broadcast stations could people with access to radio surreptitiously listen to? Reports compiled by OWI on listen-in occupied areas show variable and contradictory answers. Two Chinese men interviewed in early 1944 had worked for several years at the American Globe Wireless Company in Shanghai and so were well placed to report on radio conditions there. They indicated that Chongqing's medium-wave station XGOA could be "heard fairly regularly from 11 p.m. (Tokyo time) onwards." This suggests that owners of legal medium-wave radios were able to hear Chongqing at times when it was not being jammed (as it was during certain critical campaigns). It is possible, however, that these men listened on an illegally souped-up radio, and the average person could not tune in.[75] Ye Xianglong wrote that he could receive San Francisco more easily than Chongqing. "The Chinese Central Radio Station is intercepted and disturbed day and night. Never once have we received any news from them."[76] But in a letter some three months later, Ye reported that the only Allied stations he could receive were San Francisco and Chongqing's shortwave station XGOY. "All the rest of the stations are completely Japanese controlled or those equally as bad satellite stations."[77] Another Shanghai informant reported that he could only receive Chongqing, San Francisco, and Moscow on his shortwave, "although I can receive many Japanese stations such as Radio Tokyo, Peiping, Hongkong, etc." One of the few areas of agreement among the informants was that they could receive far more Japanese stations than Allied. Such contradictions are perhaps natural, as reception varies according to the power of the receiver, geography, hour of the day, and even weather. On stormy days, Ye had trouble receiving his usual stations, and reception everywhere was always better at night.

One station that Shanghai residents had no trouble hearing was the

Soviet broadcast station XRVN. In one of the stranger juxtapositions of wartime China, the USSR never closed its radio station in the International Concession. As the Soviets and Japanese lived in a cold detente until August 1945, there was little excuse to shut the station down. It only discussed the European front in its broadcasts but otherwise transmitted its TASS news reports freely. "Its report on Soviet-German warfare goes into the last details," Ye wrote to the Americans. "Its electric current is very powerful. Anybody with an ordinary radio in the [Jiangsu] and [Zhejiang] area can receive its broadcasts."[78]

The fact that most OWI informants came from Shanghai gives us a somewhat distorted view of listening nationwide. Reception of Chongqing's broadcasts became easier as one went west. Moreover, listeners in Shanghai were surrounded by powerful stations—Shanghai's own municipal station, the puppet government station in Nanjing, and several nearby Japanese-run provincial stations blotted out more-distant broadcasters. Regions outside of Shanghai and the lower Yangtze River region did not have to confront these challenges. The situation seems to have been much better in Hong Kong and the South more generally. A 1944 report tells us that despite a strict prohibition on listening to Allied radio, Hong Kongers still received broadcasts "from Changsha, Sao-Kwan, Chungking, Chengdu, Kunming, and Kweilin. The broadcasts from New Delhi and the Chinese rebroadcasts of the San Francisco station programs are very clear. So important and accurate news from both China and abroad can still be received in Hongkong [sic]. When the people receive victory news from various fronts, they relay it from one person to another, which makes the news even more encouraging and exciting. What can our poor enemy do to stop us!"[79] In Hong Kong, the Japanese efforts to forbid illicit listening had mirrored other parts of the country. They had registered all receivers, confiscated any set with more than five tubes, told people they could only listen to the local station, and threatened severe punishment for clandestine listening. It did not work.

What did listeners desire to hear? Many, like Ye Xianglong, were desperate for news programming and information about the war effort—just like the rural audiences encountered by the radio newspapermen. Ye liked American music—"very nice and easy going," he said—but did not find it martial enough: "At a period when we are struggling desperately for victory, it will be much more effective to release a few patriotic stirring marching songs."[80] Lucky for him, Chongqing only played patriotic songs when it did broadcast music. He had quite a few other suggestions for effective broadcasting, mostly about increasing war coverage: "People in the conquered territory in China like myself regret that your news is too brief and insufficient. They do not satisfy our demand, especially the news concerning Free China and the Pacific Front."[81] Chongqing broadcast all sorts of specialized content—programs for women, activists, children, and students. But these were of little use when most people did not—could not—spend all day listening. Regardless of the danger from the Japanese or collaborators, the lack of electricity and batteries precluded tuning in at leisure. Concise, repeated newscasts with headlines, war updates, and the most important speeches and lectures were, therefore, more important.

The effects of propaganda are notoriously difficult to measure. The British, through the BBC, kept tabs on domestic morale through polling and sample interviews. So did the Japanese and German national broadcasters. There is no evidence that Chinese Nationalists engaged in any similar sort of program. The letters sent to Central Broadcasting by listeners like Ye Xianglong would provide an informative substitute. If these letters still exist, they are locked away in Nanjing, unavailable to researchers. In the absence of opinion surveys and what must have been a voluminous correspondence to Chinese radio stations, we are left with only a few of the words from the people who experienced wartime radio. A well-connected Chinese journalist who left Shanghai for Chongqing in the early spring of 1944 left one of the fullest descriptions of the importance of broadcasting to occupied populations. He summarized the effect of radio:

> The fact that Allied sentiment is strong can be revealed by the constant spreading of reports and rumors favorable to the Allies by various people in the city, who take a great risk in doing so as the city is chock-full of petty informers in the pay of the Gendarmerie [Kempeitai], the navy, the police and the Nanking Regime. Apart from KGEI (and KWID) a fair number of people listen in to Chungking and New Delhi by long wave, and it is in this way that news comes to Shanghai.
>
> These Allied news reports are important. They serve to bolster the hopes of a lot of people whose daily life is otherwise tormented by terri-

ble economic conditions, the activities of Japanese spies and the general strain of uncertainty and desire to see the war end as soon as possible. Unless the people can expect some action of a military nature they have to satisfy themselves with whatever news they can get of Allied action against German cities, but they are elated by the American successes in the South Pacific.[82]

The theme of hope and consolation was echoed in a more personal manner by Ye Xianglong in a letter sent to the San Francisco Chinese language radio station KGEI in May 1943:

> When I listen to your commentaries on the short-wave radio, I feel much consoled—the kind of consolation to my desolate and withered spirit, just as if I were placed on the other shore of the Pacific in your honorable country.... I have to continue to seek consolation by listening to the reports of your honorable station every day, except days when bad weather prevents my doing so. Although, in Shanghai, the movement of youth is watched within the spread of the dangerous Japanese police network, I still listen to your news every day and report to my comrades, good friends and relatives to reveal the lies of the Axis news and awake my countrymen. It is quite true that after the victories in North Africa, our people were awakened from their nightmares. They seem to feel that light would soon fall on the darkness of Shanghai.[83]

Radio connected an occupied people with a remote, isolated, nearly defeated—yet still defiant—rump state. Eight years, an extraordinarily long time for a modern total war, passed with little evidence of Chinese military progress. A grand counteroffensive never came. What countered despair? News from Chongqing, and allies fighting abroad, reminded the Chinese people of what had been lost. Most did not need much convincing that that was worth fighting for. Though the Nationalists had a very effective propaganda machine, especially in the early years of the war, the Chinese didn't *need* especially good propaganda. Japanese behavior supplied that in spades. The "Free Chinese" just needed to exist; radio proved their continued existence and continued struggle. It offered a glimpse into an alternate reality where the unthinkable hadn't happened, where the nightmare of occupation was not real—or less real, at any rate. It opened up a world where fighting back was possible and sometimes even fruitful. In this way, radio broadcasts did not have to inspire people to take up arms to undermine the monolithic fact of the Japanese occupation. It blurred and undermined the regime through the transformative act of its being.

Herein lay a fundamental asymmetry between the Japanese and Chinese camps. The Chinese just had to survive. The Japanese needed to have superb propaganda, an ideological program for the ages. They believed they had such a program, and they built themselves the communication infrastructure through which to propagate it.

Collaboration

In 1950, the communist radio station in Dalian sent a recently rediscovered packet of materials to Beijing. The broadcasters had not known what to do with it. The cache was dangerous—and beautiful in its way. It was a collection of Japanese-period propaganda: bright water-colored postcards, wondrous poster illustrations, and a children's book, all encouraging radio listening. A patriotic woman listens to her set; a family safely flees a burning city. A young girl daydreams of the character she hears on the radio; a mother and child sit in a blacked-out room listening to news reports. Happy children eat, exercise, and sing along; a soldier goes to war. The children's book closes with a banner headline, in bold modernist font, "Without Radio There Is No Nation."[84]

The materials were produced by the Manchurian Telegraph and Telephone Company (MTTC), which controlled broadcasting in Manchuria. But the Japanese and their collaborators carried out similar campaigns to popularize radio listening in China proper. Occupation governments expanded the broadcast network, encouraged radio listening, and increased ownership through "suasion" campaigns and the sale of inexpensive, Japanese-made receivers. These receivers had a limited range of use; theoretically, they could only pick up nearby medium-wave stations, whose broadcasts travel shorter distances than shortwave transmissions. Thus, these mass-produced radios generally received only local broadcasts from Japanese-controlled stations (though, as we have seen, Chongqing's transmissions sometimes got through). The suppression of shortwave radio sets and the communal nature of most listening—tuning in to an illegal broadcast in front of a crowd in a society rife with informants could be suicidally foolish—intensified Axis domination of the airwaves. In short, collaboration governments rapidly and successfully promoted the infrastructures and practices of news.

As the Japanese army passed south, then west, into the urban heartland of China, they took charge of radio stations. Sometimes the departing Chinese had removed, sabotaged, or destroyed the facilities, as in Nanjing, but sometimes, as in Shanghai and Beijing, former Nationalist broadcasting stations passed into enemy hands intact. Still, the Japanese were not satisfied with their inheritance. They set about organizing two new broadcast networks in addition to the one they had already built in Manchuria. The first, the North China Broadcast Association (*Huabei guangbo xiehui*), based in Beijing, controlled Tianjin, Jinan, Taiyuan, and other broadcast stations in North China.[85] The second, based in Nanjing, purported to be the successor to the Nationalist broadcasting system and called itself China Central Broadcasting. It oversaw Shanghai, Wuhan, Suzhou, Hangzhou, and a number of smaller stations in central China. Though ostensibly independent, the radio networks in Manchuria, North China, and Central China, in fact, worked closely with NHK (*Nippon Hōsō Kyōkai*: Japan Broadcasting Corporation) officials to coordinate broadcasting and programming.

NHK conformed to a European model: a monopolistic corporation with a hierarchical network of broadcasting stations publicly funded and closely aligned with the government. To complement the broadcasting network, the Japanese government carried out campaigns to popularize radio, which I will describe in more detail later. In 1928, there were half a million radios registered in Japan.[86] In 1937, there were three million. By the height of the war in 1943, even that number had more than doubled to seven million.[87] Japanese "advisers" transplanted this model to Manchuria after their conquest of the region. On September 1, 1934, the Manchukuo government founded the MTTC to oversee telecommunications and broadcasting. All the existing stations—Dairen, Mukden, Changchun, and Harbin—were placed within the new administrative structure, and campaigns were undertaken to increase listenership. The number of (registered) radio receivers in Manchuria at the time of the 1931 occupation was estimated at a paltry six thousand. In 1945, there were roughly seven hundred thousand.[88] Thus, MTTC became invaluable to the state-building efforts in Manchuria. Indeed, Janice Mimura has argued that "through [MTTC's] control of radio broadcasting, it became the state's main instrument for mobilizing support abroad and in Manchukuo."[89]

After the successful military occupation of North China, the Japanese army handed control over broadcasting to NHK, which deftly exploited the stations in conjunction with MTTC in Manchuria. The Dalian station broadcast news live from the frontlines to Japanese residents in China. The freshly acquired Tianjin station turned its frequencies to jamming incoming broadcasts from the Chinese capital, Nanjing.[90] The Japanese national broadcaster sent radio experts to the former capital of Beijing to build a high-powered station there. Opening in January 1938, it had an initial power of 50 kw—five times more powerful than the Nationalist transmitter concurrently operating in Wuhan. In May 1940, new equipment doubled the power to 100 kw, making it the most powerful broadcasting station in China and the second most powerful in Asia, after Bangkok.[91] This Japanese-built station would eventually broadcast Mao Zedong's announcement of the founding of the People's Republic. In addition to large-scale infrastructure projects like Beijing, the Japanese established numerous local medium-wave stations. Sometimes, as in Xiamen, they converted old wireless telegraph stations to broadcast and installed street-corner loudspeakers.[92] Other times, they constructed brand-new stations, as in Jinan, capital of Shandong Province.[93]

Japan's East Asian radio empire was administered out of NHK headquarters in Tokyo by the East Asia Broadcasting Council from April 1939 until 1943. The council, which, despite its name, consisted almost entirely of Japanese officials, governed "programming policies but also facility planning, adjustments of frequency allocations, counter-jamming measures, and procurement of materials for receiving sets."[94] It had first met in 1939 in Changchun. Present at the creation were representatives of Japan's radio empire: NHK, MTTC, and Korean and Taiwanese broadcast stations. The North and Central China networks joined some months later. At the inaugural meeting, the council issued a series of recommendations urging networks to "bolster our guiding spirits of development on the Asian continent and to complete the fusion of Japanese, Manchurian, and Chinese cultures, by coordinating broadcasting policies in East Asia with our national policy."[95]

Putting this pan-Asian philosophy into practice, the East Asia Broadcasting Council organized "exchange" or "relay" broadcasting so that programming from one station could be rebroadcast from another or, indeed, throughout the empire. The flow of programming was not unidirectional, from Tokyo outward. Rather, relays frequently included broadcasts from China and Manchuria to the Japanese home islands or between client states.[96]

All Japanese-controlled radio stations became a tightly integrated network. By 1944, a listener in the southeastern province of Fujian could pick up five stations under the North China Broadcasting Association, which almost always carried the same programs in its distinctive northern dialect. "The network is if anything more tightly unified than the Japanese one," wrote an OWI official. "Occasionally when Peiping has an especially good program (a fine ch'in [zither] recital, for instance) it is rebroadcast by several stations not ordinarily in the network, including Tientsin, Nanking, Hankow, and Canton. In a way, of course, this is a greater feat of organization than that of constant network operation, since it must demand special arrangements each time."[97] The links between Manchuria's MTTC and the North China Broadcasting Association were especially strong, owing to geographical and linguistic proximity, as well as personnel overlaps between the respective corporations. The massive Beijing station transmitted Chinese opera to the waiting ears of northeastern Chinese. The flow of programming in the opposite direction, from Manchuria to North China, mostly consisted of propaganda, "inform[ing] the people of the sound development" of Manchuria under Japanese control.[98]

The Beijing station stood out for its professionalism and influence within this broadcasting network. It relayed general programming to the rest of the China networks, including, according to one listener, "the best Peiping opera, and the best Chinese instrumental music that can be heard on the Axis radio." In terms of content, he reported, Beijing "sticks almost exclusively to native Chinese entertainment, though it carries Japanese newscasts and occasionally lets some Western classical music in. It devotes roughly 80% of its time to music, 20% to news and commentaries."[99] Local news programs were "confined to the report of local social and economic activities, while the general news concerned not only current events of the world but also political activities in the four provinces of North China—Hebei, Shandong, Henan, and Shanxi. Military activities, especially punitive campaigns in the four provinces, were included in the general news." The commentary broadcasts included minute treatment of rationing and blackout regulations, among other wartime admonitions. It even sometimes carried stories about local corruption in the rationing system—presumably to deflect blame for food shortages.[100] In addition, it relayed special programs for children, women, and young people to stations throughout China. There were Japanese lessons, both beginner and advanced. Unlike in Central China and the South, Beijing gave newscasts in Mandarin only.[101]

Radio in central China, under the Reformed Government of Wang Jingwei, never became as tightly managed as in the North.[102] "Except for a few items such as general news, every individual station runs a separate program of its own," recorded a listener.[103] Operating from March 1941 to the end of the war, a new "Chinese" national broadcaster in the old capital was, in fact, managed by a team of MTTC and NHK veterans under the ever-watchful eye of the military. Fittingly, this Nanjing station relayed a substantial amount of programming from Tokyo.[104] Shanghai, as before, remained substantially commercial. Private radio stations were allowed to crowd the airwaves of the city, though with strict supervision over content.

Despite the disjunctures in Central China, the collaborationist broadcasting stations, as a whole, were far more efficient and timely with breaking news than was Chongqing. By the time news of a battle or conference had been processed by the Nationalists censors, the story would have broken over Japanese-affiliate airwaves hours or even days before. This could make it seem as though Chongqing were the one spinning or rebutting Tokyo's honest news story. One of the most egregious examples comes from the Cairo Conference, held in November 1943. The meeting between Chiang Kai-shek, Roosevelt, and Churchill resulted in some of the most momentous news in Chinese history: the formal abolition of the extraterritoriality, foreign concessions, and the unequal treaties. It was the *de juris* end of what is called today "China's Century of Humiliation." It should have been a propaganda coup for Chongqing. Instead, Tokyo scooped Chongqing, which did not make a formal disclosure of the pact for twelve crucial hours, during which it missed both morning and afternoon newspapers.[105] Though people in Chongqing may have been unaware of the news, that was not necessarily the case farther east, where Japanese propagandists preempted the news with their own spin. A report from Fujian shows just how Chongqing's generally slow transmission of news hurt the Chinese cause on the ground:

> There is still tremendous Japanese influence and much Japanese propaganda throughout [Fujian], but particularly in the coastal areas. There are constant rumors of United Nations defeats, stories that Russia and Germany are about to make a separate peace, that American and British soldiers are fighting, that the 14th Airforce is bombing Chinese coastal villages. I have already noted in other reports that the small coastal newspapers help pass on this type of stuff by reprinting rumors they pick up from word of mouth or over the air. Several local papers seem to rely on Japanese Mandarin news broadcasts for their news. Some of the larger papers with radio facilities will also use Axis materials largely because Central News Agency is so slow in getting the news on big stories to them. I was in [Fuqing] when the Cairo Conference was first announced and it was 48 hours before the local Chinese press got the news from Central. This of course facilitates the dissemination of Japanese propaganda.[106]

Many city newspapers in gray-area occupation zones still operated, as the one in Fuqing did, as businesses. To compete, they needed to turn out the latest news, not two- or three-day-old stories approved and broadcast by the Nationalist's news agency. The front pages, therefore, would be filled with stories sanctioned and rinsed by Tokyo. Their readers, presumably, did not care, as long as there was news. Indeed, throughout China, Nationalist-friendly newspapers and even party-sponsored news boards used radio broadcasts from Japanese stations as sources.[107] In addition to being slow, Chongqing's broadcasts contained confused and contradictory reporting. Though such problems would perhaps have been a benefit to the collaboration broadcasting regimes, Chongqing had no such luxury. OWI employee Liu Tsan-ch'i reported one such instance involving a dogfight above Hengyang. Despite large numbers of witnesses, news broadcasts and the corresponding newspaper reports repeatedly bungled the story. It took Chongqing four days to settle on the official line that eleven enemy planes had been shot down. "It does not make much difference for people to read a communique several days later of certain air combat that occurred hundreds of miles away, but does dishearten people who actually saw it," wrote Liu.[108]

Expansion of Ownership

Investment in an efficient, wide-reaching broadcasting network could only be one leg of Japan's radio propaganda program. It also had to encourage the growth of an audience to hear its announcements, admonitions, and exhortations. It had to cultivate the behaviors of a mass society. So, during the course of the war, Japanese and collaboration governments made every effort to popularize radio listening. Rather than eliminating all receivers, they forbid only shortwave sets capable of listening to Chongqing, San Francisco, or Delhi. Rather than restrict ownership to a select class of Chinese worthies, they flooded the market with inexpensive, mass-produced radio sets. Rather than suppress group listening, occupation governments offered prizes to institutions and social groups that installed sets for their members. They spared no effort to achieve a ubiquitous radio propaganda network—in other words, a highly developed newsscape to help them achieve their mass media and mass politics. Warfare, the Japanese recognized, required an advanced technopolitical process.

In carrying out a wartime campaign to universalize radio, collaboration governments followed the example of Japan itself. There, radio registration had more than doubled over the six years from 1937 to 1943, from three million units to seven million units.[109] In China, the outbreak of hostilities in 1931 and 1937 had caused runs on radios as people scrambled to receive the latest information. Interest and concern with the war drove parallel increases in Japan.[110] The NHK's official history recounts that people rushed to purchase radios or have sets repaired, "so great was the desire to listen to news." Indeed, Japan had seen war accelerate the technopolitical process several times before. The Sino-Japanese War, the Russo-Japanese War, and World War I all drove an explosion in print media. But this time, "the auditory sense of the entire nation concentrated on the radio, and it was the general tendency to confirm what was heard on the radio with newspaper articles."[111]

Even more than before the war, the Japanese government emphasized the close ties between broadcasting and the state. In the home islands, posters with variations of the phrase "National Defense and Radio" appeared in public places like railway stations, post offices, and city, town, and village offices. The campaign for radio listening included exhibitions, films, pamphlets, and "radio dissemination organizations" in rural areas. Following the example of Nazi Germany's *volksemfanger* mass-market radio, Japan standardized receiver models to bring down production costs. Manufacturers could essentially make only one type of radio, all of which had interchangeable parts regardless of the brand.[112]

These same mass-produced receivers were exported to occupied China, where collaboration governments, under Japanese instruction, carried out campaigns to persuade people to purchase, install, and listen to them. These concerted attempts to advance the technopolitical process aimed to mobilize the entire population for civil defense, especially against air raids, to spread anti-Communist propaganda, and increase the efficiency of government. In form, they mirrored the "moral suasion" campaigns carried out in Japan since the early years of the twentieth century, especially in that the efforts at radio distribution were channeled through local institutions and organizations.[113] The 1942 radio popularization campaign (*puji shouyinji yundong*) in Beijing, for instance, enlisted local associations to persuade residents to invest in a radio. The municipal police department ordered neighborhood police stations to "direct every institution under your jurisdiction to persuade city residents to quickly install radios, to improve the effectiveness of air defenses and the reception of propaganda."[114] Like in the successful Japanese suasion campaigns, the radio-popularization movement explicitly relied on encouragement rather than force. As the same police order from 1942 claims, "This investigation has no intention to use compulsion to persuade people. . . . It is feared that the implementers will not know the truth and act in a way . . . that disturbs the people. Attention should also be paid to the co-operation of copying and distributing radio brochures . . . to assist in persuading them to comply with this order."[115]

Evidently, though, the 1942 campaign did not meet with total success. As a result, at least one further campaign began the following year, its preamble clearly laying out the stakes involved in advancing the dialectic of mass communications and mass politics:

> In order to strengthen air defense power and to deal with other urgent matters during the Japanese Upright Holy War for East Asia, we are now working hard to complete improvements to the broadcasting equipment, but the work of popularizing radios does not seem to have been able to advance alongside. If there is a situation, this will affect the communication of orders and the unity of the people's thoughts. It is not inconsequential; it is the mission of completing our air defenses, and protecting the lives and property of the people.[116]

This popularization campaign, carried out over November and December 1943, had the goal of reaching at least one hundred thousand listening households, about 30 percent of Beijing. At the time of the campaign order's promulgation, there were seventy-four thousand households with registered radios in the municipality.[117] (Crystal receivers were exempt from registration, so the actual number of radios was substantially higher.) During the campaign, the police department reached out to ensure that school, office, and neighborhood associations tuned in regularly. Additionally, police were instructed to investigate the situation of every household without radio equipment and to persuade families lacking a receiver to install one. Any household desiring (and presumably able to afford) a radio could have one provided by the North China Broadcasting Association, which would even provide free installation. To incentivize efficiency, the Broadcasting Association offered to "give a small reward through the Police Department's relevant division to each police branch that has success in distributing radios and convincing listening households who have not registered to do so according to regulations."[118]

The scale of these campaigns seems to have been smaller under Central China's Wang Jingwei regime, although the lower Yangtze region around Shanghai started from a far higher base level of radio ownership. Wang Jingwei himself encouraged the sale of standardized Japanese-made radios "at the lowest prices ... in order to spread (*puji*) broadcast propaganda." Distributed at cost by the government, these three- and four-tube receivers were further discounted for early purchasers.[119] In addition to subsidizing the sale of radios to the public, collaboration governments in Beijing and Shanghai also repaired broken sets for "the cheapest possible charge" in order to guarantee the continuation of reception.[120] The repair of old radios became especially urgent from 1943 on, as new radio parts and supplies became rarer. The bombing of Japanese radio factories, general materiel shortages, and the American blockade of shipping lanes meant that almost no new radios appeared in China in the last two years of the war. Thus the Beijing station advised that since tubes "are difficult to obtain because of the times," radios should be turned off when there are no broadcasts. But since a turned-off radio cannot receive warning of an air raid, the neighborhood should plan to use their radios alternately so that one is always listening. Do not set your receivers in a high place, the broadcaster warned, lest they be shaken down by explosions and broken. In this way, "one of the most valuable articles in wartime" could be preserved as long as possible.[121]

Despite the later challenges, the subsidized imports from Japan and coordinated radio-popularization campaigns managed to multiply the number of radio receivers in occupied areas over the course of the war. The number of radios in Taiyuan grew tenfold, from around three hundred to more than three thousand—three- and four-tube standard Japanese radios made up the majority of the increase.[122] In Jinan, Shandong, a radio purchaser could pay for a radio up-front at the station but have the cost reimbursed in installments after the set was officially registered. Radios were given for free to schools, marketplaces, and movie theaters. By 1945, there were more than seventy-three hundred registered sets in Jinan, in addition to some three thousand unregistered crystal and direct-current sets. Of registered radios, 80 percent were located in stores and thus available for public listening.[123] The city of Jinan, just a provincial capital, had almost as many radios as all of Nationalist-held China combined.

In short, the Japanese-collaborationist push to popularize radio reshaped the listening infrastructure of the nation, bringing radio ownership to unprecedented heights. At war's end, the majority of Beijing's many thousands of radios were restricted-band Japanese models; the city's hundreds of thousands of average radio listeners had, for years, only been able to hear domestic—meaning collaborationist—stations.[124] What were they listening to?

Japanese Propaganda

Japan's propaganda strategy involved two pillars: ideology and misinformation. The ideological aspect of Japanese propaganda during the Second World War is well known.[125] It involved ideas like the Greater East Asia Coprosperity Sphere, which argued that Japan's empire was not an empire at all but a kind of economic union. Japan was fighting the corruption of the Nationalists, the disunion of the Warlords, and the threat of Communism. In a similar vein, imperial propaganda argued that its forces were battling Western imperialism, especially Britain and the United States, not conquering an empire for Tokyo. These Japanese ideologies had some purchase among the Chinese public; certainly, anti-Western propaganda could draw on deep wells of anti-British feelings in China.[126] But the Chinese as a whole were not persuaded. Many could see through the empty Japanese slogans, especially as the war progressed and the promised peace, prosperity, and independence never materialized. Therefore, the more important aspect of Japanese wartime propaganda in China was misinformation. The propagandists at the broadcasting houses of Beijing, Shanghai, and Nanjing became expert at spin, at misdirection, at having just enough truth in every lie to make it believable. They tried to make the Chinese listeners confused, apathetic, and resigned. To a significant degree, they succeeded.

This nihilistic strategy was made much easier by the fact that Japan dominated the airwaves in occupied areas. It was simply much easier for the Chinese citizenry to hear them. For those for whom radio was the primary source of news (either directly or indirectly through local newspapers), Japanese propaganda became impossible to tune out. People would certainly not stop listening or reading. The impulse to seek news was too strong. In 1938, the Zhejiang Provincial Department of Education wrote in a report:

> Since the start of the War of Resistance every provincial or privately established station has been affected by the war; they have either stopped broadcasting or reduced power. Listeners in every locale suddenly feel the difficulty of the situation, and the powerful stations of the traitors and enemy devils are all around. These stations on the one hand wantonly interfere with our side's broadcasts, and on the other hand confuse

the average person when they hear the counter-propaganda. Because our broadcast station is difficult to hear, one still must turn to the enemy transmissions as a means of disproving them, and subsequently spread falsehoods and rumors.[127]

The "falsehoods and rumors" produced by listening to Japanese radio were quite intentional. Broadcasts and newspaper articles used cherry-picked rather than false statistics and misleading rather than fantastical reporting. A Shanghai newspaperman, who recently escaped to Chongqing, reported in the spring of 1944 that "Japanese and especially German propaganda seems to have become more skillful in Shanghai. Quite often Allied news appears in the local papers in doctored form, and manages to sound convincing except to the initiated who know that they are stories of defeat."[128] Ye Xianglong discussed the practice of covering up the truth with half-truths, saying that Japanese-approved media quoted Allied communiques but only reported enemy losses.[129] Similarly, the Japanese radio played up vague but legitimate-sounding reports of disagreements between the Soviet Union and its allies, as well as strongly pro-German propaganda. This tact was meant to reassure the Chinese that the USSR was not much of a threat to the Japanese imperial project. As long as Russia still existed, the Greater East Asia project would be imperiled. "A weak Russia, pulverized by German might, could hardly be expected to disrupt GEA plans in the areas of occupied China," in the words of American intelligence.[130]

In some instances, Japanese puppet radio issued wholly falsified speeches by prominent Chinese leaders. The fake communications by Chiang and Mao, as well as well-known Nationalist and Communist generals, misdirected and confused the audience.[131] More commonly, the Japanese tried to split the Chinese by rebroadcasting unadulterated Communist propaganda, especially in areas like Beijing, where Communist influence was strong. They particularly enjoyed relaying CCP attacks on Chiang Kai-shek, the Guomindang bureaucracy, and "anti-democratic elements among the Western powers."[132] The imperial Japanese broadcasting network transmitted these messages from the Chinese Communist Party, without comment, for the last two and a half years of the war.

Chinese sources often claim that they do not believe a word of Japanese propaganda but know many other people who do. Negativity about the war prospects was pervasive, and understandably so, after years of momentous losses. For instance, in a 1939 report on underground propaganda work in Beijing, an activist claims that Japanese propaganda "cannot have the courage of conviction, and therefore has already lost." If it has already lost on arrival, why do the Chinese still need to put in so much effort to squash the enemy narrative? "Our secret newspapers in Beijing and Tianjin have developed quickly," he continues, "and never fail to shatter the false propaganda and ridiculous theories of the enemy newspapers."[133] It is doubtful whether the clever Japanese news reports could be "shattered," especially given the continued Japanese success in this period of the war. Ye Xianglong also indicated the strength of Japanese propaganda by complaining of the degree to which it influenced his friends and family: "Japanese propaganda in China is always very strong. . . . There are also many weaklings under Japanese pressure who often contend obstinately that the Allies will surely be defeated. British and Americans are powerless and China is finished. The reason for uttering such defeatist dejected remarks is simply because they are deceived and confounded by the Japanese and Wang [Jingwei]'s propaganda."[134]

Ye tended to associate susceptibility to Japanese propaganda with class and education, though the fact that he refers to his friends as among the "deceived" belies this assumption. He asserted that "We must understand that their propaganda has much attraction, especially for the 'lesser' intellectual class," and "our brethren in the conquered territories may be fooled, especially those who are semi-intellectual."[135] The young student looked down on those who would be persuaded by Japanese and collaborationist propaganda, but he recognized that there were many such people. Indeed, substantial numbers of occupied Chinese believed the collaboration government worked in their interests. This was one major aim of collaborationist propaganda: to form a sympathetic image of collaboration leadership among the Chinese public. The benefit of this line clearly flowed to the Japanese occupiers; it made people more apathetic toward the situation in their country and less likely to actively oppose puppet rule. According to two Chinese men who left Shanghai for Chongqing in December 1943, the idea of the collaborationist leadership as sympathetic characters was widespread. The men had worked for several years at the American Globe Wireless Co. in Shanghai before transferring to the *Shanghai Evening Post* and *Mercury*. As journalists under the collaboration regime, they were well-positioned to observe public opinion.

> The public on the whole is anti-Japanese, but not in any way actively. Puppet officials "admit" to friends that they sympathize wholly with the Allies and are sure they will win. Even in the newspapers there is little anti-Chungkingism. Generalissimo is referred to as "Mr. Chiang" in papers. On the other hand the public generally sympathizes with Wang Ching-wei, thinks he is a good fellow doing his best to protect them and China's interests. They say it would be much worse without him. . . . Wang is NOT regarded as a traitor at all.[136]

All this is not to imply that Japanese and collaborationist propaganda work was without fault. The Japanese language stations in Manchuria and China often struck a deeply arrogant note. The stations were "often very frank in assuming that the Japanese in their areas are a superior and privileged minority. . . . They sometimes announce rationing regulations that apply strictly to Japanese residents and not to the native population. They sometimes speak slightingly of Chinese reliability, capacity for organizing things, etc. . . . [though] they sometimes rebuke the Japanese officials and civilians in their localities for greediness and poor war spirit."[137] Moreover, the Chinese language broadcasters often put in a half-hearted performance: "The lectures are all given by traitors. At first the collaborationist government sent high level officials to speak, but afterward they had nothing to say, and they all avoided it. Soon, they started to use hired help to read, which they paid 10 yuan for each speech. Now they only pay half wages, 5 yuan." The news announcers were low-rent, but the entertainment was sometimes worse. The networks broadcast rough comedies featuring insulting Chinese characters like Idiot Zhang and Blind Tang, which caused the people to become "disgusted with the enemy broadcasting," according to a report published in a Nationalist area.[138]

More important, the Japanese propaganda program held the unenviable position of having to dissemble and lie. This was its greatest weakness; it could not fool people forever with slogans. Coprosperity became a bitter refrain to cold and starving people. Deprivation clarified foggy senses.

Deprivation drove a Shanghai haberdasher to travel through a warzone to Chongqing bearing bundles of cloth for sale, a journey that took three months and bribes for Communist guerrillas (four times), and Japanese soldiers (twice). Arriving in May 1945, he gave a report on the newsscape in the occupied territories: "War news is always printed on the front page of every newspaper. Even so, nobody will care to read it. They are tired of hearing this nonsense and they believe it no more. They are fond of gossiping over news which their friends have obtained from short-wave radio sets."[139] Japanese propaganda no longer held purchase, and people turned to radio for a sense of how the war would end.

Endgame

By the spring of 1945, Chongqing turned to preparing occupied areas for liberation. Specifically, it warned people in coastal areas to expect the landing of American troops and encouraged the population to aid them. One of these broadcasts came from a prominent citizen of Shanghai. It opened nostalgically:

> I am Wang Hsiao-lai. Those who know me will recognize me by my voice. It is exactly seven years since I left Shanghai in the winter of 1937.... All these long years seemed to me as one day.... Rumors were scattered around soon after the outbreak of the Pacific War. Some people said that I died and others said I was wounded. Well, here I am still alive. Since I left Shanghai, I have been traveling a lot.... Though my hair is all gray now, my steps are as quick and long as ever.... [I] wish to bring you the message that all your friends here are well.[140]

Wang went on to warn the audience to expect the fall of Germany shortly ("the distance between the nearest allied troops and Berlin is equal to the distance from Shanghai to Kuanshan [sic]"). Chinese liberation would also soon be at hand. In the end, the American army never needed to fight a land campaign in China. On August 10, 1945, the Japanese cabinet notified the Swiss minister of its wish to surrender. The China News Agency immediately sent word to the Chongqing radio station. Announcers broadcast news of victory continuously until 5 a.m. the next day. Station employees even forced a Japanese soldier from a nearby prisoner of war camp to announce, through his tears, the surrender over and over again in Japanese. Great crowds gathered outside the radio station, where a loudspeaker had been installed, to listen and celebrate.[141]

The Japanese broadcast networks did not transmit the unofficial indications of surrender, though one young broadcast technician named Tan Baolin risked his life to spread the word. When he heard the news of August 10, Tan switched the transmission feed so that the collaborationist station began relaying Chongqing's announcements of the intended surrender. He ran home and saw friends and neighbors in the streets and shops hurriedly discussing the news. The next day, Japanese militarists plastered notices denouncing the rumors all over the city, but Tan was never arrested.[142]

Wang Jiaqi was not so lucky. He spent the last week of the war in jail. A graduate of Tokyo Normal University's English department, he worked for a trade union in factories that made cotton yarn and socks in Manchuria. On the night of August 1, 1945, two Chinese members of the Public Security Bureau arrested him on charges of corruption and anti-Japanese thought. (In his memoirs, Wang admitted that both were true.) Brought to an underground prison cell, Wang was threatened with torture by a Japanese interrogator. Luckily, a Chinese guard intervened and helped him to his cell. Prisoner and guard became friends of a sort. On August 10, the sympathetic guard whispered to Wang that the Soviets had declared war against Japan. "Look at the faces of the Japanese, they are all panicked. At the police station Japanese men of around forty years old are being drafted as soldiers. And no one has brought your case before the court." The morning of August 15, just after breakfast, the guard returned:

> "The Japanese didn't come to work today," he said. "I heard that there will be a major Japanese broadcast at noon. I will come to you at that time. We will listen in the guard room, and find out what will happen in the end." At 12 o'clock, I turned on the radio in the guard room. The announcer was broadcasting in Japanese. It was the Emperor of Japan reading the imperial rescript of surrender to the Allies. My blood immediately boiled, my body was hot, my heart was about to jump out, and I didn't say a word more. I only nodded to my friend, and I rushed out and went straight to the corridor, cells arrayed along both sides. I

shouted "The Japanese devils have surrendered!" Then the prisoners began shouting as one, knocking on the doors and windows, stomping on the wooden floors, and calling to each other, in a wave of raging anger, that immediately merged into a symphony of joyful celebration.[143]

The emperor of Japan famously broadcast the recorded speech ordering surrender to his entire empire. Unlike the August 10 cabinet decision, the imperial surrender speech reached the breadth of Japan's territory, from Jakarta and Singapore to Harbin and Chongqing. The Japanese had thoroughly developed radio-listening habits and infrastructures across occupied territories, encouraging the conflation of the state and communications. Their success at building broadcasting and radio-listening networks meant that roughly four out of every ten radios in China resided in Manchuria, and fully one-half of Chinese radios were of Japanese manufacture at the close of the war.[144] The Japanese-collaborationist regimes had reshaped the newsscape of China, a legacy for the technopolitical process that was, ironically, constructive rather than destructive. Taking advantage of the insatiable demand for information, the Japanese inculcated widespread changes in behavior that had already begun to coalesce into a mass society, at least in urban areas.

But they would not reap the rewards. In Chongqing, Chiang Kai-shek listened to the surrender speech, said a prayer, and instructed his secretary to get in touch with Mao Zedong.[145] Over the past eight years, the Communist forces he led had been struggling to find their own foothold in the newsscape. After the war, these guerrillas' choices about how to develop the technopolitical process would have consequences for all of China.

SIX

Red News and Red Women, 1937–1949

In September 1937, a sixteen-year-old schoolgirl, soon to adopt the nom de guerre Ding Yilan, left her home in Tianjin's French Concession wearing a satin qipao, embroidered shoes, and black glasses—dressed to appear bourgeois and unsuspicious as she passed Japanese checkpoints on the bridge into the Chinese district. Attempting to escape the city for a second time, she headed to the seaport. The train station had been bombed a few days before, cutting off the possibility of flight by rail. It was the mid-autumn festival, but the crowds of refugees barely noted the holiday as they pushed toward ticket booths and quays. After several delays, she finally boarded the ship arranged for her by the underground and descended into a packed, windowless cabin. "As we headed far out to sea, we sang songs of national salvation, which gave us a great deal of comfort," she remembered.[1]

It would take only two months for her to reach the Communist capital of Yan'an, but it would be several years before Ding became one of the most influential broadcasters in China. From that moment of departure, however, she joined a generational exodus, a movement of young urban residents away from occupied areas and established social conventions, toward a guerrilla resistance and a revolutionary social movement. Many, like Ding

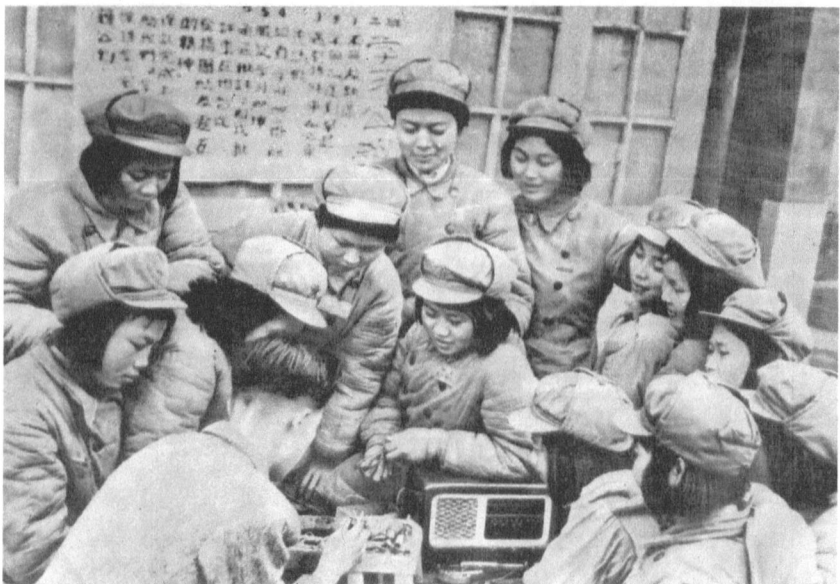

FIGURE 6.1 A group of radio monitors studying radio repair during the Korean War. Source: *Jiefangjun Huabao* [Liberation Army Pictorial], no. 11, Feb. 1952. Author's personal collection.

Yilan, would utilize their technical knowledge, education, and enthusiasm to construct a place for the Communist cause in the fluid wartime newsscape.

From 1937 to 1949, the Communist Party faced a dilemma familiar to its rivals and predecessors: how to communicate news, information, and propaganda across vast areas to millions of people who often lacked even basic infrastructure such as roads, telephones, newspapers, or electricity. Their response to the problem was also familiar. They copied practices pioneered in the previous decade such as using trained radio monitors to transcribe material and publishing small-scale news fliers. They appropriated equipment from the Nationalists they displaced and the Japanese they fought and, wherever possible, plugged themselves into the existing newsscape—surreptitiously taking over newspapers, using GMD broadcast news to sell CCP fliers, and commandeering the Japanese-built radio network after 1945. The Communist newsscape, like the movement itself, was creative, resourceful, and ambitious even under the most extreme and adverse conditions. Throughout this process, women played a prominent role, especially in the realm of radio broadcasting. Though a strong female presence had always made Chinese radio practices stand out from global peers, the wartime CCP took this tendency to the logical extreme. Founded in 1941, Yan'an radio did not hire a single male announcer until 1946.

This chapter uses the experiences of these early radio announcers to locate the seemingly unusual technological gendering of Communist broadcasting within the larger context of Chinese and world radio history. Just as significant, the chapter's analysis of the wartime newsscape offers new perspectives on the relationship between the Chinese Communist Party and the Japanese-collaborationist regimes. Official CCP history has long emphasized its role in opposing both Japanese and collaborationist armies; indeed, the historical record bears out the overwhelming role irregular Communist-allied forces played in the rearguard resistance. But the CCP also relied on Japanese-manufactured equipment (radio parts), infrastructure (Japan's vast Chinese broadcast network), and expertise (Japanese engineers who stayed after the war to help establish Communist broadcasting). Indeed the CCP's civil war period (1945–49) propaganda offensives, which helped win them control of China, operated almost entirely through Japanese infrastructure. Much previous work on the propaganda of the wartime CCP has focused on its comparison to and divergence from the Soviet model. For instance, Timothy Cheek's study of the propaganda system in a Communist base area notes that the CCP began building a "propaganda state" (a term taken from historians of Soviet Russia) with nested hierarchies of committees overseeing the major propaganda organs, in this case a large regional newspaper.[2] Taking in the full breadth of the Communist-allied newsscape, however, one in which major newspapers were but a small part, allows for a slightly different picture to emerge. Much of the CCP news and propaganda machine, at least as experienced by nonelite actors, slotted into the patterns, infrastructure, and practices of the prewar Nationalist state rather than any Soviet model. Small guerrilla papers with smatterings of news from various sources, including Nationalist radio stations, and little oversight from party officialdom predominated. The Sovietization of the Chinese Communist newsscape would wait until after 1949.

The Communist Radio Newspaper Network

As I described in chapter 4, in the 1930s the Nationalist Party organized a system of radio newspapers—small, roughly printed tabloids—to bring news to rural areas. Trained radio monitors transcribed news, information, and government instructions and delivered these to the local government while also publishing these materials and distributing them through their district. After the Japanese invasion began in earnest in 1937, Communist-allied forces occupied rural areas that the Guomindang had effectively abandoned in their forced retreat west. In these new "base-areas," Communist activists and sympathizers built an analogous newspaper network on the scaffolding of the previous Nationalist system.

One such sympathizer, Wang Xijian, a student from the coastal city of Qingdao, joined the cohort of exiled urban youths early in the war. He headed inland on foot with a motley guerrilla band, mostly laborers and college students from his hometown, about thirty persons in all. They retreated into the mountains, continually harassing the Japanese and being driven back in turn. Though Wang estimated that only one third of the group were committed Communists, they were all steadfast in their determination to fight the Japanese and so reorganized themselves as a Communist-allied unit.[3] Such largely circumstantial alliances were common—even Ding Yilan had spent time in Nationalist youth corps training in Nanjing before departing for Yan'an. As CCP associates made up the majority of irregular forces close to major occupied population centers, geography, perhaps as much as doctrine, made the Communists the default option for Chinese youth looking to quickly join the fight.

The irregular Communist forces, for their part, were in dire need of the labor and skills of the young exiles—especially when it came to communications. The Northeastern Army, for example, had no propaganda organization early in the war, so Wang Xijian and his comrades became the propaganda team. Besides organizing underground Communist activity in the army and region, they put on plays, sang songs, and published a mimeographed newspaper called *In the Line of Fire*. Wang was responsible for listening to broadcasts every night and writing down the news he heard. He and another comrade traced those items onto mimeograph paper, along with dispatches about army life, a few editorials, and some socialist theory. Between 1938 and 1940, when rampant disease and exhaustion finally caused publication to cease, they put out more than three hundred editions.[4]

Wang Xijian's Communist-affiliated newspaper was just one of the hundreds that cropped up over the course of the war.[5] In 1944, the Yan'an radio station (reporting in English Morse code) stated that each liberated district had a lead type, mimeograph, or lithograph newspaper issued by the county government or another public organization.[6] In the period after 1938, the Central Hebei base area, a region with fifty-odd counties and seven million inhabitants, contained thirty or forty newspapers: a main daily (the voice of the base area's party organs), in addition to numerous local news fliers.[7] In areas more securely under Communist control, such as the region that contained Yan'an, media proliferated even further. Radio news fliers had their circulation amplified by party-organized reading groups so that, according to their own estimations, each copy had about ten readers. Furthermore, distribution networks sent copies farther afield so that it was even possible to obtain some Communist papers in occupied cities—a border area daily was reportedly available in Beijing for $5.[8] In the Yan'an region, local residents also managed more than six hundred blackboard newspapers, which reported radio news as well as local affairs. "Forming a kind of social column, they are widely read by peasants who are often very concerned about who will get into the blackboard," reported Xinhua, the Communist news agency. "Those who are criticized are mentioned as having 'crawled into the blackboard.'" Such local stories were sourced from voluntary correspondents. Those who could not write gave their reports orally; in one locale, 79 out of the 114 stories received over the course of roughly two weeks were spoken.[9] These illiterate farmers had their words mingled with reports from national and international bulletins taken from the airwaves, prefiguring the mix of bottom-up and top-down broadcasting that would become a hallmark of the PRC's newsscape in the coming decades.

In base areas, however, conditions remained spartan. In one region, the main paper, the *Central Hebei Leader*, used lead type, a gift to the People's Self-Defense Forces by the local Catholic Church, to print its four pages. The county-level tabloids almost universally used mimeograph machines to issue their mixture of information about local guerrilla victories and outside reports—usually every three days. Sourcing international and domestic news, however, remained difficult. Most counties lacked the skilled personnel or codebooks to transcribe Morse code transmissions from Yan'an; the CCP capital had yet to acquire voice broadcasting in the first years of the war. According to one party member active in the underground propaganda work in Central Hebei, the great majority of small papers, therefore, used their radios to receive and transcribe the Guomindang Central Broadcasting Station's recorded news (*jilu xinwen*). Though Xinhua News Agency was founded in 1937, it was not until 1940 that the Central Hebei base area news bureau started issuing a mimeographed bulletin containing Xinhua's Morse-transmitted scripts. Only after this did county-level news fliers begin to incorporate Xinhua's information into their regular reportage.[10] Other party activists similarly recall depending on listening to the nightly Guomindang broadcasts to fill their newspapers in the first years of the war.[11]

In 1941, however, the installation of a more powerful transmitter provided by the Soviets improved the ability of Communist propaganda to penetrate base areas behind enemy lines. Lu Tianhong, a student raised and educated in Shanghai, experienced this shift firsthand. In 1942, Lu was active in youth work in the Tonghai District of central Jiangsu, north of the Yangtze River. He often mobilized local high school student unions to run newspaper walls (*bi bao*), even writing some articles of his own. Still, most of the regular newspapers circulating in the area were enemy and "puppet" (collaborationist) papers. The Red Army veteran who ran the administrative office, therefore, felt a strong need to establish a newspaper of their own and suggested that Lu become the editor in chief. "I hesitated about this," Lu remembered, "because without a radio, there was no source of information. There was no printing equipment, and I feared that the mimeograph was not up to the task." But the old cadre reassured him that the district administrative office, which possessed a receiver, could send copies of the Xinhua News Agency's communications and a lithography studio (formerly used to print notices for the old Guomindang "yamen") could publish the newspapers.[12]

Information about CCP agricultural and land-reform movements now passed from Yan'an over the airwaves to the base area receiver. The newspaper then distributed their political program throughout the area. In the winter of 1942, for instance, the New Year's edition of Lu's paper called on the army and civilians in the base areas to carry out winter training, winter cultural learning, and winter plowing—an agricultural program that presaged many of the movements that radio would carry nationwide after 1949. While the paper did much to advance the Communist cause, the new newsscape remained subject to the geographical, material, and linguistic conditions of the area. The radio receiving, editing, and printing divisions of Lu's paper each operated out of a different location so that if the enemy attacked, they would not suffer a complete loss. The administrative office prepared transcripts of broadcast news every two days, but Lu had to drive dozens of miles to reach its often changing camp. These telecommunications consisted mainly of reports on the victories of the Soviet-German battlefields (likely from the Soviet Station in Shanghai) and the North China battlefield, along with some editorials from Xinhua News Agency. The number of words copied each time was not very large, however, so Lu also reprinted local news from the base area's two main newspapers. Having gathered a sufficient number of words, Lu edited them to fit the small format. The first page consisted of updates on the wider antifascist war. The second and third pages reported on the battlefields in North and Central China. The fourth and final page gave the Xinhua News Agency's commentary. A woodblock-printed masthead featured the calligraphic brushwork of the base-captain himself.

The challenges extended beyond obtaining news to the printing process itself. The unskilled lithograph engravers could not write fonts in a very standardized manner. The message, therefore, had to be kept as short as possible, placed within a loose, well-balanced layout with clear (not overly beautiful or stylized) handwriting. In this way, the lithograph could print four to five hundred copies per issue. The complexity of the process meant, however, that they could only publish about once every ten days or so. It was worth the immense effort. Lu recalled that "the publication of the inaugural issue immediately caused a sensation and had a great influence on local propaganda and agitation in enemy-occupied areas. . . . It was a great victory in the guerrilla zone."[13]

Just as in the Nationalist system, linguistic problems emerged as well. As one might expect, some listeners had difficulty with the dialect spoken in the broadcast studio. Understanding the speech of, say, a Tianjin-raised girl speaking a constructed standard dialect could be difficult for local gentry outside her home region, never mind illiterate farmers and laborers. The fact that the majority of broadcast news was transcribed, reproduced, read, and repeated orally would seem to obviate the problem of dialect. But as one propaganda worker noted, farmers in northern Jiangsu had issues understanding the texts transmitted from Yan'an even when they were read aloud to them in their own pronunciation because the vocabulary and syntax of "standard" Mandarin were so foreign. The person reading the communications, therefore, had to both read and translate at the same time. The essays, dispatches, and notices were written in the so-called plain speech (*baihuawen*), but in reality, the distance between that and the worker or peasant's vernacular was enormous. The cadres in the Yanfu District of northern Jiangsu attempted to remedy this by producing two forms of their newspaper, one of which was translated into the local dialect. It was hoped that people with low levels of literacy could read it and that fully illiterate people could understand it when read aloud. According to their own surveys, illiterate or semi-illiterate people made up 80 percent or more of the residents of the district. Therefore, the editors developed a method of "writing like spoken speech" (*xie hua*), which the cadres promoted through county training programs and a wall newspaper. Each edition of the spoken speech newspaper contained more than thirty items of news and information, each individual item rather short, often translated and shortened from radio broadcasts or the editorials of larger papers.[14]

The CCP not only ran its own newspapers but also infiltrated and exerted control over nominally Guomindang-affiliated papers, inflecting these news sources with Communist spin. Fangcheng County, Henan, hosted one such paper. In November 1938, Nationalist county authorities started printing the *Resistance Daily* in the *jiaoyuguan* (education center), whose director also acted as head of the paper. He supervised an editor in chief, a publisher, and a *shouyinyuan* (radio monitor), who transcribed broadcasts through the paper's only piece of equipment, a five-tube radio. Later, the county's business bureau purchased a lithograph machine that published around three hundred copies a day. These were pasted on street corners and affixed to the four gates of the town. Each government agency, village, town, and school received a copy, along with a few private subscribers. Every week the paper reported its activities to the Guomindang Henan Party Office and Education

Bureau so that the bureaucracies could both monitor the paper for political deviation and reference it when studying local conditions.

The front of the single-sheet flier contained important local news, and the Guomindang Government's Central Broadcasting Station recorded news (*jilu xinwen*). The information copied from the radio overnight consisted primarily of reports about the invading Japanese army's military operations. In contrast, the local news was often colorless. The Nationalist Party, in order to pretend that everything was going well (*fen shi tai ping*), forbid the publication of stories about local robberies, murders, or enemy victories, no matter how insignificant. The reverse page contained longer-form material like government decrees from Chongqing, which had to be serialized over many days.

One day in 1939, the editor asked local high schoolers to submit essays to make up for the lack of literary and artistic material. Lu Kaitai, an eighteen-year-old underground Communist Party youth corps member, began writing editorials and helping with publication work, along with a number of classmates. When the non-Communist editor in chief left, the whole paper secretly fell under CCP control through these student employees. Following Communist Party directives, the *Resistance Daily* advocated resistance against the Japanese and reprinted articles from members of the Guomindang "anti-Japan faction." "We were supposed to appeal to the suffering of the people and gradually expose the darkness of the place," Lu Kaitai remembered. "The masses were under the rule of the GMD government, and there was nowhere to talk openly. Hoping that the newspaper would become their mouthpiece, people—daring not to submit articles publicly—would secretly deposit manuscripts [critical of the current situation] at the door of the newspaper." Such subterfuge continued for a number of years. After 1943, however, controls on speech became more stringent. To control the newspaper, GMD authorities not only sent party members into the newspaper office but also implemented the system of submitting manuscripts for inspection. Later, Nationalist secret services moved into Fangcheng to investigate the newspaper, criticizing a number of incorrect articles. By the second half of 1944, the paper was virtually suspended, and many of the comrades were forced to flee the county.[15]

Though the Nationalist forces seemed strict, they sometimes chose not to completely uproot Communist-affiliated propaganda infrastructure, merely to obstruct it. In Wuyang County, also in Henan Province, young Communists published an underground newspaper called *Self-Defense Daily* (*Zi Wei Ribao*) under the direct supervision of the county CCP leaders, who used it to disseminate anti-Japanese propaganda and news of victories in the resistance. The paper applied several times to the GMD party office for a News Agency License (*xinwen chuban xukezheng*) but was refused. "One Sunday morning, before we got up, the reactionary county magistrate (whose name I forget) took his guards to the newspaper office to carry out a surprise attack," one of the Communist newspapermen remembered:

> As soon as they entered the door, they turned everything over. But our county party committee had issued early instructions: be vigilant and strictly confidential. The documents and books were properly handled, and as a result, nothing was found. Comrade Li Jinglun rushed to deal with the intruders. The county magistrate found a piece of *Xinhua Daily* on the newspaper rack and asked, "Why do you subscribe to the Communist Party's newspaper?" Comrade Li Jinglun casually pointed to the newspaper rack and said, "We subscribe to all the newspapers in the country. You can read *Ta Kung Pao, Shenbao, Central Daily, Henan Daily* and so on. It is a necessary reference for our journalists!" After listening to Comrade Li Jinglun's words, the county magistrate left with his escort. Although the search was a false alarm, it sounded an alarm for us. We should not be careless in the struggle against the enemy, and we should be vigilant at all times.[16]

The Nationalist magistrate had one more card up his sleeve, however. The paper's news and information originated from the county government's *shouyinyuan* transcriptions, which had been regularly sent to the editors for publication. After this incident, the radio monitor would not send the transcriptions on time, waiting days to deliver them. "We could not publish on time, and were really angry," the activist recalled. "There wasn't any other option, so Comrade Li Jinglun and I went every night to the army's radio post outside the city walls and transcribed the news we heard there." Later that year, the activist heard that there was a wireless training program accepting students from county-level self-defense groups. The CCP county committee agreed to send him to receive instruction in sending and receiving telegrams, as well as maintaining and repairing radio receivers and transmitters. "I studied with all my might, in order that our small paper could receive and transcribe news, and to avoid their [the GMD county office] frustrations. Thereafter, the speed of transcribing and publishing increased greatly."[17] By training this young Communist, the CCP began to move away from its fundamental dependence on Nationalist infrastructure (the county *shouyinyuan* and his radio) and goodwill. It did not, however, move substantially away from the Nationalist-designed newsscape defined by hundreds of small, almost amateurish, mimeographed papers.

The Materiality of News

The wartime Communist newsscape was greatly affected by the material conditions of news; the scarcity, weight, and fragility of materials shaped who could get news and how. As we saw in the previous chapter, radio equipment was especially limited in Nationalist- and Communist-held areas because they lacked the seaports to import it and the manufacturing base to make radios domestically. But other sorts of equipment were also precious to the radio newspaper teams, especially paper and printing technologies. Lu Tianhong, when he became a guerrilla newspaperman just outside Shanghai in the last years of the war, especially prized his lightweight portable mimeograph, which he kept constantly on the move, sometimes lodged in remote temples, sometimes in the bow of a boat.[18] Lithographic printing, which in its most archaic form necessitated the application of wax and oil to a heavy porous stone, could be more cumbersome. A cadre in Henan recalled the work of this process: "First the newspaper [content] was cut into a wax tablet, then transferred to the lithograph stone, then painted with oily ink. It was a lot of hard work. In summer the sweat ran down your forehead. In winter your hands became stiff with cold."[19] Mimeographing required a simpler but far more delicate process. The day's paper would be cut onto a carbon or wax paper stencil, which could then be attached to an inked roller. The fragility of the stencil limited the number of copies a mimeograph could produce. Lu Tianhong recalled that "everyone assisted in the delicate task of mimeographing, gently pushing the ink roller, to extend the life of the wax stencil to two or three hundred copies." Lu's operation sourced their wax stencil paper, white newsprint, and ink from local schools, but they were lucky.[20] Paper became expensive as the war dragged on, severely circumscribing the capabilities of underground newspaper units, reducing both the number of copies put out and the quality of printing. The *Jinchaji* base area set up primitive foundries for type and paper factories for newsprint in the mountains to provide for some of the district's needs.[21] Still, one activist remembered that by 1942, the *Jinchaji Ribao* had been reduced to one small sheet because of the Japanese offensive. Usually, her husband liked to read the newspaper, but it had been "so poorly printed that he could not make out many of the Chinese characters."[22]

Unlike paper, radio equipment could not be supplemented by a rough homemade substitute; it had to be shipped, pilfered, or smuggled from occupied areas where radio parts, especially vacuum tubes, could still be purchased for reasonable prices. In one Shandong village, Communist Party members collected contributions totaling four hundred yuan to purchase a radio from the occupied coastal city of Yantai. They quickly started putting out a paper, which they posted in the usual places and also distributed to villagers to take home after market days.[23] As before the war, the great majority of these receivers and radio parts were imported from Japan. They became almost exclusively so after the outbreak of the Pacific War in 1941.

How did this equipment get from occupied Chinese cities to communist guerrillas operating in the countryside? Exiles leaving the occupation brought some of it in their luggage, but friendly smugglers were a more reliable source. One of the most significant smugglers was a British academic, Michael Lindsay, and his Chinese wife, Hsiao Li, both of whom left behind memoirs of their exploits.[24] Yenching University, a comprehensive American missionary university located on the grounds of today's Peking University, had hired him just after the outbreak of the war in 1937 to establish an Oxbridge-style tutorial system for its students. He spent much of his spare time, however, constructing radio sets for the guerrillas he had first made contact with during his summer vacations traveling around the region outside of Beijing. Electronics littered his living room, but Lindsay felt relatively secure from arrest. He was both a foreign national and living on Yenching's extraterritorialized campus (the university was officially incorporated in

New York State). Like many who ended up working for the party during the war, Lindsay was no committed Communist, at least not at first. His wife recalled, "Once I asked Michael why he only worked for the Communist armies. He said that they were the only Chinese army he knew that were fighting against the Japanese anywhere near Beijing so they were the people he had the most contact with. . . . He said that at one time he had tried to transmit messages through to Chongqing to a Kuomintang government organization but had not been able to make contact." It had been a simple problem—the wavelength he had been told to use was too short to get through at night—but it confirmed Lindsay in his support for the guerrilla and not Nationalist cause.[25]

Soon after making contact with the resistance, Lindsay put his privileged position to use buying radio parts. The Japanese did not wish to interfere with Westerners and possibly cause an international incident before they were prepared for war, so they did not search foreigners at the gates of Beijing like they did most Chinese residents. When Lindsay traveled to Hong Kong and Shanghai, he bought certain vacuum tubes and other radio parts that were not available in Beijing. He constructed one radio with this particular stash of equipment and sent along the parts for building nine others.[26] At times, he passed on his illicit materials to contacts in the city, but other times he borrowed the university president (future US ambassador Leighton Stuart)'s car to drive to Beijing's Western Hills. His wife, Hsiao Li, remembered that plainly dressed guerrillas emerged to meet them there. "They took the radio equipment out of the car and then saluted and moved off without a word back thorough the field towards the mountains."[27]

Safety would not last long. Early in the morning on Monday, December 8, Hsiao Li got up and turned on the radio to hear the Chinese news bulletin from the British station in Shanghai. There was no news bulletin, no signal whatsoever. She woke her husband and told him something must be wrong with the radio. Lindsay tried to get the station himself but only succeeded in getting the German station out of Shanghai, which announced the state of war between the United States and Japan. He correctly deduced that Japanese military police would already be on their way to seal off Yenching's campus and arrest its foreign faculty. They threw as much equipment as they could into the president's car, picked up a friend, and drove out the campus's

West Gate ten minutes ahead of the Kempeitai, narrowly avoiding the years of internment (or worse) faced by the faculty who stayed behind.[28]

Heading back into the Western Hills, they abandoned the car and made contact with villagers who promised to pass them on to the resistance. Hiking west, always fearful of an enemy search party, they periodically stopped to set up a radio, hanging antennas from trees, and so followed the news of Japanese assaults on Western colonial possessions across East Asia. When they left, they made sure to hide all evidence that radio equipment had been erected—any sign of such communication equipment would all but ensure the village's destruction if discovered by Japanese patrols.[29] After many weeks of deprivation, the small train of refugees finally arrived at the headquarters of the Jinchaji base area. Named after the three provinces it straddled—Shanxi, Hebei, and Chahar—Jinchaji contained a nominal Nationalist presence, though the CCP government of the area was recognized as the wartime provincial government. Indeed, Chongqing had more or less abandoned its representatives in this crucial guerrilla area, a large zone close to the ancient capital of Beiping (Beijing), the port of Tianjin, and the vital railroad arteries of North China. Hsiao Li recalled visiting GMD headquarters a few miles away from where they were staying. The Nationalists there had published their own newspaper until a year before. Though they still possessed a radio to communicate with Chongqing and receive news, newsprint had become very expensive, and Chongqing had refused money to support publication. The newspaper ceased printing, defeated by fiscal and material constraints.[30]

The Communist authorities could therefore monopolize the propaganda space if they could only overcome the infrastructural geography of Jinchaji. Only one mountain road, more of a dirt track, crossed the 180-mile-wide base area. Lindsay, traveling more than two hundred miles through the region in 1939, saw only one wheeled vehicle.[31] This isolation protected the base area, but it also inhibited its administration. Logically, radio was the most viable means of communication, crucial for both coordinating military activity and mobilizing the population through propaganda. It could reconstruct the war-torn newsscape of the guerrilla-held areas. Michael Lindsay, therefore, became indispensable. Visiting the radio station, he found dedicated people pitifully short of fundamental supplies; they even lacked a hand drill. Luckily, Lindsay had brought with him a test meter and a slide ruler, which, it turned out, were the only ones in the base area. Seeing an opportunity, the Communist general soon requested Lindsay's assistance repairing the receivers and overhauling the radio communications systems.[32]

Lindsay threw himself into the work. He rebuilt old equipment, transforming the old bulky oscillators (about two foot square by one) into smaller boxes, "though it was seldom possible to build two sets exactly the same." The receivers he built were battery-operated, while the transmitters mostly operated by a hand crank. "Some components," he recalled, "came from captured Japanese equipment, but most were smuggled out from Japanese occupied cities by merchants who would take the risk for sufficiently high prices."[33] Lindsay also began giving classes in radio engineering to the army technicians, starting from the first principles of electrical theory. Building a better transmitter required high-level mathematical calculations—"the solution of three simultaneous cubic equations"—that strained the computational resources of the area. Here Lindsay's slide ruler was again invaluable.[34]

Hsiao Li, for her part, started teaching English to the radio operators. They needed English because Chinese could only be sent in numerical code, with the attendant coding and decoding difficulties. It was far easier to send messages in English-language Morse code.[35] She also sometimes transcribed news from international broadcasts to pass on to officials, likely for later publication and intelligence purposes.[36] They could have used such assistance earlier in the war when the primary paper of the base area, the *Jinchaji Daily*, had trouble sourcing news items. They struggled to translate the Morse they heard from Yan'an and had to rely on regular voice broadcasts from Chongqing and the Soviet Union. The editors sifted through these "like searching for gold in sand," making corrections for political stances of those sources before publication. The paper, which in 1940 reached a circulation of twenty-one thousand copies, also disseminated this news through the usual secondary channels such as reading groups, private schools, wall newspapers, blackboards, and loudspeakers.[37]

Women on the Radio

On June 24, 1941, a three-thousand-character article appeared in the *Jinchaji Daily* describing the murder of a young activist by the victim's husband and father-in-law; the men believed that her participation in the cause of the resistance brought shame on the family. The author, a "women's work" (*funü gongzuo*) activist in Pingshan County, denounced these men with eloquent indignation. The editor in chief of the paper, an intellectual (and acquaintance of Michael Lindsay) named Deng Tuo, took an immediate interest in the young *tongxunyuan* or amateur correspondent, penning letters and poetry to her. Ding Yilan, the girl who had fled Tianjin four years before, had now grown into a mature activist and received the letters with interest.[38] Deng and Ding soon embarked on a decades-long union near the heart of the CCP establishment that would marry intellectualism and mass propaganda, radio broadcasting and newspaper editorializing.[39] In 1941, however, Ding was still a lowly amateur correspondent, not yet a famous broadcaster and the head of the Beijing People's Radio Station. Still, even then, hers was not the only female voice on the page of her husband's daily. Ding's polemic appeared alongside the words of other women, the broadcasters of Yan'an, whose news, opinions, and updates the Jinchaji base area received on a daily basis.[40]

In fact, from 1941 to 1946, all of the voices on Yan'an radio belonged to women. While the dominance of women in the radio profession in Yan'an may seem striking, the Communist Party hardly pioneered the field. In 1927, the Tianjin branch of the Manchurian broadcast network hired its first announcer, a Beijing-accented woman, "to recite the broadcast programming and announce the domestic news."[41] The Nationalists continued the pattern, placing women in prominent roles throughout their radio system in the 1930s. In China, the voice of authority was often female, marking it as distinctive among radio cultures, such that foreigners often commented on the phenomenon. Writing in the 1960s, MIT media researcher Alan P. Liu wrote that in both Nationalist and Communist China, "it almost seems that women dominate the field of radio announcing," something he ascribed to "the fact that Mandarin is the main language used in broadcasting; it is more distinct, more graceful, and perhaps more pleasant when it is spoken by a female than a male."[42] Certainly, it is hard to claim that China possessed a stronger feminist tradition than other countries. According to classical mores, Chinese women were expected to be silent in public, neither officially participating in public affairs nor speaking on behalf of themselves, their families, or communities. (Of course, this was simply a patriarchal ideal, which was far less true in practice.) Women only began to be seen and heard on their own terms after 1911, when as David Strand has described, they began giving political speeches, playing the female roles in the theater, and attending school in increasing numbers.[43] Hsiao Li Lindsay and Ding Yilan were part of this first, privileged generation of educated women—a cohort not limited to China. Other countries also experienced similar liberatory processes at the beginning of the twentieth century. Women in Britain and the United States won the right to vote and increasingly worked outside the home. Women in Russia and Germany helped carry out revolutions and labored to sustain newly industrialized societies. In colonies like India, women worked to subvert the political order through economic and political acts, like Gandhi's home spinning. None of these places, however, chose to have women symbolize their technological modernity by placing their voices so prominently on the air. Why did China?

Time and again, commentators came back to how women's voices *sounded* on the radio. Pleasant, strong, and, above all, clear. This is not to say that some did not adopt the all-too-familiar misogynistic complaint of women sounding "cloyingly sweet" or "delicate and helpless" (*jiaodidi*), the Chinese equivalent of strident or "shrill" vocabulary used in English to denigrate women's voices. The ubiquity of such complaints in history and, indeed, in our contemporary society has been the subject of much public comment in recent years. The classicist Mary Beard has argued that the tie between authority and the male voice has been embedded in Western culture for millennia, implicitly and explicitly denying the femininity, and certainly the pleasantness, of women who make their voices heard.[44] Others wading into this debate have taken a more technologically centered approach. Tina Tallon has argued that the design of recording and broadcasting technology itself has limited women's voices by cutting off the higher pitches of the human voice spectrum. Citing studies from the 1920s and 1930s, she argues that women's voices were circumscribed not only by the vociferous and subjective opinions of men but by the technology those men designed—especially the limitation of transmission signals to between three hundred and thirty-four hundred hertz ("the voiceband"), which supposedly cut off the upper pitches of the human voice range disproportionately present in female speech.[45]

Tallon's point is well-taken. But I would critique this primarily technological reading of the problem of radio, sound recording, and women's voices in two ways. First, women were not entirely absent from early radio broadcasting in the West and, at times, played a prominent and unproblematic role in broadcasting. Michele Hilmes has shown that, in fact, women's absence from the history of early American radio is more a perception than a reality—that early female broadcasters, especially of the 1920s, were written out of history, not nonexistent. Still, Hilmes does not dispute that by the 1930s, women had largely been excluded from the "hard news" and main announcer positions on the major networks, to be largely confined to "women's content" like fashion, homemaking, or children's programming.[46]

Such an exclusion never happened in China, which leads to my second critique of the technological explanation for the denigration of women's voices. In other cultures, the same technology, with the same limitations, led to the opposite result—a *preference* for women's voices. Throughout the 1930s, the same period women were being driven out of broadcasting booths in America because of supposed technological constraints, the most famous broadcasters at Nanjing's Central Station were women—for instance, Liu Junying, the Warbler of Nanjing (the most celebrated radio women were all nicknamed after birds, a nod toward their pleasant voices). Though Wu Daoyi, the station chief, tried to keep women from announcing the most important news broadcasts, he had to fight the tide of opinion that demanded more of them.[47] This preference seems, on the surface, difficult to explain. It may seem hard to sustain the position that China was more conscientiously feminist than other nations; indeed, it remained a deeply oppressive patriarchal society for many more decades. But both the Nationalists and Communists, as modernizing national Leninist parties, embraced the theoretical emancipation of women as a sign of their break with the recent (backward and oppressed) Chinese past. Indeed, a 1934 advertisement for female announcers for a new GMD-run broadcasting station stated that it would exclusively hire women "in order to advance the professionalization of [that sex]." More than 130 women meeting the qualifications—a high school graduate or equivalent, fluent and eloquent (*kou chi lingli*), and well-conversant in the national speech (*shan jiang guoyu*)—put their names forward for the positions.[48]

If listeners cared particularly about the advancement of women, however, they left little record of it. What they wanted was a broadcast they could hear, and over and over again, they indicate in the sources that they could hear women better, that their voices were "clear" (*qing*).[49] One radio listener in Kunming in 1945 expressed his dislike of a male announcer: "his mandarin is not correct and his voice is so low that sometimes one cannot understand what he is saying."[50] If the consensus in the West, both now and in the early days of radio, was that women's voices performed poorly through sound-reproducing technologies, why did Chinese listeners report such a different experience? The first conjecture I would like to put forward centers on the tonal nature of the Chinese language. According to generalizations proposed by linguistic studies, women speaking Mandarin tend to have a greater range of pitch in their tones when speaking than men. That is to say, men tend to speak with a flatter affect.[51] In a language like Chinese, which depends on reception and recognition of a spoken tone to convey the meaning of many otherwise homophonous phonemes, this difference can matter a lot when one hears distorted or electronically mediated sound—like, for instance, the rough, static-filled airwaves of the early radio. This would explain why women are often preferred for their clarity rather than any perceived gravitas or personal attributes.

A second gendered-speech phenomenon may also be at play here: a female inclination toward accent standardization. Anthropologists have sometimes noted that men feel more pressure to preserve their local dialect (and hence identity), whereas women can be quicker to adopt a perceived norm of their environment (in particular urban or upper-class accents). These are, of course, broad generalizations but may have some purchase here, as the announcers at all the stations were generally well-educated (high school or more) upper-middle-class women from coastal Chinese cities, precisely the kind of people with malleable, upwardly mobile accents.[52] As has been discussed in recent works by Gina Tam and Janet Chen, however, the process of language standardization was very much in flux before 1949.[53] There was really no thoroughly established spoken standard for women to aspire to;

in fact, the Yan'an announcers simply used their peers as a sounding board and referenced a beat-up old dictionary to settle "proper" pronunciations.[54] Regardless of technological and linguistic factors, however, the choice of women to be the voice of radio in China was, more than anything, just that: a choice. In the end, this culturally informed decision, made by many actors in China over the early twentieth century, including listeners, technicians, and administrators, says more about the cultures that chose not to follow this path than the few that did.

Women were the only broadcasters at the first Yan'an Xinhua radio station broadcast from December 30, 1940, to March 8, 1943, when equipment failures forced its closure. The Soviet Comintern had gifted its transmitter to Zhou Enlai on a visit to Moscow in 1940. To transport it back to Yan'an, Zhou had to break the equipment down into its component parts and run a gauntlet of Nationalist checkpoints in Urumqi, Lanzhou, and Xi'an. The CCP purchased additional equipment from Nationalist-controlled areas and Hong Kong. Combined with materiel inherited from Zhang Xueliang's pre-Xi'an Incident Shaanxi administration and captured Japanese parts, the assembled transmitter was a minor miracle. Finding the city of Yan'an both vulnerable to bombing and unconducive to broadcasting (it was situated in a deep valley), the radio station was placed in a small village twelve miles west.[55] There was no gasoline to run the electric generators, so the radio employees converted two large gasoline barrels into stoves, burning charcoal to produce gas and using that gas to drive a car engine that powered a generator. "The technical cadres became 'boiler workers' with burning smoke and soot all over their faces and hands," one announcer recalled. "Comrade Xu Lu, a female mechanic, was knocked down by the gas and carried to the outside of the kiln cave to save her. When she woke up, she got up and went on to work."[56]

Who were the women who broadcast from this station? Four were students at universities in Yan'an, and one was the wife of the communications school president, a Japanese woman who conducted the broadcasts in her native language.[57] None had any professional training for the job, though at least one, Xu Ruizhang, had a background in theater. It was not until 1946 that the station had its first male announcer.[58] Xu, along with another of the girls, was from Jiangsu Province. Just eighteen years old when she began, she

FIGURE 6.2 A group of radio monitors practicing the transcription of news from a simulated broadcast. As in most of China, the announcer and the majority of monitors are women. Source: *Jiefangjun Huabao* [Liberation Army Pictorial], no. 11, Feb. 1952. Author's personal collection.

was assigned to the radio station with a classmate after only a semester at Yan'an Women's University. Broadcasting conditions were rudimentary, a simple two-room cave dug into the hillside. The outer workroom contained a small wooden table and a telephone to contact the control room, along with a wooden bed for those on duty. Thick sheep's wool blankets covered the walls and floor of the inner studio to dampen sounds. A small wooden table with a microphone, a lantern, a gramophone, and some twenty records completed the scene. The announcer sat at a wooden bench, working in this humid and stuffy environment. Xu remembered often hearing the sounds of goats, roosters, and donkeys from the countryside in the studio. The wool blankets were rather poor sound insulators. Her only technical instruction came from the director and an older comrade who patiently taught her to read the *Jiefang Daily* like an announcer, practicing it over and over again until the performance was satisfactory. "We marked the cadence and speed on the manuscript, and discussed with each other. Fortunately, I had a life experience of being a drama actor. It was easier to learn."[59] The others were not so lucky, and the pressure was immense. Since there was no recording equipment in the radio station at that time, the announcer could only broadcast directly into the microphone. Mistakes, once made, could not be retrieved. The announcers carefully prepared before broadcasting, however, reading through the manuscript several times until there were few mistakes. One announcer claimed that in more than two years of broadcasting work, she never missed a word.[60]

The Newsscape of a Communist Base Area at the Surrender of Japan

On August 11, 1945, at 8:00 p.m. the *Jinchaji Daily* received a wireless telegram from Yan'an Xinhua News Agency warning listeners to be on alert for an urgent flash message. The comrades at the newspaper all waited anxiously around the "telephone receiver" (*dianhuaji*: a handset rather than a loudspeaker) for the important news. At midnight, the electric signal finally came through. Everyone anxiously waited for the message to be translated from Morse into plain text: "The Japanese Government has announced its unconditional surrender." (Of course, this was the first tentative indication of surrender, passed to the Swiss Embassy, rather than the official Imperial surrender, which occurred four days later.) Ding Yilan, who had sacrificed seven years of her life for the resistance, let out a cheer. The operator ran out into the courtyard shouting, "Japan has surrendered." All the newspaper workers chanted, "Long live the victory of the Anti-Japanese War" and "long live the Communist Party" with tears in their eyes.[61] Soon, the Xinhua News Agency also transmitted a statement issued by Chairman Mao and an order by Zhu De for all enemy and puppet troops in the liberated areas to surrender to Communist forces immediately. If the enemy and puppet troops refused to surrender, "they should be resolutely eliminated." Ding and her colleagues immediately got to work printing an emergency evening edition of the paper to convey the news and the orders of the Party Central Committee to the army and the people of the whole border region, a single-page bulletin with the blaring headlines in gigantic font.

In 1945, after so many years of war, the border areas still lacked radio-broadcast posts or loudspeaker networks. Nor did they possess a regular post or telecommunications system. The transmission of information, as Ding recalled, usually followed a "village to village" pattern where the first settlement passed on oral news or fliers to a second settlement, which in turn passed it on to a third. News of momentous import might incorporate more archaic and spectacular practices. On the night that the news of surrender arrived, the *Jinchaji Daily* was quartered in Leibao village, Fuping County, surrounded by mountains on three sides. When villagers heard the cries of victory from the newspaper office, they spontaneously lit torches to navigate the dark trails above. Singing along to the clatter of drums and gongs, they spread the news to all the surrounding villages. "At the beginning, I saw only a few beams of fire flowing along the side of the mountain paths," Ding remembered, "but gradually more and more people went by, and red lights were flashing all over the hills and fields."[62] Her husband, Deng Tuo, ever the poet, wrote several lines to evoke the scene:

> The sound of gongs and drums in the sky, the red flag dances
>   on the ground,
> the long march of the revolution, and the hardships of the
>   world.
> Today, people all over the earth are singing with one voice,
> hailing the success of the Anti-Japanese war.
> I'm glad to see the mountains and the fields covered with fire.
> The stars and moons are bright.[63]

Inspired, the newspaper organized a team of thirty or forty torch-bearing people to march with official word to the district's party offices, army units, and people's organizations. Ding Yilan, responsible for reading out the news of the victory and the statements of Chairman Mao and Commander Zhu, recalled the night:

> Every time the propaganda team arrived in a village, the gongs and drums team beat hard, and everyone yelled in unison: *the Japanese devils have surrendered!* and *the Anti-Japanese war has been won!* Villagers woke

up from their dreams and quickly ran out around us. Some of the people also lit torches and held them high, standing beside us. The sound of gongs and drums stopped. I stood on the high stool and read aloud. The red fire was shining on my manuscript and my whole body. At the end of each reading, the sound of gongs and drums and cheers shook the sky and earth![64]

Hsiao Li Lindsay similarly learned of the surrender through these traditional forms of noise and celebration:

All of the sudden, in the silent night I heard the sound of the Yangge dance music played at festival times. It gradually became more distinct and finally I could not restrain myself from waking up Michael. We both got up quickly and went to the edge of the grounds in front of our rooms to see what was happening. We were not the only ones getting up. Down below there were quite a few people wandering about trying to figure out what was happening. Then we saw a procession of people with torches marching down from the caves on the sides of the valley and shouting slogans. "Long live the victory over Japan!"[65]

The scene of gongs, drums, and fire conveying news again evokes Alain Corbin's French village church bells—a traditional sound, thickly encoded with religious, social, and political meaning, repurposed for a new era.[66] Here, then, the newsscape offers a curious juxtaposition: the sound of gongs and drums, used for centuries as part of religious worship, rousing villagers to announce information received through coded electromagnetic waves in the air. New and old, side by side. Close analysis of this newsscape reveals a snapshot of the technopolitical process in time. It reveals that, for all that was new and useful, communication technologies and political structures had not yet created a mass society in these rural areas. Many more people had tasted the practices of mass communications, but the infrastructure was not steady enough to sustain it. Stability—that is, peace—would be needed for the technopolitical process to deepen outside of major cities.

Ding Yilan and her team, for their part, did not return to the newspaper office until dawn. "Although we didn't sleep all night, no one felt tired. Full of joy in completing the task, we talked about our work and life after our victory." That morning, as Ding recalled, some of her comrades commented on her good *putonghua* (standard Mandarin) and the loudness of her voice. In the future, when we enter the cities, we will have a radio station. You can become an announcer, they said. "I didn't expect the joke to come true," Ding later remembered.[67]

The Scramble for China

Over the last days of the war, the Yan'an station repeatedly transmitted orders for puppet and Japanese forces to surrender to the People's Liberation Army, in direct contradiction to the instructions emanating from the Chongqing station.[68] These dueling broadcasts set off a mad scramble to assert control over the infrastructure, economy, and resources of the newly liberated territories. The race in many ways mirrored that occurring in Western Europe in the same period—the urgent competition between the US and the USSR to obtain German scientists and advanced industry for their own side. Similarly, in East Asia, former allies competed to control the assets of a fallen enemy, though in China the situation was complicated by the fact that the enemy had not been defeated in the field. A million or more Japanese troops remained, complemented by many more collaborationist forces. The Soviet army had overrun Manchuria. Forces formerly loyal to the puppet governments had to decide who to support and how. Irregular forces had to decide whether they really wanted to fall under the formal CCP structure. Displaced civilians, including the many Japanese who had settled in Chinese cities over the last half century of semicolonialism and conquest, had to decide where and how to resettle.

It was clear to leaders in both parties that a great civil war was coming. The CCP in 1945 was still an almost entirely rural party, with a presence only in the suburbs of major cities. The Jinchaji base area party committee recognized this as an urgent problem, especially as it related to propaganda. "At present," they wrote, "the GMD uses its radio stations and unified leading organs in the city to rapidly spread its deceptive propaganda, while our propaganda materials rely on transportation, which is very slow." On July 25, 1945, the committee, therefore, issued a directive to increase efforts to smuggle books and documents into cities.[69] But radio offered a faster and more efficient way to transmit the party's message, if only they had the broadcasting power to reliably reach receivers. The remote and jerry-rigged transmitter at Yan'an barely reached North China; its weakness and isolation would also make it easier to jam. There was a great urgency, then, to lay claim to other broadcasting facilities that stood exclusively in territory previously controlled by the Japanese.

These so-called puppet stations had ceased distributing overt propaganda since the surrender, but they had hardly ceased broadcasting. Instead, most of the stations obeyed the orders sent from Chongqing to stand by until the central government authorities could take possession, confining broadcasts to the information on the local situation and police regulations. Some of the puppet radio stations, like Beijing, contacted Chongqing directly and began relaying news from the GMD's Central News Agency. According to a US intelligence report, the Shanghai radio network also notified Nationalist radio officials of "its readiness to make transfer arrangements with the authorized central government representative" given the "the present 'confused' situation in which many unauthorized [broadcasters] were attempting to take over control of the Shanghai stations."[70] These "unauthorized" groups refer to Communist affiliated radio units, making clear the urgency behind Chongqing's repeated broadcast instructions to the Japanese and their allies forbidding their negotiation with any forces other than Chiang Kai-shek's representatives. The collaboration between Chongqing and Japanese-collaborationist forces in this immediate postwar period is clear enough. They worked together, as often alleged by the CCP, to squeeze Communist-allied forces out. This long-told story is, in reality, far more complex, however, for the CCP also had its own collaboration with Japanese and puppet forces after the war, though on a smaller scale.

This association with the former enemy began close to home. Zhangjiakou City, in today's Inner Mongolia, faced profound questions about its identity, potential culpability, and future in the days and weeks following the end of the war. The city had been integral to the Japanese imperial project as the capital of the nominally autonomous Mongolian border state led by Mongolian aristocrat Demchugdongrub. On August 23, 1945, the remaining garrison decided to surrender to the Communist Eighth Route Army rather than holding out until Nationalist forces arrived. This decision probably had as much to do with geography as with ideology: Zhangjiakou was the closest occupied city to the Yan'an base area. Regardless, the Japanese-built station that the garrison handed over to the Communist forces would prove crucial to the CCP's cause. Moving quickly, the PLA put Deng Tuo and his *Jinchaji Daily* colleagues in charge of the broadcasting station and its powerful 10 kw transmitter. The newly rechristened Zhangjiakou Xinhua radio station immediately superseded all other CCP transmitters in power and importance; it could be heard clearly across most of China and all the way to Japan and Southeast Asia. In October 1945, Ding Yilan was appointed its primary announcer, inaugurating her forty years in news broadcasting.[71] The radio station remained in Zhangjiakou for two years, during which time Ding cemented her reputation as a reliable announcer with a "sharp and clear" (*qingcui*) voice. When, in 1946, she presided over the live broadcast of war-crime and treason trials, everyone praised the emotions she poured into the transmitter, a colleague recalled.[72] In October, the station and all its equipment fled a Nationalist offensive back to their old holdout, Fuping County, Hebei. The sturdy Japanese transmitter continued to broadcast from that location until 1948, when it was merged with the Yan'an North Shaanxi Station.[73]

Though one of the earliest broadcast facilities to be occupied by Communist forces, the radio station at Zhangjiakou was not alone. The other stations were all located in northeastern China, the former territory of the Marshals Zhang and the puppet state of Manchukuo, now occupied by the Soviet army. The most important radio station, located in the erstwhile capital Changchun, possessed an immensely powerful transmitter capable of both domestic and international broadcasting. The Soviets had captured the city in the last days of the war. Soon afterward, a plane landed at the local airport, carrying a small contingent of Soviet-trained activists, among them Comrade Wang Yizhi, a young woman who had been studying in the USSR since 1941. She went immediately to the station to inform the editing, broadcasting, and maintenance personnel that they should continue working as usual so as to maintain broadcasting capabilities. Only twenty people actually showed up to meet her. Some were hostile, some were prevaricating, and some were favorably disposed toward the Communist cause. "At the first meeting," she remembered, "I explained the situation to them, assigned work and announced discipline. They all looked at me with astonishment as a uniformed female cadre of the Anti-Japanese United Army." But being a local girl, Wang understood the insults suffered by the people of Northeast China. "Therefore, those people who were in a complex mood but eager, felt that I was warm and kind. Soon, some people became close to us and introduced inside information to us, so that we had a clearer idea of our work." With the assistance of these former Manchukuo radio officials, the station began broadcasting as the Changchun People's Radio Station. Though the Nationalists later took possession of Changchun, it was the Communist Party's voice that was heard first over the airwaves in northeastern China.[74]

Wang Yizhi's memoirs do not indicate the nationality of the engineers and other radio workers who agreed to work with the new Communist regime she represented. We know from other sources, however, that a substantial number of Japanese engineers ended up working for the CCP's new radio network. The motivations of these individuals are difficult to parse. All must have been, as Wang observed, in an at least somewhat "complex mood"—fearing the fates that awaited them under the new administration. Tens of thousands of Japanese soldiers were being deported to Siberia, where many died. Perhaps, then, working with the CCP to run radio stations could represent a way out of this fate. Equally, however, some may have had sympathy for the CCP; a surprising number of Japanese intraimperial emigrants had a leftist bent, and indeed some individuals remained in China after the founding of the People's Republic, as late as the early 1950s.[75]

Dalian, the former Japanese colony and naval port on the Liaodong Peninsula, saw more Japanese radio personnel stay on than any other city. Like the rest of Manchuria, the Soviet army had occupied it in the August sweep of northeast Asia, but unlike other cities, Dalian never reverted to *de jure* Nationalist control. A GMD-Soviet pact had granted the USSR control over Dalian, the erstwhile Port Arthur, retroceding the territory Moscow had lost in the Russo-Japanese War (1904–5) and making the district an island of safety in the accelerating civil war. The preservation and control of the radio station was critical, then. The week after Japan's surrender, from August 15 to 22, some Japanese staff members at the Dalian Central Broadcasting Bureau burned documents, scripts, technical drawings, and account books. Luckily, the Soviets "liberated" the city before any real harm could be done to the broadcasting and telecommunications equipment stored in a warehouse. By December, when the Soviets turned the station over to Chinese Communist control, the one kilowatt broadcasting facility was up and running. On December 19, the last Japanese director of the Dalian Central Broadcasting Bureau formally examined the radio facilities, properties, and supplies, registering them in a stamped handover agreement. Kang Minzhuang, the new station chief, received the document and began the difficult task of managing the facility.

The greatest difficulty, Kang recalled, was the shortage of personnel.[76] The Communist Party, because of the terms of the Soviet-Guomindang treaty, could not operate openly in Dalian. Moreover, the population heavily favored the Guomindang, a fact that Bai Quanwu, the Japanese-language translator, attributed to "the long-term enslavement education of Japanese imperialism."[77] Kang quickly realized that to organize an effective broadcasting team, they could not rely on the same patterns as in the base areas but had to adapt to the urban and diverse conditions. Therefore, they put careful effort into persuading and educating the holdovers from the Japanese broadcasting bureau, aiming to "solve the problem of unity between foreign cadres and local cadres. . . . The Japanese who stayed in the radio station were guaranteed their lives and personal safety. With strict discipline requirements, they could work conscientiously, so as to repair the broadcasting machine as soon as possible and ensure the return of normal broadcasting."[78]

Bai Quanwu, a fluent Japanese speaker, was central to the effort to retain the personnel at the radio station. He later wrote, "The party and the government required us to treat the Japanese in a correct way." The party's official position was that the Japanese people were also victims of the capitalist-imperialist system, so cadres should strive to educate Japanese individuals, especially those who might contribute their expertise to China's reconstruction. At the time of the handover, Bai asked all the Japanese from the original broadcasting bureau to gather in the big office downstairs. He told them in Japanese about the decision of the Dalian municipal government to accept control over the Broadcasting Bureau and gave a detailed account of the postwar situation and the spirit of the policies regarding Japanese subjects. He urged them to cooperate in organizing the Dalian Radio Station of the Chinese people's democratic government before returning home. "At that time, in the Dalian area, our party had not gone public," he remembered. "Although I was very tactful and did not mention the word *Communist Party* some Japanese still guessed that we were sent by the Communist Party of China, and there was a tacit understanding. After I spoke, they offered to stay and work with us."[79]

The station retained more than thirty Japanese staff, mainly engineers. A few of these sweated as generator-maintenance technicians, but most were engaged in transmission and sound engineering. Some editors and broadcasters also stayed, including the reporter Aso Taro, the female announcer Saito Yoshiko, and male announcer Harada. By January 16, 1946, when Dalian Radio Station began broadcasting under the leadership of the Communist Party, there were more than fifty staff in the station, thirty of which were Japanese.

Comrade Bai often held study sessions with these personnel, studying Mao Zedong's works and the speeches of the Japanese Communist Nosaka Sanzō. Dalian's government soon established the Japanese workers' labor union, which the radio station's management actively encouraged the retained staff to join. Several of them even joined the party's Dalian Japanese Communist Alliance, an affiliated branch of the Japanese Communist Party. At the same time, the Japanese technicians began training new Chinese students and advising on repairs and improvements, assisting in the design, assemblage, and installation of new shortwave transmitters as part of a plan to expand overseas propaganda in 1947. This new transmitter doubled the power of the station. "At that time," Bai remembered, "there was a shortage of materials and no ready-made drawings. However, the Chinese and Japanese radio technicians worked together for many days and nights. Finally, on January 16, 1948, the second anniversary of Dalian Radio Station's first broadcast, the trial transmission was successful."[80]

Dalian began to repatriate overseas Japanese in December 1946; by April 1947, more than 185,000 Japanese had returned.[81] After this initial repatriation, fewer than ten Japanese engineers remained. When Bai Quanwu left the Dalian Radio Station in October 1949, five or six Japanese were still working there. He later heard that some of them did not return home until 1953 and that the announcer Saito Yoshiko never left but stayed to marry a Chinese man. Though most eventually departed under varying degrees of duress, some Japanese colleagues retained a degree of affection and nostalgia for their time building Dalian's Communist radio system, even returning to be honored with a special banquet after the Reform and Opening of the 1980s.[82]

Content and Listening

The fact that the CCP's civil war (1945–49) propaganda offensives, which helped win it the control of China, operated almost entirely through Japanese infrastructure has been little noted. The general importance of broadcasting in civil war was frequently noticed, however, and long-remembered. Recollections of the war recall surreptitious listening to news that the Nationalists suppressed, and the influence of the personal testimonials of officials and military men who had crossed to the CCP side. The intimacy of radio brought these confessions-cum-exhortations to the influential bourgeois classes, who could more easily listen in and whose support would turn the tide. The regular denunciations of Nationalist corruption and incompetence carried over the airwaves made their way into newspapers nationwide for those who could not listen. A relentless propaganda machine moved south with the People's Liberation Army as it continued to capture city after city and radio station after radio station.

Kang Minzhuang recalled the strategy of "relying on the party and the masses" they used during this period to create rich, persuasive content. Bearing in mind the slogan "everyone running broadcasting" (*dajia ban guanbo*), they organized party and government departments, mass organizations, and regular people to contribute to broadcasting. "At that time, many cadres and members of the masses were very interested in radio, which was no less popular than today's television. They mainly listened to the news and information on victories for the sake of 'being quickest in the know' (*xian zhi wei kuai*)."[83] This interest, along with the relatively large number of receivers in Dalian (a legacy of Japanese administration: there were an estimated sixty thousand sets in the city), helped integrate radio broadcasting into people's everyday lives. For instance, after a snowfall, the Dalian municipal government issued a general order to shovel the streets. "Even the old ladies and housewives who couldn't read knew the order all at once," and they helped clear the snow. Radio officials encouraged cadres at all levels to use the Dalian station as a powerful tool for propagandizing, organizing the masses, and implementing decrees. They urged individuals to come to the radio station to give reports and lectures so as to make Dalian radio an on-air university for the people. In 1946, regular people and model workers spoke at the radio station 102 times.[84]

Outside of the Northeast, where radio receivers were less common, surreptitious radio listening still occurred, though it was less integrated into everyday life. The Guomindang certainly noticed the problem of its citizens listening to stations like Ding Yilan's Hebei People's Station, Yan'an, or Dalian. The municipality of Beiping (Beijing) and its North China Bandit Suppression Headquarters banned listening to "traitor bandit propaganda" and demanded that every radio be registered and pay for a licenses.[85] The municipality of Chongqing planned similar restrictions on listening to Maoist bandits (*Mao fei*),[86] as did the Hubei Provincial Government.[87] The Education Ministry wrote worried memorials to the central government about students at Fudan University in Shanghai listening to Yan'an broadcasts.[88] There was a clear and justified concern that the CCP message was getting through to young people, leftists, and urban populations through radio. At the same time, the Nationalist government placed strict limits on radio imports as part of a suite of protectionist measures ostensibly designed to encourage the growth of a domestic electronics industry.[89] Still, many people desired the forbidden fruit of illicit information, especially when the official news they received was obviously lacking. One resident of Beijing, a comedian working at a private radio station, listened to the CCP radio news furtively at night. The broadcasts confirmed what the sound of gunfire coming from the outskirts of the city indicated: the "Great Victory" predicted by Nationalist stations was a mirage.[90]

Far away in the southwest, in Yunnan, student activists listened to the broadcasts from the captured northeastern stations and used their content to carry out an effective pressure campaign against the local government. In March 1946, students from Yunnan University began secretly transcribing Xinhua News Agency bulletins and publishing them as a weekly paper that reached up to three hundred copies per issue. To avoid suspicion, the news was made to look like it emanated from Hong Kong. The paper printed important instructions, proclamations, and orders of the CCP Central Committee and exposed supposed Guomindang plots. Mostly, however, it spread news of PLA victories and GMD defections.

Underground party members and progressive students also frequently copied down important news from Xinhua onto big-character posters. They called these posters wall newspapers, giving them the names *The People* and *Democracy*. Later, the Xinhua radio station began broadcasting lists of war criminals among the Guomindang leadership. Activists pasted the names in big characters onto walls near street corners, those ancient cruxes of the newsscape, where many people would see them. They printed copies of letters urging enemy generals to surrender, along with the "war criminal" lists, and sent these items to the families of Guomindang military and political officials—acts that could not fail to be read as threats.[91] Broadcasts also appealed to the families of rank-and-file Nationalist soldiers to oppose their participation in the war. A piece that aired in May 1947 began by recounting how even Chiang Kai-shek's crack troops were being decimated. "Dear listeners," the announcer pleaded, "if your fathers, husbands, brothers, relatives, friends, fellow townsmen, or school mates are now serving in Chiang Kai-shek's forces, please inform them that even the 74th Division was annihilated. The other troops were much inferior to the 74th Division, so why carry on the fighting. If they insist on fighting, they will simply send themselves to their death. Why sacrifice lives for Chiang Kai-shek and American imperialism? What good will it do those who become widows and orphans?"[92]

By December 1948, most of the newspapers in Kunming had underground party members or members of the party's associate organizations embedded in them. These underground members used their newspapers' receivers to copy the news from Xinhua and publish it, uncredited. One paper got away with publishing the list of war criminals by citing a foreign radio station as its source; merely reporting on other people's reporting is a time-honored way of publishing borderline material. In the end, this propaganda campaign succeeded spectacularly. Large parts of the leadership of Yunnan defected to the Communists in 1949, yielding the populous southwestern province, which Chiang Kai-shek had intended to use as the heart of his final redoubt, without a fight. Activists in Yunnan had transformed radio news into concrete influence.[93]

The infiltration of Communist agents into nominally Nationalist or independent newspapers was common toward the end of the civil war. In 1948 one such agent, Hu Guanzhong, arrived in Xiamen on the southeast coast with orders to insinuate himself into the newspapers there. The city had four major papers, two privately owned and two GMD-affiliated, along with many small tabloid newssheets. Hu soon discovered that one of the major private papers, *Bright Star*, which he had joined, in addition to publishing the Nationalist Central News Agency wires also put out regular "Exclusive Communiqués" without attribution. Asking around their newsroom, particularly the radio operators, he discovered that *Bright Star* collected these pieces from all over the airwaves and put them under the false heading of "Exclusive" to attract customers. This gave him an idea. If the newspaper was already lying, he could, too. Following their pattern, he began to label the Xinhua's communiqués he transcribed "Exclusives" and placed them in *Bright Star* to spread his party's message, in his mind, "laying bare the GMD Central News Agency's lies."[94] Everyone wanted to know the source of the news, so Hu spread the word of Xinhua's wavelength and call sign. Xinhua news appeared in other newspapers. Though it was illegal to transcribe these broadcasts, Hu received the secret support of his bureau chief at the paper. Throughout the peace talks of January 1949, *Bright Star* continued to give the official Communist side of the story; the sheer amount of news meant that Hu had to double the amount of transcribing he did. As troops moved south through China, accounts of the destruction of GMD troops and refutations of GMD spin appeared in the Xiamen newspaper. Eventually, in April, a front-page order from Mao Zedong and Zhu De outed the underground activists, and Hu was forced to leave town.[95]

Ding Yilan's long war ended on March 25, 1949, when the North Shaanxi Station, following the central government, entered Beijing. They moved into the facilities built earlier in the decade for the collaborationist North China government and began broadcasting as the Beiping New China Broadcasting Station. The preceding twelve years had been a long struggle for Ding. Worst of all, she had lost an infant son, one she had not even had time to name, a month before her marching finally stopped. Yet her determination shone as bright as ever. She had begun the year by reading Chairman Mao's New Year's speech, "We Will Carry Out the Revolution to the End," on the radio. Ding Yilan, the scared schoolgirl in the embroidered shoes, had become the voice of Mao, and Mao had become the master of all China. When, that fall, she broadcast the founding of the People's Republic from atop Tiananmen, the newspaperman Hu Guanzhong was listening in Fujian, his eyes filled with tears. Still moved from the broadcast, he attended a celebratory banquet where each comrade was given a few small pieces of pork. After a decade of struggle, long nights of transcription, editing and placing news stories, inserting Communist spin into the structure of the existing newsscape at great danger to himself, this was his reward. "It was just a few pieces of pork . . . so small in value. Yet the feelings I felt were ineffable. It was a symbol of our victorious revolution."[96]

SEVEN

Socialized Media, 1949–1958

On October 1, 1949, Mao Zedong and a bevy of leftist grandees climbed up the worn steps to the top of Tiananmen Gate, where they could stand perched overlooking a banner-waving sea. Technicians had been prepping the platform for nearly a month. Wires from eight microphones snaked in several directions. One led to a magnetic wire recorder that inscribed the day's events for posterity. Another led to the loudspeakers, which allowed the crowd in the square to hear the incantations of national foundation being spoken on top of the old imperial wall. Still other wires led, roundabout, to the Japanese-built radio station. Two announcers, including the redoubtable Ding Yilan, stood by, ready to narrate the proceedings. At three in the afternoon, the live broadcast began. Mao, speaking in his thick Hunanese accent, announced the formation of the People's Republic.[1]

The founding-day broadcast should have reached all corners of the country—theoretically. In reality, China's radio infrastructure was in a parlous state. Most of the broadcasting stations had been built under Japanese occupation, or even earlier, and had changed hands two or three times in the intervening chaos. The majority of listening posts had been abandoned or destroyed in twelve years of war. The number of receivers had not grown

FIGURE 7.1 A crowd of men, women, and children gather to listen to a radio receiver in a rural area. Source: *Wuxiandian* [Radio], no. 1, 1956. Author's personal collection.

in more than a decade. The Communist government, therefore, faced the necessary task and rare opportunity of remaking China's newsscape according to their own design. Ideology, technological limitations, and fiscal constraints largely precluded the individualist consumer model embraced by the post-Stalinist USSR, Japan, Taiwan, and the West.[2] Instead, the government settled on a socialized media infrastructure as a way forward in 1949. This path shaped the technological and social practices of news for decades, profoundly affecting the implementation and experience of socialism. Indeed, China's socialized model of mass communications—in the form of both wired and wireless broadcasting—would help enact party rule while also amplifying the damage of party caprice.

A New Socialized Media

Writing from exile in the late 1930s and 1940s, the German philosopher Theodor Adorno sought to explain the disaster of fascism, in part, through the rise of mass media (what he called the "culture industry") and radio in particular. In so doing, he set an enduring paradigm for understanding authoritarian newsscapes. Drawing on the work of his friend Walter Benjamin, he theorized that the "sameness" produced by mass technologies betrayed a tyranny inherent within those techniques of reproduction. The effects of phenomena like radio or film on individuals and societies were thus convergent regardless of those bodies' overt political formation—citizens of dictatorships and democracies both became media-directed automata.[3] In Germany, he wrote, "radio becomes the universal mouthpiece of the Fuhrer; in the loudspeakers on the street his voice merges with the howl of sirens proclaiming panic." Indeed, "the Fuhrer's metaphysical charisma ... turn[s] out finally to be merely the omnipresence of his radio addresses."[4] Sound reproduction technologies, in other words, are themselves the appeal of fascism. Thus, the audience is transformed into a passive vessel for the leader's agenda. "Not only do the listening subjects lose, along with the freedom of choice and responsibility, the capacity for conscious perception," he proposed, "but they stubbornly reject the possibility of such perception."[5]

Adorno attributes this so-called regression to the "authoritarian voice," which arises from the intimacy of the broadcast form and the social practices of listening. "The authority of radio becomes greater the more it addresses the listener in his privacy," whereas "an organized mass of listeners might feel their own strength and even rise to a sort of opposition."[6] This conjecture has had an enormous influence on the popular conception of media in the twentieth and twenty-first centuries, though Kate Lacey and Carolyn Birdsall have thoroughly demolished his case for the passivity of Germany's "listening public."[7] Still, some of the theories seem borne out in the case of China. The sensations of immediacy, connection, and contemporaneity inherent in radio—Adorno describes it as the "illusion" of closeness—certainly surfaces in Chinese listener accounts. The philosopher would also recognize the process by which radio more directly connected the Communist Party center at the top and the masses at the bottom, minimizing other

FIGURE 7.2 A utopian image of progress under the worker's state features advanced technologies like vacuum tubes and radio broadcast towers. Source: *Wuxiandian* [Radio], no. 10, 1958. Author's personal collection.

centers of information or power. At times, radio networks gave Beijing the ability to synchronize the nation, keeping everyone apace of each campaign, of every political advancement and reversal. Like a metronome, it paced their start, their crescendo, their finale. Local variation, while still present and important, became more difficult as some lower level officials lost the ability to throttle the revolutionary process.[8]

But the practice and experience of news in China also diverged significantly from Adorno's theories of authoritarian media. This should not be surprising. Adorno sought to understand the consequences of mass media technologies within the context of bourgeois capitalism—in both its liberal and fascist forms. But what does mass media look like if it is not bourgeois at all? What if it is largely rural and socialized? What if the leader of an authoritarian society does not himself speak? What if radio is not used to structure "free time" but organize work? What if listening is not a passive exercise but an expression of active participation?

Examining the unexpected conjunctures of the 1950s Chinese newsscape, we can begin to challenge some of the assumptions made by the Western media studies tradition. First, mass media emerges as a substantially rural phenomenon. Second, the power of communication techniques can increase with their publicness, not their intimacy or individuality. Instead of radio being the voice of a stranger in a cloistered room, it can be the voice of a neighbor heard in a crowd. Third, passivity need not be the hallmark of the mass-consumption newsscape. In 1950s China, voluntary and forced engagement gave the newsscape its power, as both the ownership of the means of communication and the act of receiving information became socialized and participatory. Radio listening transformed into a community activity required of all citizens. People had always listened in groups, but now they analyzed and responded to the programming together as well.

At the same time, media also became local and "contributory." Wired broadcasting networks, systems run over wires serving a single county or commune, shared news and information (though highly doctored) between putative peers, avoiding the appearance of a top-down directive. Adorno claims that one of the most significant and tyrannical facts about radio is that "no mechanism of reply has been developed," but this was not true in the early People's Republic. As the country moved into the ultimately disastrous Great Leap Forward, up to half of all news accounts in some rural areas came from local contributors sharing stories within their own communities, a novel arrangement that increased enthusiasm and credulity. This theme of active engagement with local, peer-to-peer news technologies (as well as the unwavering faith some people placed in their reportage) certainly parallels developments in our postmillennial news environment, making the affective consequences of the 1950s newsscape easier to conjure.

In fact, the trust created by these systems was central to the power structures of the early Communist revolution. Yang Jisheng writes in his monolithic elegy *Tombstone* that, during the Maoist period, "the government's monopoly on information gave it a monopoly on truth. As the center of power, the Party Center was also the heart of truth and information."[9] While theoretically true, to most rural Chinese in the late 1950s, it did not *seem* that the party was the source of all information. The newsscape was different at the level of experience. The wired broadcasting network, local and intimate, gave the impression that some news and information was coming from one's peers—the members of one's own small community. If you did not take something on the word of Beijing and the party, you very well might take the word of someone from the next village over when they testified to a policy or methodology's virtue. In this way, the very structure of the radio network (local and social) reinforced propaganda, allowing that state's preoccupations and policies to strengthen and travel to extreme and tragic ends.

Thus, the story of the early PRC's newsscape illustrates radio's capacity not merely as a tool of persuasion but also as a tool of governance. To use Sun Yat-sen's metaphor: the communications network became the brain that "controls the limbs and digits."[10] Broadcasting was indispensable to the administration and execution of government policies, especially in the countryside. Integrating rural areas into the mass media and mass politics of the era, radio brought the technopolitical process to new heights, though the results ran counter to the utopia imagined decades earlier. Most broadly, then, this chapter addresses the question of *how* the new government carried out its agenda. Until now, most explanations have focused on the political thesis of the technopolitical dialectic. Established accounts constitute some combination of three factors: propaganda, party, and war. The propaganda portion refers to the most obvious forms of persuasion: public trials, theater, criticism sessions, posters, news, songs, and books. Highly structured campaigns carried this information and material through nested hierarchies of party organization—the "party" leg of most explanations.[11] Finally, government programs and policies found a populace conditioned by years of war to accept militarized forms of social mobilization.[12]

The study of news infrastructures and practices complements these existing narratives. By identifying the technological substratum of the CCP's cultural, legal, economic, and political programs, this chapter highlights two facets of the growth in governance. First, it was voluntary to a significant degree. Many people, especially in rural areas, welcomed the increased flow of information that the new structures allowed and actively reorganized their communities and individual behavior around them. This increased the state's capacity. Second, these infrastructures pervasively influenced daily existence. Jeremy Brown and Matthew Johnson have argued that high socialist China "came to life at the grassroots in the form of class status labels, grain rationing booklets, propaganda written on the neighborhood blackboards, loudspeaker broadcasts, mandatory evening meetings, and anxious interactions with local officials who wielded arbitrary authority."[13] Of their six exemplary characteristics of Maoism, two are direct extensions of radio: the loudspeakers and the blackboards maintained by *shouyinyuan* (radio monitors) and *guangboyuan* (broadcast station workers). The remainder of the characteristics, as this chapter will show, also fell under radio's influence. Though the communication technology did not *give* local figures their arbitrary authority, it did direct them to use their power in capricious ways. Though not every evening meeting included a broadcast, information and practices heard by cadres over the airwaves (or printed in a radio newspaper) often suggested their content. Though radio did not create rationing and food shortages, it delivered the specious outside harvest figures that encouraged local party secretaries to confiscate yet more grain. In obvious ways and in invisible ways, radio helped manufacture the everyday experience of high socialism.

Establishing Control

In 1949, Communist radio officials throughout the country moved into the facilities built by their many predecessors. In Shanghai, they moved into the Peace Hotel on Nanjing Road, whose broadcasting facilities the Shanghai Municipal Government had constructed in 1932. Tao Baichuan had led the defense of Shanghai from the same rooms in 1937. The Japanese had stormed them in 1938, eliminating the Nationalist's last formal presence in the city. Passed, in succession, to collaborationists and then back to the GMD, the communications suite at the Peace Hotel changed hands in the summer of

FIGURE 7.3   Rockets soar and Sputnik flies, while Tiananmen's radio waves echo around the world. Source: *Wuxiandian* [Radio], no. 11, 1958. Author's personal collection.

1949 for a fourth time. That heady season saw many leading Communists attending cocktail parties in the hotel lounge. Armored cars idled outside, guards paced the roof, and the bar boys sipped every cocktail served to test for poison. Upstairs, the East China Radio Station began preaching communism to tens of millions of people.[14]

At the Peace Hotel, the radio station occupied the entire second floor and soon swallowed much of the first. Finding even this insufficient, the radio administration quickly incorporated other Shanghai stations into the Communist propaganda system. The main facilities on Nanjing Road focused on long-wave transmissions to the wider region, while the old German station broadcast physical training programs, military songs, and propaganda stories through the metropolitan loudspeaker system.[15] In this new environment, private stations came under pressure to go off-air or become public. Strong-arm tactics were common. Announcers were vetted for political reliability and registered. Stations that violated regulations, for instance by accepting money to transmit commercial messages, had their managers and employees arrested, their client list interrogated.[16] For appearance's sake, the party seized stations outright only in rare cases. Instead, the municipal committee purchased them, though no doubt at heavily discounted prices.[17] By 1953, the thirty-year history of private broadcasting in Shanghai, dating back to E. G. Osborn's first station in 1923, had ended.[18] In theory, the state now directed all radio work, though the reality of broadcasting and the experience of radio listening were far more complicated than a statist model would suggest.

Though the new government could control broadcasting stations on the mainland, it could not prevent radio waves emanating from overseas. Guomindang broadcasts from Taipei, Voice of America, and US Armed Forces Radio from Tokyo all regularly reached Communist China. Like the Japanese occupiers before them, the CCP moved to suppress shortwave receivers because they could pick up broadcasts from around the world. No exception was made even for equipment at foreign embassies and consulates.[19] In Shanghai, the new People's Government moved with warlike urgency; the city was still being bombed by Nationalist forces, and many expected an imminent counteroffensive by the exiles. The possibility of communication with GMD forces had to be foreclosed. Control over radio listening would help stamp out Nationalist cells by making operations behind Communist lines much more difficult. Just as important, preventing Nationalist propaganda would cauterize sympathy for the lost cause among a still unreliable population. Communist authorities referred to this as "cutting out the enemy's tongue."[20]

The new administration estimated that there were 250,000 to 300,000 receivers in the city, of which roughly 40 percent were judged capable of long-wave reception. With tens of thousands of shortwave radios to eliminate, the task seemed enormous. Luckily for CCP radio officials, the postwar Shanghai radio industry had rebounded. The years 1945 to 1949 had witnessed an increase in small-scale private radio stations. Set dealers had returned after the wartime privations. According to the wireless electronics trade association, five or six hundred people in the city could be employed as radio repairmen. Of those, only one hundred were technologically capable of inspecting and dismantling shortwave receivers, though not all were politically trustworthy—some had received electronics and communications training during the war through their service in Nationalist or collaborationist regimes.[21] Still, Shanghai authorities estimated that eight hundred people, mostly employees from the telecommunications bureaus and radio stations, could be mobilized to do the work of reducing shortwave reception capacity. One person could dismantle seven radio sets per day. The city would be shortwave free within one month, at least according to Communist numerology.[22]

The government mobilized all sorts of authorities in their anti-shortwave project: public security, scientific associations, garrison and PLA communications divisions, news publishers, women's associations, and trade unions. Newspapers published editorials, and stations broadcast pithy orders: Oppose listening to Voice of America! Turn in your shortwave radios! Unions and the Communist Youth Corps urged their members to turn in those people secretly harboring a shortwave radio or listening to enemy broadcasts.[23] In a pamphlet, government propagandists described the Voice of America with typical hyperbole as "an instrument 10,000 times more destructive than the atom bomb. The bomb can do damage to only a limited area, while the Voice of America corrupts every person who listens to its enticing broadcasts."[24]

The concern was justified. Not everyone turned in their radios, and enough people tuned in to Voice of America (VOA) or Taipei to make these stations the most important source of "rumor"—that is, the sharing of any news or information inimical to the CCP.[25] Moreover, there was a strong desire to listen, since information and stimulation were lacking elsewhere. All the newspapers were the same—"to read one was to have read them all," said one Shanghai resident.[26] It was not uncommon for the front page stories to be identical in all major journals, so some people listened out of boredom or simply "because they were curious to know what was supposed to be bad for them."[27] A former radio operator visiting Hong Kong reported that, though power shortages were curtailing the use of receivers, "many Shanghai residents saved their electricity allotment to listen to the VOA, and used candles for illumination." Since people would choose news over light, "VOA news carried rapidly by word of mouth despite listening handicaps."[28]

The extent of the anti-VOA campaign—which resembled many of the early PRC campaigns in form—reflected the perceived risk. In Nanjing, shop employees gathered in a meeting to pledge not to listen to the "poisonous" Voice of America. "It is a political, aggressive tool of the enemy and the source of rumors of secret agents," they exclaimed.[29] Though the government never legally forbid listening to the Voice of America, Radio Hong Kong, or any foreign station, to be caught doing so was to risk imprisonment or worse. One former resident of Harbin noted that an "apprehended listener of VOA would likely be placed in detention for from one to two weeks under reduced rations and released with a warning not to listen again. . . . The poor prison food coupled with questioning and daily lectures was sufficient to deter offenders from further listening."[30] But in Nanjing, "rumormongers" who spread gossip about VOA news were arrested, tried, and made to confess in front of mass rallies.[31]

In the face of such threats, people began policing themselves and one another. The employees of one business in Qingdao discussed the matter and turned in the office shortwave set for their own security.[32] In Guangdong, close to dangerous radio stations in Hong Kong and Taiwan, all radio owners were required to provide two guarantors that they would not listen to VOA. In case of violation, both the owner and the guarantors would be punished.[33] Neighbors were enlisted to monitor neighbors. People could still hear Voice of America, sometimes even on their medium-wave sets. But they could only listen secretly at home. The wealthy and foreign educated, especially keen listeners to stations like US Armed Forces Radio from Tokyo, VOA, or BBC, could not trust other members of the household, especially domestic help. One businessman from Hangzhou reflected that his peers particularly feared the servants turning them in to Public Security for listening and thus being sent to a labor camp for "education."[34] Even children could not be fully trusted. Schoolteachers encouraged students as young as four to reveal their parents' listening habits.[35] This mutual surveillance was easily exploited, as the government itself realized. Con artists posed as government agents to confiscate radios or to demand registration fees. Eavesdroppers blackmailed

FIGURE 7.4 The installation of wired radios signals the prosperity of a happy rural home. Chairman Mao's image looks out from an altar-like space inside. Source: Huang Entao, "Hongse laba jiajia xiang" (Red loudspeakers sound in every home), 1972. Author's personal collection.

large sums from people observed listening to overseas broadcasts.³⁶ Illicit radio represented a threat to lives and livelihoods. It invited danger from all directions, so, eventually, the number of illicit listeners was whittled down to only the most ardent and secretive. Even building one's own set—long the resort of furtive, curious listeners—became difficult since all electronic parts were registered at retailers and sales had to be preapproved.³⁷ Though no system could be perfect, in the 1950s, before the advent of inexpensive transistor radios, people's self-policing behaviors came close to granting the CCP a monopoly on the radio newsscape.

While officials suppressed certain types of radio (shortwave in particular), receivers did not disappear completely from urban markets. Radio sets came from new Soviet bloc countries: Hungary, Czechoslovakia, East Germany, even Romania (these last being "not so good").³⁸ None of them compared to the quality of prewar British or American receivers. One Shanghai resident recalled a factory worker who purchased a Czech radio through his union on an installment plan. After a few months, and long before the instrument had been paid for, a tube and a coil burned out. The worker "rewound the coil himself but, although he scoured the city, he was unable to find a new radio tube."³⁹ Inferior quality led to inferior results. Thus, while the government denounced all things American, it still sought American exports. In 1950, party representatives approached the RCA Victor distributor in Shanghai for a quote on radio tube components with the idea of manufacturing the tubes themselves. The American distributor reported that the inquirers had no idea of the technological complexities of making tubes, nor did they have any interest in patent rights. After the CCP succeeded in driving foreign business from Shanghai, the trade moved to Hong Kong, which acted as an entrepôt and launderer for electronics destined to help build New China. Radio tubes and loudspeakers from Chicago, potentiometers from Indianapolis, aerials from Cleveland, and capacitors from Washington, DC, all made their way in quantity to the island city to satisfy Beijing's demand.⁴⁰ Some Hong Kong importers, the US State Department estimated, reshipped 90 percent of their radio tubes to Communist China.⁴¹ In this small way, American technology helped construct New China's newsscape.

China had never succeeded in creating its own domestic electronics manufacturing industry. Attempts by the Nationalist government and Shanghai entrepreneurs had been thwarted by circumstance, and until the 1950s, what little "manufacturing" went on was, in reality, only assembly using American or Japanese parts. The Communist government, therefore, invested heavily in electronics research at places like Shanghai Jiaotong (Communications) University. By 1952, tubes and amplifiers had been successfully test-manufactured, and the electrical appliance factories could prepare for mass production.⁴² By 1954, the Chinese were able to produce their own radio receivers and transmitter equipment in Nanjing, the old Nationalist electronics hub. Meanwhile, the production of other equipment remained in the hands of small enterprises, like that owned by a half-British, half-Cantonese electronics engineer in Shanghai. Interviewed in the summer of 1954, the engineer described his company's manufacture of loudspeakers for public address systems, along with radio valves (tubes) and tape recorders for broadcasts. Though problems with the state-directed economy were already apparent, the expertise of the Shanghai electronics enterprises compensated for some of the systemic issues—for instance, the man's company sometimes rebuilt equipment that had arrived from the USSR with missing screws and ill-fitting parts.⁴³ Despite such setbacks, the combination of imports and domestic manufacturing enabled China to embark on an ambitious construction and installation program. By 1955, the total wattage of all transmitters in the county was 9.5 times that of 1949. Beijing and fifty-seven additional stations served hundreds of millions of people for more than twenty hours a day.⁴⁴ Each week a variety of programming—political and domestic, tranquil and violent, entertaining and mobilizing—filled these airwaves. It is to this content, and to the experience of listening, that we now turn.

One Week on the Radio

At seven in the morning on Sunday, April 22, 1951, radio waves across China awoke in the usual way, with the announcement of the time and fifteen minutes of news relayed from Beijing. Provincial news followed. At nine, listeners in Shanghai, those lucky enough to have the day off, enjoyed the Sunday music program—a cello sonata followed by Edvard Grieg's Piano Concerto in A minor. All stations then carried up to two hours of news and propaganda material for radio newspapers and blackboards, as well as instructions for petty officials, laying out directives for the week. These broadcasts were addressed, at dictation speed, not to the average listener but to "governments at all levels."⁴⁵ Children's hour reached kids basking in the late spring weekend at different times, depending on the region of the country. In Tianjin, children heard the story of comrade militiaman Wang Qi's heroic sacrifice just after eleven. Peasant choir songs sung by the Second District Elementary and a musical called "Do a Good Job" followed. Chongqing's youth program featured contributions from eight local schools, whose students sang folk songs, performed plays, and recited poems. In Shanghai that weekend, the family hour broached an important topic, "So, Mama Has Had a Little Brother," before the adult fare began in the evening. This particular Sunday, the grown-ups' entertainment program kicked off a long celebration of International Labor Day with music by the Police-Government Ensemble, a violin solo by a famous musician, and tunes from the movies. A political segment discussed the exemplary achievements of an agricultural production mutual aid group. In Chongqing, an arts and literature program featured comic cross-talk dialogue and thought reform; Tianjin broadcast a play called *The Enlightenment of a Porter*. The day wound down with more news, editorials, and dispatches from Beijing, relayed through every wavelength nationwide, before all stations signed off around 10 p.m.⁴⁶

Nothing about this schedule would be unusual for listeners over the coming decade, except perhaps the lack of a broadcast exercise routine, which later became part of the morning ritual across China, and the relative dearth of Russian-language lessons.⁴⁷ Like much of the world, Chinese radio programming contained news and lectures, music, and general interest programs. But the broadcasts also delivered content particular to the Communist bloc. For instance, the radio repeated incessantly that it would take fifteen years to achieve socialism, after which everyone would have a job, two rooms with good furniture, electric light, a telephone, and a radio.⁴⁸ Some of the methods for delivering these messages were particular as well. Broadcasts played through radios and loudspeakers strung up along streets, in parks and apartment blocks, in schools and offices. Each work unit (in Maoist China, every urban resident was assigned to a *danwei*, or work unit) was supposed to have a radio receiver, loudspeaker, and an individual (usually an "activist" or group leader) to organize listening. Whether they wished to or not, everyone was supposed to hear what the Communist Party wanted them to hear. From the 1930s, many shops and offices in China had communal radio receivers, used for attracting customers, keeping up with markets and news, or simply entertaining employees. But in the Communist period, radio listening in the workplace or street became largely divorced from commerce and entertainment. The purpose now was to encourage production and devotion to the party and nation while discouraging deviation from government policies.

These propaganda goals pervaded the regular programming aired that Sunday. Enforced communal listening instilled and fixed the message. Mass "radio meetings" (*guangbo dahui*) emerged from experiments in northern cities like Tianjin in the first full year of the People's Republic, 1950, and quickly spread to the rest of the country. At an appointed time, large crowds would gather to listen to the broadcast of a live rally at their school, office, factory, or neighborhood meeting place. The most persuasive and reliable propaganda theater, which previously might have only been witnessed by thousands, thus became replicable for millions. This broadcasting practice increased the numbers of participants in the mass social rituals of news by several orders of magnitude.[49] Realizing the immense potential of listening groups, the government instructed every factory, association, institution, and apartment building to organize permanent ones.[50]

On Monday and Tuesday, these groups came in handy as activists and propaganda officers set about preparing factories, shops, and neighborhood alleys for radio meetings on Wednesday, Thursday, and Friday. The East China Shanghai Resist America Support Korea Association, official organizer of the rally, instructed every party, government, and civilian organization to underscore the importance of the meeting as part of the upcoming Labor Day celebrations. The responsible individuals at each institution would have been familiar with the protocol. This was not Shanghai's first such meeting; that had occurred the previous December. Indeed, before this last week of April, eleven broadcasting stations in East China had carried out a total of nineteen mass-radio meetings. The Southern Jiangsu "Oppose America's Rearming of Japan" events had attracted half a million direct listeners. In Nanjing, 50 percent of the population had heard a "suppress-counterrevolutionaries" meeting. Their efficacy seemed undeniable. "This kind of broadcast mass meeting uses the least expense, the shortest amount of time, and fewest personnel to educate" hundreds of thousands of the masses, announced one editorial.[51]

Like previous meetings, instructions on how to prepare went out in the days leading up to the event. Organizers set up equipment at appropriate listening sites, ensuring that not too many people crowded around each radio or loudspeaker. Associations that governed each industry made sure that their members tuned in. During a similar broadcast meeting the following week in Chongqing, for instance, the Finance Industry Association issued careful listening guidelines. If they had a radio, the leaders of each banking house would gather employees around the receiver to hear broadcasts welcoming home volunteer soldiers from the Korean War. Presumably the stories of heroism at the front would cause an upsurge of patriotic emotion, after which leaders would call a vote. By show of hands the employees would demonstrate their opposition to American imperialism and the rearmament of Japan. A secretary would record the number of people in attendance and how they voted. (Results of the vote would be reported at a meeting later in the week.) Those businesses without a receiver could either join up with another bank to listen or else instruct their employees to listen together with their own apartment complexes.[52] The Pharmacist Association similarly asked every shop with a street-facing loudspeaker (the kind run outside Chinese stores for decades) to organize small listening groups to enable the masses on the streets to hear the returned soldiers.[53] During the broadcasts, a responsible person from each location recorded listener reactions live and reported these by telephone, telegraph, or letter. This individual would also explain parts of broadcast content that the audience did not understand and facilitate discussion afterward. If there were technical malfunctions, they would direct the audience to another listening site.[54]

Corralling people to participate in such activities was not always easy. In early 1951, the workers of the Huaxin Flour Mill in Wuxi, Jiangsu, sat through a meeting broadcast by the regional radio station. The program urged listeners to remember the harms inflicted by the Japanese devils, to recognize the sinister conspiracy of American imperialism, and to combat the American rearmament of Japan by raising production. Prior to the meeting, a few of the cadres worried that "if there's no one on the stage, only a few radios," no one will listen. Indeed, some of the workers did not want to sit through the four-hour broadcast or participate in the accompanying denunciations. One worker complained that the Japanese had burned down houses and killed people. "What use, then, is denunciation?" To overcome such sentiments, the activists at the flour mill talked through the content with the workers, discussing the issues at stake in small groups after the broadcast. They claimed to have made a breakthrough but also set a policy of "no arriving late, no leaving early, no absence without excuse, listen attentively on the factory floor." The minority of backward workers who did not want to participate in listening soon disappeared.[55]

By Wednesday, April 25, listening stations across East China were ready. At around 6 p.m., factory workers and students filled cafeterias, auditoriums, and recreation clubs. The broadcast began at 6:30 with the national anthem and the *Internationale*. The secretary of the meeting announced the circumstances of the broadcast: tough battles were being fought by Chinese armies on the Korean peninsula and the campaign needed to deepen the spirit of the Resist America and Aid Korea campaign on the home front. Listeners heard recently returned army "volunteers" enumerate the brave actions of the Chinese and North Korean foot soldiers, while highlighting the cruelties of the American-led coalition. The speakers led the audience in chanting slogans and urged them to sign peace petitions. Songs like "In the East a Red Sun Is Rising" punctuated the ceremonies.[56] Fifteen People's Radio stations in East China relayed the program, along with the twenty-one remaining semiprivate stations in Shanghai. Three million people listened.[57] Though some factories had to provide translators for those workers who did not understand *putonghua*, most of the audience could feel an unmediated connection with the events. As one student reported: "When I shouted the slogans, I felt that 30 million people in East China were shouting with the same voice, and I shouted even more powerfully."[58]

The movement's name might lead one to conclude that denunciations of America always occupied center stage in the Resist America Aid Korea campaign, but in practice many of the denunciations focused on Japan, the far easier target. To the extent that the average person thought of America at all, until just a few short years before, it would have been as an ally. Enmity with Japan, however, was now more than a generation old. Memories of violence were still painfully fresh and nearly universal. The so-called rearmament of Japan was, therefore, the perfect topic to engage the attention of an audience, to indoctrinate them into the practices and habits of denunciations, to build anger and enthusiasm. Just the previous week, four million residents across Zhejiang tuned in to a broadcast, mostly in the Zhejiang dialect, during which people discussed atrocities they had experienced at the hands of the Japanese. A sixty-year-old woman named Wu, speaking her native Ningbo dialect, sobbed at the microphone as she described how the Japanese had killed her son and husband. Listening groups telephoned the station to express indignation at what they had heard on air. Three additional telephones had to be installed to handle the flood of calls. "The people must turn their hatred into strength," said the announcer, seeking to redirect the outpouring of emotion.[59] Luckily, the listeners had someone close-at-hand to redirect this energy in politically useful ways. In each workplace, school, or neighborhood listening station, an activist or cadre led listeners in postbroadcast discussion and facilitated the shift from passive to active listening. These cadre-overseers expected people to interact with the information given to them in the broadcast through written responses, confessions, accusations, production pledges, manifesto signatures, and hand-raise votes, all seen and supervised by the group. Cadres could make clarifications and observe adverse reactions like boredom. Criticism sessions could follow, imitating the format and language of the broadcasts and reinforcing the political lesson.

An example of such active interaction and analysis can be found in the broadcast-response letter of one Chongqing bank from the following week. Twenty-four bank employees attended the listening session, along with six of their family members. Everyone rushed to finish the day's work before the broadcast started; any tasks that could not be completed were left until after listening. Attention had to be undivided. Everyone milled around "in a mood of unlimited excitement," as the broadcast began. (Some of the most effusive phrases, such as this, were added to the report in another hand. Evidently, the bank employees were still mastering the art of speaking like Communists and had to be made sufficiently enthusiastic retrospectively.)[60] As a soldier told of his experiences in Korea, the listeners raised hands to show support for resisting American imperialism and opposing the rearmament of Japan. "The mood was running high the whole time we were listening. When they reported that the American beasts had massacred peaceful people, *every breast filled with hatred*. Some spoke through tears, some ground their teeth in anger and vowed to exterminate these bandits.... *Hearing of victories... made us spontaneously cheer.*" In the end, everyone came away with a deeper understanding of the war, and was satisfied with the meeting, according to the report. Each comrade-employee promised to write a condolence letter to families of soldiers. Along with intangible emotions like pride and anger, the broadcast also produced tangible fiscal results. The bank raised two million yuan in cash to donate to the cause, and every employee individually donated cash or other valuables.[61] While it is easy to be cynical about the reported effects of group listening exercises, such records are best read with a grain of trust as well as salt. The success of the Korean War was transformational for Chinese self-esteem and the new regime's prestige. China had fought the world's greatest power to a standstill, pushed an "imperialist" away from its borders in a war that was, for the first time in fifty years, fought on foreign soil. Discussing the war effort and denouncing America and Japan, as the East China Radio special did throughout Wednesday, was a good way to elicit support, sympathy, and participation from the audience.

Thursday's propaganda task was far tougher. The four hours of programming that evening shifted focus to domestic enemies, the threat of spies, evil landlords, capitalists, or saboteurs. The audience, reportedly 4.5 million people across East China, rose in feverish excitement to shout slogans: "Raise up the iron fist of the people, and ferociously strike the counterrevolutionaries!" Having listened to the program, students from Shanghai Datong University personally reported certain supposed special agents to the police.[62] The next day, Friday, April 27, the focus on internal enemies continued as listeners made public pledges to support the strict suppression of counterrevolutionaries. "The audiences expressed their determination to aid the government in mopping up enemy agents," read the headline in Shanghai's largest paper.[63] Having elicited shows of public support through participation, having solicited accusations, and having garnered enthusiastic consent for the punishment of enemies from the millions of listeners in East China, the four-hour long broadcast ended at 10:30 p.m. At midnight, the arrests began. Overnight raids netted 8,359 accused counterrevolutionaries "in response to the popular demand of the people." The detainees stood accused of being US-Chiang special agents, bandit leaders, members of reactionary religious sects, and despotic landlords. Those arrested included middle-school principals and the employee of a tram company. They had committed numerous supposed crimes, carried out sabotage, disseminated rumors, robbed the people, and plotted armed riot.[64] This effort to consolidate the revolutionary order on the home front by removing the threat of possible fifth columnists must have been long-planned, as was the broadcast propaganda theater surrounding it.

On the afternoon of Sunday, April 29, events culminated in an unannounced finale to the week's mass radio meetings. The usual relaxed Sunday fare was superseded by a special broadcast from the Yu Yuan dog track in the former French Concession of Shanghai. The Joint People's Representative Conference publicly tried several of the accused at that appropriately humiliating venue while 2.8 million people, including more than half of Shanghai's population, listened live. In every corner of the city, people gathered next to the radio. In Changning District, the audience reached 31,668 people, not counting high school and university students or teachers. More than thirty thousand residents listened to the broadcast in the alleyways by the North Station. In Yimiao District, almost fifty thousand people listened, including an old woman who was seen hurrying to her relative's house to be early to hear the program. In Alleyway 904 of Liyang Road, more than two hundred residents gathered, even though it was raining. They paid close attention underneath their umbrellas; not a single person left in the middle of the broadcast. In factories, the majority of workers showed up even though it was Sunday. At the Number 6 National Cotton Factory, more than two thousand workers arrived early before lunch to sit in the dining hall to listen. Students came to school and took notes, almost like they were in class. The Guangming Cantonese opera house stopped performing and relayed the conference to neighboring residents. Churches halted services so their congregations could experience this epochal event.[65]

By all accounts the people of the region were filled with emotion and tension as they listened to the trial, understanding the stakes both for the defendants and the still-new political order. "Now we will try the despicable enemy agent, Chen Xiaomao," announced the trial's officiant. When the workers at one factory heard this, they burst into applause, the sound of which merged with the sound of clapping coming from the radio. This sense of immediacy and participation continued through the many hours of testimony. "When everyone listened to Liu Xiuying's denunciation, they spontaneously said, 'This is Liu Xiuying's voice!' Suddenly, Liu Xiuying's voice became their own voice, Liu Xiuying's denunciation became exactly what they desired to say. When Liu Xiuying cried, they also cried."[66] At another factory, the workers heard some of their former colleagues stand trial; they were now determined to have been "special agents" who had been "concealed" within the company. The workers shouted without restraint, "Shoot them!" The listening groups all moved to this inevitable conclusion under the watchful eye of their activists and cadres. "Shoot them!" came the shouts from the alleyways. "Shoot them!" schools and factories demanded over the phone. At the close of the broadcast, in a spontaneous, choreographed denouement, nine of the defendants at the dog track lined up to receive their sentence: bullets to the head. Two-hundred eighty-five more of the accused followed the next day.[67]

The audience across the city of Shanghai evinced a full spectrum of human responses: they were emboldened or cowed, shocked or righteously happy, fearful or optimistic. A student at Zhongxi Girls High School listening to the broadcast suddenly ran up to an image of Chairman Mao and pledged an oath of allegiance. She turned to her classmates and denounced her father, who had been arrested the night before. The former dean of a law college, he had expelled more than one hundred progressive students before the liberation. "At first, I was very sad," the girl told her classmates, "but today I promise you: as a Youth League member, I will firmly stand my ground and with all my heart and mind endorse the government in arresting him!" In Hongkou District, two individuals unexpectedly attacked the radio receiver as residents were listening to the broadcast, attempting to destroy it. Bravely and foolishly registering their disgust, they shouted at their neighbors: "You have all listened to this. How will you sleep at night?!" A wrathful crowd seized the two and delivered them to the police station as counterrevolutionaries.[68]

The music, news, trials, denunciations, and slogans broadcast over the course of this last week of April constituted just one small part in the longer Resist

America and Aid Korea campaign. By its conclusion, an estimated one hundred million Chinese had listened in alongside the alleyway residents of Shanghai, the schoolchildren of Tianjin, and the employees of that small Chongqing bank. Xinhua judged that three quarters of the populations of Tianjin, Nanjing, Wuhan, and Tangshan heard regular radio propaganda during the movement. A year and a half after the founding of the People's Republic, radio already brought political campaigns, news, and other programming to an unprecedentedly wide audience in an unprecedentedly united newsscape.[69] Effective and efficient, mass-listening campaigns proliferated and lengthened over the course of the 1950s.[70] Storefront radios and corner loudspeakers carried the sound of news, music, and trials throughout urban areas so that even if one avoided it at work or school or the apartment block, one could not escape the denunciations. Loudspeakers and the unavoidable sound of broadcasting pervaded life as campaigns were organized for every major policy and purge.[71] The technology was particularly useful in inculcating new vocabularies and behaviors through group listening, voting, slogan shouting, and radio-led denunciations. Indeed, in the eyes of the officials and in the eyes of one's peers, the very act of participation in these activities meant assent to and acceptance of the new sociopolitical order. Radio was a technique of propaganda but also of control.

Radio's effectiveness as an administrative tool here reflected, in part, the advanced stage of the technopolitical process in cities. In other words, its mobilizing potential in urban areas mirrored a highly structured society. People were already closely organized into residential blocks, schools, and *danwei* that could be directed to ensure individual participation. More important, newspapers and telephone systems already assisted mass coordination in ways that were not possible in the countryside, where the great majority of Chinese lived. In rural areas, therefore, the radio-inflected newsscape took on a different valence. It ensured that campaigns were carried out on time and to specification by delivering instructions. It enabled mass mobilization in ways never before possible. The morning after the great trial in Shanghai, the Beijing national station transmitted an eager follow-up to cadres throughout the country. Broadcast at dictation speed so it could be easily copied down, the item asked propaganda officials to reflect on what had been successful so far in the Resist America campaign and what had failed. What methods were most welcomed by the people? What methods were worthy of retention and elaboration? Urging reflection, striving for improvement, the radio paced and encouraged the movement. In counties across China, rural radio monitors carefully took down the instructions.⁷²

Rural Listening Posts, 1949–1955

In Yiyuan County, Shandong Province, there were no roads passable by car, just a few simple tracks built during the Resistance and civil war for military use. In 1950, the cadres sent there had to carry their luggage on their backs or push it in a wheelbarrow as they made their way on foot. Newspapers still crossed by post over the mountains in baskets suspended over the shoulder on a pole or else on horseback. The papers arrived three to five days after appearing in the provincial capital. When it rained, they were even later.⁷³ The situation had not improved since the 1930s; maybe it had even become worse through decades of warfare. And Shandong, on China's eastern seaboard, was a relatively accessible place. Other locations, like Yunnan's far west, were even more remote. There, daily newspapers "seemed as if they were weeklies or monthlies." There was no telephone to speak of. Therefore, according to the recollections of one resident, the county government had to rely on the listening post to understand the party directives and policies for the suppression of bandits (anti-Communists) and land-reform work. The radio-listening post became the county government's "right hand man," its indispensable tool for administration, mobilization, synchronization, coordination, and compliance.⁷⁴ In a vast country where transportation was so consistently dire, where newspapers were so few, and where illiteracy was so high, radio became rural China's most important information technology—a communication lifeline but also a facilitator of tragedy.

The rural radio-monitoring network (*guangbo shouyinwang*) had its origins in an April 1950 Central Committee decision. The six large military districts that governed the country began implementing the plan that summer. They ordered every locale to set up radio-monitoring posts (*shouyinzhan*) "to increase efficiency and speed of government orders, control mass movements [*zhangwo qunzhong yundong*] and educate cadres and the masses."⁷⁵ In a now familiar pattern, the radio monitors (*shouyinyuan*) would be responsible for listening and transcribing news and government directives from national and provincial radio stations for use in blackboards, mimeographed newspapers, and broadcast to small listening groups. In form and structure, the program was identical to the radio-monitor program built by the Guomindang starting in the early 1930s. But whatever facilities the Guomindang *shouyinyuan* left behind were in shambles (*lantanzi*); radios were damaged and parts were missing.[76] In fact, most memoirs and contemporary records of New China's *shouyinyuan* make no mention of the Nationalist predecessor program.[77] Instead, the Communist regime, never one worried about historical amnesia, largely pretended to work from a *tabula rasa* when it came to radio communications. As a result, the uncommemorated Nationalist antecedents faded from view as the CCP listening-post network grew.

In the 1950s, the new system benefited from the absence of war and a continuing decline in the cost of radio equipment. Within a year and a half of the nation's founding, more than eighteen hundred of China's roughly two thousand counties had a listening post and a *shouyinyuan*.[78] At first, there were only one or two *shouyinyuan* per county, usually chosen from among the local school teachers, government employees, and eager party cadres. Sometimes they had a background or interest in radio, but just as often they did not. They were reliable, young, functionally literate, and often female.[79] Once selected, local officials submitted their names, geographic origin, education, and class background to the provincial authorities as a security precaution. Eventually, the trainees made their way to the provincial capital for instruction in electronics maintenance and repair, shorthand transcription, and propaganda strategies. As a class, the radio monitors exhibited a great deal of comradely *esprit de corps*. They met regularly at conferences (once yearly at the provincial level, more often locally) to exchange experiences, tactics, and battle stories. One monitor described the often brutally taxing work as "fighting a war on one's own." They labored to carry heavy equipment, including batteries (each weighing more than twenty pounds) on their backs up mountain paths and over streams to bring radio listening to the people. Sometimes *shouyinyuan* went through all this effort only to have the broadcast marred by interference, static, or poor reception because of weather. "We worked deep into the night while others slept to write down the news," recalled one monitor. "I remember a certain *shouyinyuan* who was pregnant still copying down broadcasts in the middle of the night, and because of this had a miscarriage." Another woman in Qinghai went blind from writing down the news in weak lamplight.[80] Radio work was not for the fainthearted.

The monitors returned to their home counties with a month's training and enough equipment to set up a monitoring post—a radio and batteries, a microphone, and a couple of loudspeakers. One returned with a Columbia Gramophone.[81] They set themselves up wherever they could find a set of rooms—the local land reform committee hall or inside the local Propaganda Bureau or a commandeered private home. Here they erected aerials (antennae) over roofs and made space for a mimeograph operation. In Yiyuan County, the monitor installed two loudspeakers, one on the roof and one over the door of the propaganda bureau offices. Every morning, afternoon, and evening, they broadcast the news. At night they might play a record. Regardless of the content, whenever the loudspeakers were turned on, several hundred residents, both cadres and average folk, gathered around to hear.[82]

The dissemination of news, a core duty of the *shouyinyuan*, occupied much of their time. Newspaper boards located at the head of a street, outside a school or government office, or in a marketplace—anywhere frequented and central—were a key aspect of this work. Pursuant to the central government's order establishing the monitoring network, the blackboards included "each province's important government orders and announcements, and important nationwide domestic and international news." They also reported on village work affairs, the progress of model personages, criticisms, summaries of mass opinion, as well as ordinary policies and laws.[83] In short, the boards contained the entire communist program, condensed and simplified for mass consumption, regularly refreshed, and freely available for the whole community to read.

In addition to the posted newspaper boards, the *xiaobao*, or mimeographed tabloid, format endured. Most *shouyinyuan* produced from one hundred to four hundred copies per edition—about as many as the mimeographing technology of the time could handle before copies became illegible. Given names like *Shouyin Jianbao* (condensed radio news) or *Shishi Kuaibao* (timely paper), the papers were delivered to villages, schools, work teams, cadres, and communes.[84] The *shouyinyuan* used schoolchildren to speed delivery on publication, usually every two or three days.[85] Important news warranted special editions and broadcasts through loudspeakers at the monitoring posts.

The radio monitors also facilitated radio listening by traveling into the countryside. While farmers rested from their labors at the side of their fields, the monitor would perch a loudspeaker on a piece of luggage or hang it from a tree and broadcast from a battery-powered receiver. In this way, news of collectivization policies, or enjoinders to work on water conservation projects, or to repair roads, reached even remote audiences.[86] *Shouyinyuan* did not always enjoy such treks, especially over the holidays. One year on New Year's Day, the Central Government ordered *shouyinyuan* across the country to go into the countryside to organize the peasants to listen to a special broadcast. Evidently, some must have neglected to leave their offices for this midwinter excursion because *shouyinyuan* who did head out over the holiday received a special commendation.[87] Despite the primitive and labor-intensive methods, the traveling radio-listening program yielded substantial results. In less than two months during the Aid Korea campaign, 59 radio receiving sets and four amplifiers were sent to 782 rural villages through Pingyuan Province,[88] and a total of 896,927 people heard anti-American war propaganda.[89]

Rural residents seemed to enjoy these radio-listening events. A few had never heard mechanically reproduced sound before, but most had only listened to a record player constantly playing the same tired recordings. Radio brought variety and a sense of renewal with its endlessly changing content. "Now there is the radio," said one farmer in 1950; "it has songs, it has conversation. Every day there are new speeches. It lets us understand national affairs. It really is amazing. . . . Our new society has changed completely in just a year."[90] More important for Communist officials, broadcasting brought the government respect. In fact, in the first months and years of the new regime, the radio-monitoring network helped Beijing maintain control over the countryside by continually delivering news of the administration's power. As one monitor recalled, during the "bandit suppression campaign" (to quash partisan resistance to CCP rule), the news from the radio-monitoring post brought information on CCP victories, thus "dispelling enemy rumor, encouraging the people, and quieting discontent."[91] News of victories in the Korean War delivered by radio was even more effective at quieting seditious rumors and encouraged faith in the new system. The memoir of *shouyinyuan*

Wang Helin in Fujian recounts that, during the war, "elements" in the villages spread "bandit" rumors that "the third world war is coming soon" or "the retaking of the mainland will happen soon" or "there is a concealed anticommunist army." Wang countered these rumors by spreading news of success in Korea, which he remembers quieted the naysayers. Even ignoring the content, early CCP broadcasting work could have an impact just through its existence. Hearing a nationwide broadcast could instill awe in the power of the government and offer a sense of connection to the capital. Listening to wartime news in a mountainous area of Fujian, one local exclaimed, "I never expected we hill folk in these ravines would hear the Party's voice coming from Beijing!"[92] The rural propaganda program enabled by radio thus began a feedback loop. The system delivered support and stability to the local government; the government could therefore build ever more penetrative communication infrastructure and denser newsscapes. The dialectic nature of the technopolitical process comes out clearly in this synthesis of political and technological practice.

The goodwill engendered by radio and its increasing penetration was put to use in the creation of cooperatives, the first stage of collectivization, in 1954. Ordered to broadcast ceaseless procollectivization propaganda, radio-monitor teams encouraged hundreds of peasants to gather together for listening meetings. Saturdays and Sundays had special programming that reported on the cooperative movement, repeating speeches that explained central government decisions, and conveyed positive experiences of collectivization elsewhere. When *shouyinyuan* encountered a recalcitrant village unwilling to surrender its private landholdings, they organized groups to listen to happy reports of other villages who had complied. "I loaded equipment into a car on Saturdays and drove from village to village, from field to home," remembered one monitor in Jiangsu. "One by one we organized the masses to listen to the special programs from Beijing. Before long we had our first cooperative."[93]

Radio became an important tool in executing government programs because of the confidence it generated. Hearing about the benefits of socialism through such a modern, authoritative machine instilled trust. Since the voice came from central government, many assumed its message had to be true. The news and orders were direct, ungarbled by intermediate officials. As one village cadre succinctly phrased it: "Listening to broadcasts for one night is more effective than us talking for ten nights. The peasants all believe that the policies discussed on the radio are good."[94] Even cadres evinced a similar reverent trust in the device. One day in March 1953, a *shouyinyuan* named Zhao Hanzhou was conducting his usual propaganda rounds in a rural district when he heard the news that Stalin had died. He quickly packed up the equipment and carried it on foot the twenty-five *li* back to town. There he reported the momentous news to the first secretary, who replied with an admonition: "Don't speak so carelessly, it must be a mistake." The secretary refused to believe it until he heard it himself on the radio, the only incontrovertible source of news. Finally, having confirmed the disaster, the secretary allowed the radio monitor to dash off a mimeographed notice, which the news office staff personally delivered to every district in the county.[95]

The receivers themselves also represented tangible proof of the promises of the socialist future, a concrete down payment on the assurances of future prosperity that helped convince villagers to join collectives. One *shouyinyuan* recalled a particular meeting where radio became a deciding factor for the wavering audience:

> I had the radio on top of a suit case, and the battery linked up beside it. I had carried it 10 *li* over the mountains. At that time we used vivid slogans. One went "Socialist modernization is an upstairs, a downstairs, electric lights and telephones." But there were no electric lights in the village, most people had never seen one, and didn't understand their use. Telephones? What benefit could those be to a peasant? But people were excited, amazed, quietly listening. They didn't move for an hour, listening like that. I shut down the radio to start the [discussion portion of the] meeting when a village cadre said in a loud voice, "Beijing, Nanning [the provincial capital] . . . speeches, and songs. When we hear these things in the mountains, this is socialist modernization! When we have collectivized, us villagers can very quickly have radio too, and lights."[96]

The village, having conscientiously ordered themselves to receive this information, wanted more. They decided to collectivize.

While it was capable of directly incentivizing people to reorganize their behavior, rural radio in the first five years of the People's Republic exerted its greatest influence on the *function* of government, through its power to direct local cadres. In a situation where newspapers and mail were slow, where telephone systems were in disarray, and transportation archaic, county governments had to rely on radio and the monitors that operated them to understand party directives and policies. One *shouyinyuan* recalled how he transcribed news and "study materials" (*xuexi cailiao*) from the central government onto bamboo paper, writing with hair-tipped brushes. This humbly copied material became the way the great majority of cadres and people understood the Central and Provincial parties' intentions and obtained important information. Studied by the county leadership, it became the basis for policy decisions.[97] During the land reform, lectures on the party policy broadcast from the provincial station, having been copied and printed and given to each land reform work team, became the blueprint for the movement. Some cadres in the field called these radio-produced guidelines Land Reform's "Timely Rain" (*jishi yu*) because it was a saving grace for the program.[98]

Endless mass meetings are one of the most recognizable and remembered aspects of Maoist society. Radio supplied them with form and content. The cadres themselves listened to ("attended") radio meetings that modeled formats for new campaigns. Meanwhile, *shouyinyuan* labored sixteen hours a day to transcribe materials that provided the content. A monitor in Guangxi remembered the role of the listening posts in those days. He copied down party lines, directives, and policies to give to the county leadership to study for prompt implementation. A radio program, he recalled, gave the county officials ideas on how to run the opening ceremony for mass meetings, helping to shape the form of future campaigns. The information he transcribed became the content for the gatherings. When he received news of the liberation of Xikang and Chamdo, it was read at the people's representative meeting to much applause and foot stomping. When the provincial Guangxi station broadcast ten guidelines for suppressing bandits, the leadership used this report and its content as its work directive the next day.[99]

Through the radio program "Agricultural Science" (*Nongye Keji Zhishi*), new "scientific" farming techniques (like chemical fertilizers or close-planting methods) reached local cadres. The listening station conscientiously copied down the agricultural knowledge relevant to their county and printed it in a small paper or sent the information directly to the villages and cooperatives. Sometimes they gathered the agricultural cooperatives' cadres for a big meeting to distribute the new "scientific" techniques to the attendees.[100] As in agriculture, the campaigns announced by radio came to affect all aspects of life, from childhood to marriage to burial. The government had stated in 1950 that the purpose of the listening post network was to "increase the efficiency and speed of government orders, and control mass movements." By 1955, the system could be judged a success by any measure. That year, the government claimed that the radio network served hundreds of millions of Chinese through 28,000 monitoring posts, keeping them abreast of news and information by the hour, or even the minute.[101] Gone, allegedly, was the time of waiting days or weeks for news. More important, gone was the time of waiting weeks or months for a government directive. Millions of cadres received orders, guidelines, and instructions simultaneously. The radio newsscape they built had already delivered enormous power into the hands of the government.

Wired Broadcasting

Still, this *shouyinyuan*-mediated information network failed to deliver the *sound* of radio to most people in the countryside. From 1956 onward, however, radio technicians, farmers, and laborers constructed a wired broadcasting network through which sound could reach the population directly.[102] The following year, the government began transitioning the country to a new form of broadcast propaganda, one that would allow the population to listen directly to information from the center with fewer mediating print materials. This was wired broadcasting—a system where wires connected a centrally located "broadcasting post" (*guangbozhan*) to hundreds or thousands of loudspeakers.[103] Amplifiers increased the power of signals broadcast from Beijing or a provincial radio station and then relayed the broadcast over wires to the fields, farms, canteens, and factories of the county or commune. The system was analogous to, and sometimes ran on, a network of telephone wires. This was a highly intermedial, integrated newsscape whose component parts were difficult to differentiate. Wired broadcasting and telephone connections, for instance, were often conflated.[104]

The broadcasters (*guangboyuan*) had two primary jobs. First, they continued much of the work of the radio monitors, transcribing, replicating, and transmitting broadcast information from the center—though now they included sound just as often as print. The second task represented a more substantial shift away from the *shouyinyuan* model. The broadcasting posts created their own news, collecting stories from the county or commune members to share with their neighbors. Inevitably, these contributors would recount personal experiences with production or politics; they were intended to be Stakhanovite model workers. But people saw in them stories about themselves and their community. If contributors were not people known personally to the listener, they were local enough to be relatable; and the information shared by neighbors could, in key instances, be far more believable than that coming from an outsider, on the radio or in person.

Why did the central government in Beijing undertake this shift toward wired broadcasting? First, it wanted to move toward a listening-based radio model, and wired broadcasting offered much more control than promoting personal radio sets. There was very little chance of anyone tuning in to Taiwan or Tokyo or BBC when the system was not private. Second, the wired broadcasting system was less expensive than providing the equivalent number of people with individual radio receivers. Third, the Soviet Union provided a good precedent for such a system, as with so much in Communist China. In the 1920s and 1930s, the USSR faced similar geographical and developmental conditions and turned to a wired broadcasting network as a solution.[105] Their wired broadcasting infrastructure had proved invaluable during the war against Nazi Germany. During the siege of Leningrad, the city's wired speaker system operated around the clock. Throughout the night, or when there was no regular programming, the speakers transmitted the slow beat of a metronome, testifying that the city had not fallen. Before an important announcement, the metronome quickened to let listeners know that orders or warnings would soon be coming through. Officials thus maintained direct contact with the people at all times, helping them hold out until eventual victory.[106] The party, especially Mao, now sought such constant warlike mobilization against imperialism, or nature, or internal enemies. Wired broadcasting allowed the immediacy, intimacy, and universality it desired.[107] It could manifest a mass society.

Unlike other government interventions, wired broadcasting also had the benefit of being welcome. It would allow many more people the pleasure of listening every day rather than on the rare occasion when a *shouyinyuan* happened to be in town. Villages requested installation, often covering the expenses themselves and providing the labor.[108] They hailed the arrival of the wires, sourcing and erecting the long lines of wooden poles that carried them across the landscape. One *guangboyuan*, Fu Changsheng, remembered how the cafeteria workers came out to the countryside to cook for the comrades doing the installation so they would not have to trudge back to the village for meals. In return, they placed some of the first loudspeakers in the communal canteens. Fu's team completed construction of his broadcasting post within two weeks, though initially only connected to sixty-two loudspeakers. The haste and amateurishness of this construction led to problems—a pattern repeated across Maoist China. For quite a few comrades, it was their first time climbing a telephone pole or planting one. It was a case of feeling the stones as you cross the river, Fu recalled, since there was very little knowledge or experience with technology. Sometimes, they had to run miles up into mountains at broadcast time just to see if a recently installed speaker was working.[109] In Shandong, the construction quality was such that up to 40 percent of the speakers required repair soon after installation.[110]

Still, people wanted more speakers and more programming. This was a "contradiction," to use the then-current Marxist phrase: the desire could not be fulfilled just yet. The broadcast time was limited by a sporadic electric supply, insufficient equipment, and a lack of money.[111] By April 1957, one county in Hubei had installed speakers in twenty-five villages and towns, and more than 250 communes: 59.2 percent of the towns and villages, and 59 percent of the communes. Yet in meetings and in written reports, cadres from the communes and villages without loudspeakers still protested that the network was not being put up quickly enough. "The people have a lot of complaint [about this]," one wrote.[112] Strong demand spurred the system's rapid growth. At the end of 1956, eighty Shandong counties had built wired broadcasting stations. By the close of 1958, 120 counties there contained broadcasting facilities, along with 90 percent of communes and 80 percent of large work teams.[113] Nationwide, there were 9,435 commune and 1,689 county wired stations by 1959. By 1960, 4,570,000 loudspeakers had been installed.[114] Even accounting for some exaggeration in figures in this period, it appears that wired broadcasting became pervasive throughout the country.

Who ran the system at a local level? As with the monitors, the new broadcasters were local figures like teachers, students, or minor cadres. Their amateur status was not always a boon for local broadcasting. In Guizhou, a teacher named Yin Huazhen was transferred to the Jun County People's Broadcasting Station in 1958. The new role was overwhelming. "At first I didn't put in a conscientious effort into the broadcasting work. I didn't know where to start." The former teacher in her missed her students and the classroom. The county leadership saw that she was wavering and patiently helped her improve her understanding of radio. They sent her to Wuhan to study at the Hubei People's Broadcasting Station, where she carried out copyediting for broadcasts. The experience, she recalled, improved her level of *putonghua* and her ability as a broadcaster. "In a short time of apprenticeship, I understood broadcasting skills, and met many experienced old broadcasters. It moved me immensely and increased my confidence. When I returned to my post, I was reassured and threw myself into the work."[115]

Yin's wired broadcasting station was located in a small courtyard home in the county seat. Two of the bungalows along the courtyard were renovated; one became the broadcast booth. All four walls were hung with sound insulating material. A partition inlaid with a thick glass window separated it from the control room. Both operator and announcer could easily see one another and coordinate their work. The control desk (*caozuo-tai*) started and ended transmission and controlled the volume. Next to it stood a transmitter, whose ninety wires broadcast sound to the streets, districts, villages, towns, and works teams—multitudes of people. "Though our equipment was very simple, it was brand new; this ensured our hardware for broadcasting [functioned]," Yin remembered. "Our offices were inside of the county paper's offices, and we worked together with them. We shared the small courtyard, where everyone was busy. Though old, the building was very cheerful."[116]

The arrival of such a broadcasting post was a significant moment for rural people. It was "a large event in the political life of our county," recalled Yin. "From then on Jun County people could hear their own village and locality's news, and find pleasure in arts and culture programs. They were able to immediately hear important information from the Central, Provincial, and County governments."[117] The people were enthusiastic, too. A cooperative member (*sheyuan*) said, "Broadcasting really is a treasure. You can understand all national affairs without going out your door. There is an old saying: a scholar can understand the world without leaving home. Now *we* can really understand the world without leaving home." Another person summed up their emotions by declaring, "Even in my dreams I didn't believe that broadcast could reach us here."[118] Indeed, some compared the radio to a living Bodhisattva, a saintly being, because of its potential for good.[119] Radio's good deeds were often related to the weather forecasting. For instance, on April 15, 1957, the local broadcast post in Hanyang transmitted warning of a massive rain storm approaching. The local commune mustered twelve hundred people to go in the fields to prevent flooding, minimizing losses to lives and livelihoods. Another commune turned the radio forecast into a challenge, having people work to combat flooding as long as the transmission continued. "Whenever the broadcasting went until, we worked till then." In a telling exaggeration, one collective member reported, "Whatever the radio says, we do."[120]

Of course, much of what the radio was saying (at least initially) consisted of the kind of endless litanies of production numbers beloved by communists. In a fit of honesty, one broadcaster reported that people thought radio programming was boring, reporting that a commune member said, "Every day it's *Production! Production!* There's never a bit of anything else." The *guangboyuan* were not unsympathetic to this complaint. In response, they added programming "closer to the life of the masses," stories about national reconstruction, family concerns, science and sanitation lectures, question and answer sessions, and casual conversations.[121] Some of the audience wanted more entertainment. Plays, especially by the county theater troupe, were most popular. One broadcast cadre described the audience when theater was carried over the radio: "They would rather not eat dinner and stay listening under the loudspeaker. One old man more than sixty years of age waited days to hear a play sung, and one night when it finally came said 'I've waited so many days, tonight I am hysterically happy' [*guo yi zu yi*]. When we were organizing listening groups, people said 'you only need to broadcast plays every night. Then you won't need to organize, we guarantee underneath every speaker the ground will be filled with people.'" Of course, the broadcast post could not simply conjure the much-needed entertainment. The people's stated desires and the broadcasters' need to propagandize were a contradiction, wrote the county broadcast team. There were not enough records to play an opera every night.[122]

Listeners also appreciated information about local economic conditions; such information sometimes even helped to maintain order. The inflationary pressures affecting everyday commodities posed a danger to the regime, especially given the role similar fiscal problems played in the fall of the Nationalists. Thus, in the spring of 1957, when the price of salt increased, the local wired station made sure to broadcast a piece on the reasons behind the price change. This quieted an incipient panic among the farmers. Before the broadcast, a long line of people stood waiting to buy salt at the store just outside the town gate. But the next afternoon only around ten people waited at the door of the shop.[123] Listeners also paid particular attention to health- and sanitation-related broadcasts. In one county, people transcribed the medicinal recipes from broadcasts, noting the ingredients and the conditions they could supposedly cure. Villagers began the practice of buying and wearing face masks (*kouzhao*) to avoid illness after hearing a broadcast on disease. Among the local farmers, a broadcast about the proper treatment of hog-cholera created a stir. For days afterward, they came to the station to copy the medicinal formula. A village head told the station that "the material has been extremely helpful to us. We have mastered this leading production method, and hope that you can broadcast this kind of material often in the future."[124] By curing the hogs of flu, local broadcasters won the trust of the people. Not all radio content would be so benign.

Socialized Media in the Great Leap Forward

In the earliest days of wired broadcasting, the majority of content transmitted on the local network came from outside—that is, from the provincial station, or Beijing. One broadcast post in Zhejiang only transmitted twenty to thirty minutes of self-written material a day. Initially, locally produced scripts primarily originated from the county party and government, town committees and departments, work units, the county propaganda department, or the county paper.[125] But during the buildup to the Great Leap Forward—a mass campaign for industrialization that collectivized agriculture and resulted in the worst famine in human history—news content multiplied. The increase in information from the center necessitated an expansion of broadcast hours. Initially, most stations broadcast two hours a day, but by 1958 seven or eight hours was not unheard of.[126] "There was a lot of news, editorials were long," Yin Huazhen remembered, "so we had to take turns copying down broadcasts deep into the night, with the goal of preparing the next day's editorials." The amount of locally sourced news reports also swelled. "Advanced persons came to the broadcast room to make speeches and introduce their experiences and spread their techniques," she recalled. They used every tool in their belt to advance the Great Leap: posting policy throughout the county on blackboards, broadcasting poems, songs, conversations between men and women, question and answer sessions, and short stories. The results were rather good, Yin thought. The county leaders came themselves on important occasions—there were always important occasions—to make speeches directing those below toward higher production and harder labor.[127]

Critically, though, this new form of media was not just a top-down affair. Local broadcasting posts soon introduced a new, less hierarchical form of media, where listeners became participants and contributors. The consumer also became the producer of news. Communist regimes are famous for their promotion of "model" workers used to encourage production or rectitude. These model workers, like the original Soviet Stakhanov, were often nationally famous. But inspirational figures were much more numerous at the local level. Wired broadcasting stations recruited dozens of individuals per month to contribute their own small-scale stories of success and production. In 1959, one county-level broadcast post in Shaanxi transmitted speeches from 178 individuals.[128] Daye County, one of the locales we will examine in greater detail, publicized 367 stories from their community in the first six months of 1957. This represented just under half the broadcasting post's total output. Everywhere contributions exceeded the number of "stories" actually transmitted. One county in Zhejiang received a daily average of fifty contributions from people all clambering to get heard by their peers.[129]

These contributors were called *tongxunyuan* in Chinese, which roughly translates to "reporter" but does not connote a profession; they came from everywhere and could be anyone politically correct. In the same Zhejiang district, it was the employees of the broadcasting post who developed these contributors. Competition to be selected to share a story on the radio was fierce; selection meant recognition of "advanced" status and a variety of political and economic benefits on the smallest scale (access to better food or more hours away from heavy labor, for instance). Competition to contribute also existed between districts and work groups. No one wanted to be left behind in the race to proclaim their own virtue.[130] The community enjoyed the new socialized media and paid increased attention to the broadcasts. One cadre contributor commented:

> When the station broadcast an article I wrote promoting the Abundance Commune Bumper Wheat Harvest, [our work team] was in the middle of a commune meeting. As soon as the team leader heard that they were talking about our commune, he quickly called an end to the meeting to go listen to the broadcast. Soon other commune members and nearby institutions became very eager to have their work reported over the air, and came to this reporter with reports of their experiences. In this way, our energy was not only much greater than before, but we also received the people's enthusiastic support.[131]

This system of socialized media had enormous power within the community it reached. It could multiply pride or shame. When a story about one woman's argument with her mother-in-law was broadcast to the community (as an example of conflict-resolution), she "threatened suicide and summoned members of her natal family to beat up" the woman who had shared it. "It was three years before the offended woman would speak to" the contributor again.[132] In addition to this power of shaming and extolment, the wired broadcast system brought new levels of *gamification* to campaign mobilization. Gamification is the practice of applying game-like elements to encourage participation—for instance, awarding points, status, or labels to successful participants, or creating competitions. The CCP had been using these tactics for years, but radio allowed them to operate on an unprecedented scale. Through local broadcasting posts, people and groups could be put in direct, live competition with one another. Sometimes these activities resembled a telethon but without passive observers at home on the couch. Here, everyone stood shoulder to shoulder with their neighbors and comrades, participating.

One such event took place in Huanggang County, Hubei, on the first two nights of December 1957, at 6:15. This "radio meeting" had almost one hundred thousand participants. (Pseudoscientific precision records exactly 93,789 "attendees.") The topic of the meeting was water conservancy and the great surge in wintertime production planned for the coming months. Water conservation work promised a farmer's paradise; there would be no more worry about drought or flood. In reality, most types of "conservation" work were ecologically destructive ventures riddled with unsound engineering that would be constructed in the depths of winter. By January, up to one in six people in China would be digging frozen earth in horrid conditions.[133] In Huanggang, that was secondary to the task at hand: mobilizing people for the project. It began with a notice in the county newspaper on November 27 urging everyone to listen in. Written pledges and water conservation plans were collected from select villages beforehand. Five commitments from model workers were prerecorded to avoid any possibility of error. As the radio meeting kicked off, the *guangboyuan* managed the phones and the broadcasting booth as the responses of villages and communes poured in, each competing to top the other. Village heads pledged live on the air to eliminate drought in 1958, to turn barren fields into productive ones, to work through the night to hasten the production of dams. People sat beneath loudspeakers across the county simultaneously experiencing the event, all hearing and registering the public promises. The program delivered an immense pressure to commit to more time and labor, along with a sense of accountability for one's pledges. It also delivered a participatory excitement. In places where the speakers were indoors, there was no place to sit, and people stood outside the doors of the buildings. In certain locations there were one hundred or more people gathered around every loudspeaker. "Listening to the radio meeting, the more I want to do something, the more energy I have," said one resident. Reports of reluctant participants who had come to see the light about water conservancy inevitably followed. "If people who are suffering so much can take up the cause, why can't we? It was a defect in our thought. Now of course we will take it up," said one repentant farmer.[134] Through the wired broadcasting network, only five people had organized this affair of one hundred thousand. Five people had organized the already overworked, underfed population of this county to labor outdoors at the height of winter, without compensation, without adequate shelter. Across China, untold thousands of people died from exhaustion and exposure, but the number of deaths during the Winter Production Campaign in Huanggang is unrecorded.

Radio continued to propagate and enforce new farming techniques and policies, as it had before the Great Leap Forward period. But now, instead of curing swine flu, the policies intensified the quest for rural mass production just at the moment that the ability of the local government to propagate and enforce the directives increased. It was, ultimately, an unfortunate confluence of the technopolitical process. In the spring of 1957, Daye County's radio station broadcast orders for cadres to attend a large meeting before the transplantation of seedlings, the busiest time of year for farmers. Convening in a model village, the station conveyed information on that village's collection of manure, its methods for planting the fields, and its irrigation systems to the whole county. Afterward, the broadcast post began reporting on the transformation of barren clay fields (*bainitian*) into fecund ground. The youth of two villages had shown in a trial that one could ameliorate *bainitian* through the application of chemical fertilizer. "This news greatly mobilized those commune members who were not paying attention to the elimination of fallow fields," according to the report. In Liangtang Village, there were two production teams that had more than 110 *mu* of clay fields. Originally the commune residents thought they were short of labor and were too busy to fertilize the clay fields. Undoubtedly understanding from generations of experience the futility of the exercise, they did not plan to apply fertilizer. But, having heard how other villages and communes eliminated these fallow fields, the two production teams gathered twenty-five laborers and treated more than sixty *mu* of land within ten days.[135]

Even after this concerted effort at persuasion, a few village and commune cadres still neglected agricultural production (that is, planting marginal land) and laid too much stress on doing paid side occupations (*fuye*). This was a rather prominent problem with production, according to the cadres, who feared that "if we do not prevent and overcome this reluctance, it will affect the commune's production and the completion of this year's production plan." In response, the radio station reported on yet other communes' use of fertilizer and its profitable outcomes. The station "used truth to criticize the idea that one can plant without" the heavy fertilizer that Communist agricultural ideology demanded. Fertilizer use finally surged throughout the county. For instance, in Wuqiao Village, the farmers eliminated 105 barren *mu* by collected twenty-four hundred catties of cow dung, which they burned to ash (destroying much of its nutrients) and spread on the fields.¹³⁶ All this effort had little effect on crop yields. Most of the "fertilizer" was useless; some of it even consisted of ashes of homes torn down for the very purpose of spreading the remains in the fields.¹³⁷ Soon, instead of spreading stories of successful change, the local broadcasting stations would turn to denouncing those who would doubt party policies.

In Wufeng County, Hubei, wired broadcasting was said to have "armed the cadres, and united thought." Indeed, here the listening post led the way in pursuing party policy and disciplining those who resisted—even before the Great Leap. During the 1956 Hundred Flowers Campaign, which embraced liberalized expression, the post enthusiastically broadcast sixty-two openly critical news items in accordance with the county committee under whose authority the post operated. But when the reactionary antirightist campaign came, the county bureaucracy was slow to begin carrying out its message. So the *guangboyuan* took it on themselves to begin spreading antirightist messages through the arts and culture program, kicking off the repression of those who had spoken out. The post's broadcasters batted away questions that had been raised during the Hundred Flowers with remarkable directness. They discussed topics such as whether, after the collectivization of agriculture, the lives of the peasants had improved or deteriorated. (Clearly, there was no consensus among the farmers.) "Has the leadership of the Chinese Communist Party caused the peasants to suffer? Or have the peasants emancipated (*fanshen*) themselves in politics, economics, and culture? We collected correct opinions to correct negative opinions." They denounced a man who claimed that the current tax office was no better than the Guomindang tax office. They recounted how much money, land, and clothing had been collected from the landlords in the early years of the revolution and how much it had amounted to per person. The broadcasters had residents come testify to the abuses they had experienced under the Guomindang, how they had received land, housing, and even a mattress after the revolution.

Through its control of information and communication technology, its utter domination of an intensified and unfamiliar newsscape, the Wufeng broadcasting post undermined the authority of the local county bureaucracy, which had not sought to emphasize the reaction. The county-level leadership could no longer mediate between high and low to dampen the wildly oscillating enthusiasms of the center. The old saying, "Heaven is High and the Emperor is Far Away" became less true. Subverting local hierarchies, the listening posts increased power of the party center—Beijing and the highest echelons of the party.

This process, called disintermediation by media theorists, is a common characteristic of mass society, often sold as a form of liberation. Disintermediation frees you of the middleman who might corrupt a message or tax a trade. At other times, it is sold as a method of enfranchisement. Connected to the center, every individual becomes a participant in national politics. Yet frequently, as Elihu Katz points out, this process results in even greater burdens. When Gutenberg's Bible disintermediated the priest, people became solely responsible for the salvation of their souls. When broadcasting techniques disintermediate some levels of the bureaucratic hierarchy, individuals became responsible for carrying out the revolution within themselves and within their community. Thus, Katz argues, disintermediation is a cause for concern. You may lose an "informed and reliable intermediary," a sense of community, and a "countervailing power" inherent in hierarchical social structures. In sum, "if enfranchisement is the reward of disintermediation, its 'punishment' is becoming newly dependent on direct relations with Kellogg's, God, or the medical encyclopedia, or worse, on the media that deliver their messages."[138]

In 1950s China, these losses—the so-called punishment of disintermediation—became acute. The commands issued by higher-ups were, in the counties under examination, executed regardless of the consequences. In its report, the Wufeng broadcast post could brag that everything had been put in order, despite all the questions raised by residents. "Labor productivity has been greatly increased. In grain work, besides carrying out policies, we also broadcast criticism of food speculators and of departmentalism [hoarding enough food to stay alive] in agricultural cooperatives." They condemned villages that did not participate in the common system for buying and selling grain, shaming them for selfishly hurting the poor. They asked farmers to come to the broadcasting booth to announce how much their grain intake had increased under collectivization to corral the wayward locales. The results were predictable. "Following a reduction in grain production due to natural disasters and striving for no reduction in total income in 1957, the [production] plan was revised in all townships of the county. And according to the demands put forward by various communes [it was decided] to rely on mountains and streams for sustenance [*kao shan chi shan, kao shui chi shui*]." In October 1957, rural residents were already being told to forage the wilds for food. Socialist modernization had arrived.[139]

A Tragedy in Xishui County

The Hubei Provincial Archive preserves the reports of several counties' *guangbozhan*, written between the founding of the wired broadcast stations and the eve of the Great Leap Forward. By chance, the most well-represented place among these reports, Xishui County, also happens to be the home of famine survivor and historian Yang Jisheng, who spent decades laboriously compiling evidence of the disaster.[140] Xishui offers a compelling case study in the operation of local radio broadcasting propaganda, mobilization, and coercion as the Great Leap Forward commenced. Though not necessarily representative, these documents illustrate the particular effects of the radio newsscape revolution on the lives of the people who lived in a single county over three of the most turbulent years of China's twentieth century.

Xishui County was only 100 km from Wuhan but seemed 'impossibly remote' to its residents. From the outlying areas it could take a day to reach the county seat over uneven mountain paths. Many villagers went to this town only once or twice a year. It was so isolated that some villagers still expected the return of Pu Yi, the last emperor.[141] A place more susceptible to the relentless persuasion of wired broadcasting could scarcely be imagined. The Xishui County station started transmitting on July 1, 1956, closely following the national and provincial orders establishing the system.[142] By the end of November, they had installed 1,510 speakers throughout the county. Wired broadcasting served 35 villages and 737 agricultural communes, representing 92.8 percent of the county's total communes. The great majority of the county's institutions, schools, factories, businesses, and handicraft communes had speakers.[143] Maintenance of this much wire (2,900 *li* of it, in addition to 1,063 *li* of telephone wires) posed a logistical challenge.[144] Village chiefs established committees to safeguard the lines, and unlike many other public responsibilities, this one was kept up. People liked the radio and were motivated to maintain the wires. When something malfunctioned, they came unprompted to the county seat to report it. Otherwise, they repaired equipment and replaced poles themselves.

On average days, those without special transmissions from Beijing or Wuhan about some new campaign, the station broadcast to the countryside for around two hours. The bulk of the broadcasts coincided with dinner, filling two precious nonworking hours with propaganda.[145] Cadres stood by to help interpret for those farmers who, because of their dialect, could not understand the broadcast Mandarin. About half of the pieces broadcast by the station—757 articles out of 1,467 in the period ending January 1, 1957—were transcribed from Central Broadcasting in Beijing or from the Hubei Provincial Station. The remainder the *guangboyuan* wrote themselves, mainly about the collectivization of agriculture, public health education, arts, and culture. For entertainment, the broadcasters relied on a collection of more than 420 records, which they increased daily by recording songs played from the provincial station. On weekends, an acting troupe, a school, or a rural social club would give a speech or performance, as a special treat.[146] These, too, were recorded for the future pleasure of the county.

Initially, the station relied on the county newspaper for local news—supplying the paper with provincial and national news in exchange—but soon developed its own reportage. To begin, it selected more than 130 *tongxunyuan* (contributors) from among residents with a good grasp on the political situation and an appropriate level of literacy. The station employees cultivated these contributors to improve the quality of the incoming stories, which had left much to be desired at first. They were taught to write short, punchy pieces, from two hundred to three hundred characters in length.[147] Drafts were corrected and sent back with words of encouragement. Whenever a movement began, the station promptly reported the plan to the contributors and bade them report their specific thoughts, providing the station with needed scripts. During certain campaigns, the station provided particular topics and concrete requests to manufacture a perfect range of news. By the end of 1957, they had increased the number of contributors from 130 to 413 people, among whom 70 were classed as peasants. In February 1957, they supplied 73 stories. In October, every *tongxunyuan* submitted a story. Of the 413 pieces contributed, the station broadcast 37 percent. On an average day, sixteen people shared their stories with the community. The highest day saw thirty-six people share; the lowest saw eight.[148]

Listening groups were organized throughout the county to encourage attention among the masses. Station workers sent around a mimeographed news-sheet titled "Red News" to notify the village and commune of important upcoming news, weather forecasts, technology, and work-results broadcasts. The commune then used the local address system (which also carried the broadcasts) to organize the residents. Staff from the broadcast station often went to nearby villages to observe audience reactions and sent listener opinion surveys to contributors, changing broadcasting methods and content according to the responses. For instance, formerly the station had broadcast the Radio Club program every Sunday, but the farmers did not follow the Western time pattern of a seven-day week. The station changed the Club program, which contained no politics but only music, opera, plays, and conversation, to the first and fifteenth of every lunar month, the traditional Chinese days of rest. On these days, the appreciative audience would gather under the speakers to listen to the program. A commune resident commented enthusiastically, "Radio is our social club."[149]

The station broadcast news of the Hundred Flowers movement, during which it encouraged the local population to speak out against mistakes, contradictions, and injustices, even those perpetrated by the regime. Then, in a rapid volte-face, mimicking the power politics at the top of the party like a shadow, the station urged on the antirightist campaign to help the county and people ferret out local "rightists." Denunciations soon flowed through the same wires that had moved the gullible or desperate to speak out. The radio station mirrored national patterns in other campaigns as well. In 1957, Xishui cadres and citizens wrote self-criticisms and confessions of instances where they had deviated from the party as part of a nationwide hunt for "deviationists."[150] These confessions were no doubt similar to the radio-inspired confessions recorded in a nearby county. There, people reported that certain programs about a cadre's deviation and redemption had "snatched [them] from the morass of capitalism" or "fostered belief and happiness in the socialist road." Radio was judged "an effective medicine" for capitalist thought.[151]

When the movement turned on certain cadres, the Xishui station broadcast the officials' own self-criticisms. The radio had carried the confessional model to the most local level.

The station's employees stayed in close touch with the county administration, attending every leadership conference or group. Before every major movement (like water conservation, rural socialist education, winter production, or the fortieth anniversary of the Russian Revolution), the county propaganda chief sought out the radio station to coordinate perspectives and requirements. This ensured that the station's day-to-day propaganda operations accorded with the intentions of the party. But the county committee also relied on the information held by the radio station "to examine the reported situation and research reported problems." For instance, when the time for reaping the early rice came, and it was time to plant the second crop, the first county secretary came to the station to find information about the early season rice forecasts. Presumably, he wanted information about what other areas of the country had reported back, so he could know what he was expected to report to his own superiors. The pattern continued when the propaganda department chief sought out the broadcasting station to research the reported plans during the Socialist Education and Water Conservancy Campaigns.[152] Receiving both plans from above and reactions from below, the station was the most well-informed body in the county, the heart of the social and technological infrastructure of news. This position within the newsscape gave it an informational advantage over county committee members, such that officials relied on the *guangbozhan* for instruction in a fundamentally uncertain political environment.

The station had a hand in organizing the increasingly frequent agricultural and labor movements. At the start of all Party Center Work (work projects ordered from Beijing), the county secretary spoke to the entire district via radio. He ordered every village to work, setting off the mass movement. At the same time, the broadcasting post passed on stories of exemplary work experiences to mass meetings and shamed noncompliant districts. For instance, during the campaign to increase production during winter—traditionally the farmer's only period of rest—it was discovered that a few villages were not implementing orders. The county secretary, therefore, broadcast a stern correction, and the station organized more broadcasts focused on implementing production in resistant villages.[153] The station helpfully coordinated the dates for planting, and it organized people to carry out repairs on irrigation systems—in the dead of winter. News about "successful" new agricultural practices encouraged the further spread of those practices. "A commune leader heard us broadcast about Bahe Village collecting fertilizer in field-side pits," wrote one broadcaster; "the next day he gathered the commune's committee to research Bahe Village's fertilizer collection method and put it to use." Another cadre organized eighty-three people to gather manure; they filled 112 manure pits in just a few days. Having toiled on the dung piles (their supposed use is unclear), one resident of the commune remarked: "Things on the radio have all been researched by the county committee. Of course we'll strive to do as they say."[154]

In the odd moment when the loudspeakers were not haranguing exhausted residents to greater production, the radio discussed solar and lunar eclipses, among other scientific topics. In October, it reported to the bone-weary farmers that the Soviets had launched the world's first manmade satellite.[155] Loudspeakers transmitted the party's thoughts on the importance of Sputnik to villagers knee-deep in manure pits that would not, in the end, serve any productive purpose. By 1958, many county residents were already showing signs of edema, the swelling of limbs that accompanies the body's self-cannibalization. It signals the advent of a painful, lingering death by starvation. Yang Jisheng noticed these signs in his father but did not recognize their cause. By the time he rushed home in April 1959, his father was skin-and-bones, unable to speak or move. All around the village, bark had been stripped off trees, eaten by desperate farmers who had also drained the ponds to consume the muddy, repulsive snails that lived at the bottom.[156] The villagers were indeed "relying on the mountains and streams for survival." They died surrounded by loudspeakers, hounded by news, testifying in their own terrible way to the persistence of the technopolitical process and the mass society that had, finally, arrived.

Radio and the propaganda it carried had not needed to convince everyone that the misguided policies and superfluous movements were essential. It had only needed to fully convince the lower cadres, a class vetted by the party but selected by the people from among themselves. Like the news that circulated through the broadcasting posts, this class appeared to emerge organically from the bottom up, a reflection of society's technopolitical reorganization and an embodiment of the new mass forms. These lowest level of representative held significant powers of coercion, a near monopoly on both food and legitimate violence. They were already predisposed to believe what the party told them. But the unending stories of success from around the country, and from the local community, made *everyone* more willing to overlook evidence of failure in front of their own eyes. Enough everyday citizens believed what the news had told them, such that few instances of mass resistance are recorded. There was no guarantee that this would be the outcome of radio's newsscape revolution. The disaster that resulted must primarily be blamed on ideology and the policy choices made by leaders, especially Mao, within the framework of their utopian-totalitarian ideal. Nonetheless, the changes in the newsscape that enabled those policies to become a widespread and relentlessly long-lasting tragedy cannot be ignored. News multiplied through new technological practices. A mass society reordered itself around listening posts and wired networks. Through the resulting socialized media, the technopolitical process brought the catastrophe of the Great Leap Forward within the realm of possibility.

Yet techniques of politics and communications do not change in a linear fashion. People become accustomed to formerly destabilizing information systems. They learn to manipulate new formulations of the newsscape. History teaches us that credulity, unlike the desire for news, is not infinite. The dynamics of centralized hierarchy that characterized the 1950s would not last forever.

EIGHT

The Technopolitics of Disorder

The summer of 1976 stretched long in the rural towns and villages of coastal Zhejiang. Near Taizhou, the grandchildren of the students who had spread the news of May 4, 1919, wore short-sleeve undershirts into September. At one countryside high school, a sixteen-year-old named Li Dai, who had just started his sophomore year, recalled that they conducted some classes and activities in the shade of the courtyard, where one could catch a breeze. He did not find it unusual, then, when at noon on September 9 the teachers asked him and some other boys to move the classroom benches outside, between the schoolhouse and a large tree, from which a loudspeaker hung. Nor did he find it unusual that they would hear an important broadcast there at 4 p.m. In those days, people had become accustomed to such "emergency" broadcasts—for more than ten years they had dutifully organized for such communication events in the name of the Great Proletarian Cultural Revolution—in addition to their regular listening. Thus, the students and teachers were chatting among themselves, unconcerned, when a slow, monumental dirge burst from the horn above them. After a series of long circumlocutions addressing the "whole party, whole army, and whole nation"—an audience their grandparents' generation rhetorically claimed but could never

FIGURE 8.1 A cadre uses a telephone handset and amplifier to exhort hundreds of people hoeing the earth. The telephone poles, wires, and loudspeakers at their backs carry the address. Source: *Wuxiandian* [Radio], no. 12, 1959. Author's personal collection.

realize—a male broadcaster finally revealed what some had probably already guessed. The chairman was dead.[1]

One hundred million loudspeakers, strung through wired broadcast networks, delivered the news of Mao Zedong's passing to hundreds of millions of individuals.[2] The system was unprecedented, not only in its scale but in its comprehensiveness. Many locales claimed to have one loudspeaker installed in every workplace or household.[3] Officially, 92.7 percent of production teams and 70 percent of households in China were connected to the system.[4] This did not even take into account inexpensive wireless transistor radios beginning to flood the country.[5] Before this new radio age, within every adult's living memory, regular news and information had largely been restricted to the elite, literate, and urban. Even though information often passed through radio waves as it circulated through the newsscape after 1923, the 90 percent of people who lived in the countryside primarily experienced news about the outside world through word of mouth. The broadcast network had thus radically shifted the infrastructures, geographies, and practices of news over the previous decades. By 1976, most people received information directly, on a daily basis, often through the medium of mechanically reproduced, sound-encoded information. A mass media, oriented toward and theoretically capable of universal consumption, had been realized.

The development of this new informational and technological environment was inseparable from the development of new forms of political organization, particularly the mass mobilized party-state, over the same period. More intensively organized political forms and more penetrative and reliable communication technologies grew in concert. This book has described this coproduction of an information-saturated newsscape and an all-encompassing political system as aspects of the technopolitical process, arguing that mass politics cannot be formed without the technologies and practices of mass media, and vice versa. Yet, it would be a mistake to assume that the technopolitical process has a linear relationship with state construction, stability, and centralization. Information saturation, for instance, does not necessarily lead to uniformity, greater control, and more order. In fact, the forms and structures of a highly developed newsscape can, under the right conditions, result in political fragmentation and even chaos. Examining several episodes from the first two years of the Cultural Revolution, we conclude our story of China's newsscape revolution by examining the ways in which, perhaps counterintuitively, the proliferation of news can accompany social breakdown.

Instigated by a marginalized Chairman Mao in an effort to seize back power and cement his legacy as the founder of a truly *new* China, the Cultural Revolution (1966–76) aimed to overturn a supposedly calcified party bureaucracy, as well as thousands of years of cultural heritage. Though prompted by elite politics, millions of young people took up the cause of this revolution as their own, marching to destroy all remnants of the old, oppressive order—flattening ancient monuments, burning feudal books, and attacking authority figures. Red guards, self-organized troupes of young people, began by struggling against teachers, administrators, and bureaucrats, beating some to death, and driving far more to suicide. Then, in a fashion reflecting the anarchist nature of their organization, these groups turned on each other, seizing weapons caches and fighting pitched battles in the streets while claiming to be the true defenders of Mao's liberatory ideology.

Systems and processes of communication propelled and framed this movement. Rather than aid a centralized stability, intensely developed infrastructures like the wired radio network mediated and synchronized a process of social disintegration. Factions formed according to the ideologies and orders coming over the radio. News technologies became weapons in an internal struggle; groups even maimed and killed in order to seize broadcast stations. The very structure of the communications system encouraged the strife. Wired loudspeaker broadcast networks reached hundreds of millions of people but operated from an extremely local level (perhaps a single factory or school), simultaneously allowing nationwide mass organization and close-quarter conflict over their control. This polycentric infrastructure, along with big-character posters and mimeographed news fliers, encouraged a society already primed to see enemies in every quarter to explode. Indeed, recent study has shown a strong correlation between the quality of radio reception in a given county and the number of people killed there in the period.[6] The Cultural Revolution, in other words, can be interpreted as a phenomenon within the newsscape revolution, a communications event from the beginning.

Seizing the Means of Communication

Few spaces hold such a prominent place in the history of China's twentieth-century newsscape as the east wall of Peking University's former cafeteria. Though it has since been torn down, the wall once abutted a triangular intersection where the paths between the dormitories, classrooms, and library crossed. As people moved through their day, they could not help passing by, so—as in decades or centuries past, as in many other cultures around the world—the crossroad became a space for communication in both senses of the word.[7] In 1957, during the short-lived Hundred Flowers movement, intellectuals criticized the injustices in society through essays pasted to this stretch of gray brick they named the "Democracy Wall." Those who spoke out were quickly suppressed. It came as a surprise, therefore, when at the end of May 1966, students and faculty once again began to encrust the facade in hand-written commentaries. Photographs of the time reveal individuals climbing up ladders to paste new screeds attacking this or that faction, calling for ever-increasing radicalism, demanding revolution. In one image, a large crowd mills about, standing back from the wall, reading the bold characters, and observing the posting of new texts, which begin to crowd and then obliterate the old. Indeed it was here—within the dynamic interactions of the built environment, communication technology (in this case, paper, ink, and writing), and human behavior—that the Cultural Revolution was born.

It began with a single poster, on a single day. At the instigation of Mao's agents, a committee from the Philosophy Department drafted a political manifesto—Nie Yuanze, a disgruntled assistant professor who was about to be denied promotion and fired, joined at the last moment, though her name would become most closely associated with the document. Written in characters several inches high, on broad sheets that took up most of the width of the wall, it attacked the university president and two municipal officials for insufficiently supporting Chairman Mao's effort at cultural revolution, which most people then understood as yet another run-of-the-mill political campaign, one of dozens carried out since the establishment of the People's Republic some sixteen years before. The manifesto and its author would soon become immensely famous, but when it appeared on the cafeteria wall at 2 p.m. on May 25, 1966, it was uncertain whether its daring attack would lead to a swift backlash or, just as likely, be quickly forgotten by everyone but Nie and the president she had berated. It was, in other words, highly unlikely to spark a revolution. Indeed, over the first few days, the backlash seemed to win as hundreds of posters appeared attacking professor Nie as a rightist or antiparty element.[8]

Lost in the furor, and in its subsequent fame, however, is the actual content of the manifesto. In it, Nie (as lead signatory) attacks the university

FIGURE 8.2 A broadcast worker adjusts a loudspeaker as people labor in teams to pull carts of earth uphill. The broadcast system appears to dominate both the people and the landscape. Source: *Wuxiandian* [Radio], no. 1, Jan. 1960. Author's personal collection.

administration for attempting to stifle the revolution through their "guidance" of the movement. The university "has been deathly still and quiet," she wrote. "The strong revolutionary demands of the broad masses of teachers and students have been suppressed." She denounces the methods of this suppression, the ways in which the leadership has suffocated speech on campus. In repetitive and quotation-filled prose, Nie criticizes attempts by the university to monopolize and circumscribe the techniques of communication available to people. They have controlled discussions. They have only allowed wall newspapers in *small* characters. She mentions meetings and posters—the technologies of news—a dozen times in the document, which is only three typewritten pages in translated English. She cries out in one representative passage:

> [The university leadership says] "problems cannot be completely solved by merely holding meetings," "Peking University should not post wall newspapers in large characters" "We should lead them to hold group discussion and to write wall newspapers in small characters"—why are you so afraid of wall newspapers in large characters? . . . To hold meetings and to post wall newspapers in large characters are mass militant styles in the best form. You would "lead" the masses not to hold meetings and not to post wall newspapers in large characters. You have made such regulations. Have you not oppressed mass revolution, forbidden mass revolution, and opposed mass revolution?![9]

What is the revolution that Nie and her colleagues in the Philosophy Department are calling for? They demand the ability to hold mass meetings and write big-character posters—messages that large groups of people can read at a distance, as opposed to regular printed type, with small font, which only a few people can read close-up. The difference between big and small characters, therefore, is the difference between mere theorizing and mass communication. Understood this way, the big-character poster becomes a technique of written broadcast, just as the meeting transforms into a technique of oral organization. The reason why these two social technologies lie at the heart of her manifesto thus becomes clear. The rebels are calling for a fundamental shift in the practice of news, demanding the right to use their voices and pens and paper to organize and transmit political thought. She is demanding that the people, however defined, seize the means of communication.

Mao was well aware of the revolutionary potential of such a seizure, how it would undermine the party and the state, which had kept him out of power for six years since the disaster of the Great Leap Forward had forced him into semiretirement. He also knew that China's mass communication system was highly diffuse and vulnerable to a seizure from below. A more centralized system based around a few key radio stations might have been susceptible to a coup from above, as so much twentieth-century history has shown. But Mao did not trust the loyalty of the radio administration or the military. (In the event, Mao maintained control over the main radio stations in Beijing and kept the army at bay, but this was certainly not guaranteed.) Since, however, China's newsscape consisted of many small broadcast stations connected by wire to networks of speakers, small groups of rebels could easily seize and mobilize them against the party hierarchy, initiating a thorough, society-wide destruction of the ossified old ways. On June 1, having read about the Peking University manifesto in a domestic intelligence report, Mao ordered it broadcast the same day.[10]

As the Chairman himself later said, the call to overturn a hierarchical newsscape "caused a huge uproar."[11] He was not wrong. Listeners and readers across China expressed shock as the news reached them, though they had been prepared for *something* important. Beijing had broadcast instructions ahead of time, warning institutions of an upcoming announcement and ordering them to organize people to listen in to the 8 p.m. special. One former high school student from Shanghai vividly recalled the day. Even before the announcement, he had already noticed harshly worded editorials coming through the radio and unusually large, bold characters in the newspaper headlines. "The news that evening must be big and very urgent, how could we not listen to it on time?!" he remembered thinking. At the appointed hour, then, he heard the text of Nie's big-character poster being read, along with his entire school.[12]

The momentous information also reached comrades far away from Shanghai and China's other large cities that evening. Near midnight in rural Guizhou province, a cadre settled in to listen to the radio after a long evening at village production meetings. Soon after, he recorded in his journal an editorial from the main national newspaper, the *People's Daily*, which was broadcast alongside Nie's manifesto.[13] The piece had gone from Mao's hands to the ears of a remote rural party member in less than half a day. No doubt the cadre or someone near him also transcribed the editorials for publication, as intended. Newspaper items were still broadcast like this, as content on the radio, sometimes even before their official publication, multiplying information across many media. The practice extended the geographical reach of news, helping to overcome the physical limitations of print—namely that

FIGURE 8.3 People laboring at night gather to hear encouraging news from the radio loudspeaker. Source: Liu Zhigui, "Hongse dianbo chuan xixun" (The red radio waves bring good news), 1974. Author's personal collection.

the *People's Daily* editions were published in a few large urban centers, and transportation infrastructure was nowhere near developed enough to ensure same-day delivery except in proximate suburbs. Physical geography remained a significant determinant in the newsscape, as it had been in 1919. To transmit texts with greater accuracy, a Morse numerical-code transmission of the text supplemented the manifesto's broadcast. A local radio monitor could write down the four-digit sets and know exactly which characters to use for publication, according to a widely available codebook. Indeed, the CIA's Foreign Broadcast Information Service picked up one of these Morse transmissions just after the first voice broadcast, at 8:31 p.m.[14] For many people, then, the *People's Daily* did not exist as a "newspaper," or even a "broadcast," but as a source of the information reprinted in their local newssheet after having been picked up on the radio. Such was the highly intermedial complication of the Chinese newsscape.

Still, the radio broadcast astonished those who did hear, even in Shanghai, with its large newspapers and regular flow of information. "Before we had even seen the newspaper, the important points of the editorial had been transmitted to our ears through radio waves," remembered the student, "telling all revolutionary masses nation-wide to 'sweep away the cattle, ghosts, snakes and gods.' . . . Smash the so-called bourgeois 'experts,' 'scholars,' 'authorities' and 'revered figures.' Wipe them out."[15] Such unsettling news quickly became the center of discussion as people amplified the information by word of mouth, speculating on the implications and worrying about the personal consequences. A teacher at Beijing Normal University recalled that she was still at dinner when she heard the broadcast. Colleagues from the Education Department soon found her and, having ascertained she had heard the news, asked what to do. She replied that they should learn from the antirightist struggles of the 1950s, which were surely being repeated again. She could not know that she would soon become one of the earliest victims of the movement, accused of being a capitalist roader for teaching foreign education methods.[16]

Though most who heard it were sure of the broadcast's importance, fewer were certain of its exact political meaning. One who believed that he really understood this nuance was an American-born party member and broadcast station worker who was visiting the CCP's wartime headquarters of Yan'an, in the wild north of Shaanxi Province. His reaction is telling:

> That night in the little inn, we once again tuned in to Radio Beijing. This time, I was the one who was shaken up. . . . Suppression is wrong, the *People's Daily* reported. In ringing, almost visionary language, the paper announced the beginning of a new revolution in China. *The way a country goes rotten is from within,* the broadcast continued. *There are revisionists in the party, people who would suppress, control, lead us down the wrong road, and even attack Chairman Mao himself. Don't let that happen in China! The high tide is now rising! A new day is dawning! Think for yourselves. Revolutionaries, arise! Think of the words of the Great Chairman Mao. Ignore all orders that conflict with that. It is right to rebel.* The words were electric.[17]

The liberatory language made the broadcast extraordinary for him. Suppression is wrong. Think for yourselves, put up big-character posters, and take up the technologies of communication. When he returned to his office in Beijing just a few days later, a remarkable transformation had already taken place. "The hall fluttered with posters. They were hanging from the walls, or from clotheslines strung out across the room. Some were written on sheets of unused newsprint, others on newspapers themselves. The sheets were pasted together so that the tails of the longer ones dragged on the ground. Most were painted in big, thick black Chinese brush strokes."[18] Anyone could now share their ideas with the crowds that gathered to read. For a while the conflicts and attacks contained in those messages stayed on the walls, in words and not in actions. But by early the next year, internecine struggles broke out at the broadcast administration. Criticism sessions, beatings, and suicides became common.

Pacing the Revolution

Just to the north of Yan'an, the dusty and largely barren Yulin County received Nie's speech denouncing the Beijing authorities on the evening of June 1, 1966. The morning after hearing the Peking University manifesto, eleven students from Yulin's high school took up its call to seize the right to convey their thoughts. They posted a big-character poster on the door of the Yulin Party Committee, calling it a big black shop (*da hei dian*)—essentially labeling the local party apparatus class traitors. Later the same day, June 2, forty or so workers also posted a big-character poster at the entrance of the bell tower (the town's own crossroads) calling for people to "open fire" on the county committee.[19]

As in much of the rest of the country, broadcast news from the center paced and encouraged the revolution in Yulin. We have seen how individuals and governments as disparate as Zhang Zuolin, the Nationalist Party, and Japanese-collaborationist regimes dreamed of enacting politics at a local, everyday level through a communications infrastructure. Since the founding of the People's Republic, this had become a quotidian reality. Now information arrived over the airwaves and transformed in concrete action at the local level, such that the rhythms, behaviors, and arcs of the Cultural Revolution played out in this remote place. At 4 p.m. on June 3, the Beijing radio station broadcast the decision of the newly reorganized Beijing Municipal

Committee to assign "work teams" (*gongzuo zu*) to lead the Cultural Revolution at Peking University. A *People's Daily* editorial praising the decision was transmitted early the next day.[20] Soon after, in a similar attempt to direct the early outbursts, the Yulin County party sent work teams of three to five people to supervise the writing of big-character posters in schools and in the newspaper office. They hoped to "mobilize and guide teachers, students and cadres," preferably in ways that would not implicate the establishment in any malfeasance. This was difficult, however. Accusers quickly began condemning teachers and leading cadres with landlord family backgrounds, poor interpersonal relations, or so-called "bad lifestyles." As the local history recalls, "critical articles, expository big character posters, and cartoons . . . blotted out the sky and covered the earth. Denouncing and criticizing birds came tweeting out of the woodwork, and the movement gradually escalated amid the noise."[21]

By July, the schools, government, and even the supposedly party-directed work teams had split into left, center, and right factions. For weeks, the radio had been carrying stories of revolutionary struggle, stories about brave factions facing covert counterrevolutionaries, putting up their big-character posters and striking a blow for true Mao Zedong Thought. Troublingly, however, no one knew which faction constituted the true revolutionaries and which the subversives. Dueling manifestos, each the hero of its own story, denounced rival groups with full confidence in their own correctness.[22] With no clear direction, but with every incentive to signal their own righteousness through public communication, groups quickly coalesced in every institution. In Yulin schools, teachers who had problematic backgrounds, who quarreled with others, or who had made political mistakes were classified as part of the "right" or "gangster" faction. Soon, big-character posters, cartoons, and speeches exposed their crimes. On July 8 and 9, national broadcasts once again carried stories encouraging the "broad masses" to "criticize the local monsters and freaks."[23] Accordingly, on July 11, some students painted four of the town teachers black and made them wear high conical paper dunce caps. Placards enumerating their supposed sins hung around their necks. In the two months following the June broadcast and the sudden ability of people to create their own widely read news, a total of 70 percent of the teachers in Yulin were criticized. Four teachers suffered mental crises and committed suicide under the frightening attacks. Crowds beat an additional 113 school leaders and teachers as "gangsters," "ghosts and snakes," or "three anti elements." The undesirables subsequently went to rural farm communes to perform supervised labor.[24]

This was only the beginning. In late July, Chairman Mao decided that all party oversight, which had existed mostly in theory anyway, should be removed.[25] On August 8, the Sixteen Point Manifesto announced, through a nationwide transmission, that only the masses could liberate themselves. Subsequent editorials, also transmitted over the airwaves, criticized the work teams for stifling the revolution.[26] "After that," the local history recalls, "news of rebellion against the Cultural Revolution work teams came from other places." Posters, meetings, fliers, and marches propagated tidings of the insurgency across the county. The work teams then withdrew from Yulin's schools, newspapers, and other units. With the change in national politics, the students and teachers who had been criticized earlier in the summer gained the upper hand. They "took to paper in late August, posting big character messages attacking the working groups for implementing a bourgeois reactionary line to the revolt. Then some schools seized members of the work teams to denounce them, and many people were beaten."[27] With all guardrails removed, positions and factions hardened, and people became further radicalized amid the torrents of information.

By the end of September, there were more than ten thousand Red Guards in the county. Explaining this phenomenon and the destruction it caused, the local history records that, over the last two months of summer the "trend of *destroying the Four Olds* suddenly arose." Yes, "suddenly arose," but, of course, there was nothing passive or mysteriously spontaneous about it. The Sixteen Point Manifesto had urged party activists to greater violence on August 8, saying, "Don't be afraid of disturbances. Chairman Mao has often told us that revolution cannot be so very refined, so gentle, so temperate, kind, courteous, restrained and magnanimous."[28] The sounds of the Red Guards receiving the approbation of Mao and a million comrades on Tiananmen Square went out live to the nation on August 18, encouraging youth across China to take up their banner.[29] On August 23, the central radio station carried stories and editorials approving of the destruction of the Four Olds in the streets of Beijing: the harassing of bourgeois hairdressers, the burning of foreign books and magazines, the renaming of streets, and banning of fancy foods.[30] Two days later (one day to publish the news, one day to organize), on August 25, the Red Guards of all the high schools in Yulin City marched through town, posting fliers and ordering all the shops to annihilate the Four Olds. They proceeded to destroy statues, shrines, tombs, ancient stone carvings, traditional houses, and precious paintings. They ransacked four-hundred-year-old temples. They rushed into the homes of prominent families, scholars, and any perceived counterrevolutionaries, smashing all vestiges of the old order.[31] Through the synchronization of radio, through intensely developed technopolitical infrastructure, the Red Guard groups of dusty Yulin replicated, nearly contemporaneously, the events occurring across the country.

They were not, however, a unified body. In fact, the structure of the newsscape and the encouragement of the elite politicians meant that Yulin society continually ramified over the course of the fall, as faction split from faction. Each utilized convenient means of communication—big-character posters, loudspeakers, marches, and drums—while they awaited an opportunity to seize the real prize: the wired broadcast station.

Mediating Social Collapse

As we have seen, by the time of the Cultural Revolution, many communes, farms, and factories had their own wired broadcast stations—a network of cables strung between loudspeakers and a central transmission office. These networks tolled the rhythms of everyday life. As one *guangboyuan* from a rural factory recalled, the wired broadcast station announced the start of the day at 6:25 a.m. with *The East Is Red* and the pronouncement, "This is the People's Machinery Plant Broadcasting Station. Today's first broadcast is now starting!" At 6:30 they relayed Beijing Central Broadcasting station's "News and Newspaper Summary," along with the provincial news program. "Because in those days there was no mobile phone, no TV, and even newspapers were delivered three or five days late," remembered the broadcaster, "radio waves became the fastest transmission channel for national events." The lunchtime broadcast focused on news from the factory. At 4:30 p.m., a third transmission reported important news and the daily production results as people got off work. "Surrounded by the sound of broadcasting, people shook off the tiredness of the day and went to the canteen, the baths, and the library. Of course, in the family area, under the sounds of the broadcast, cooking smoke curled upward, and every family had a fragrant smelling meal." At 8 p.m., a final broadcast relayed the Beijing Central Station's news program, followed by the Internationale and the exclamation: "The People's Machinery Factory Broadcast Station daily broadcast is now ended. See you tomorrow!"

> The sound of the radio was the wake-up number in the morning, the punch card for going to and from work, the clock for dining in the canteen, and the adviser and helper for people's daily life. The mountain people [of this remote district] had long formed the habit of listening to the radio. If the loudspeaker didn't sound for some reason, it was a big deal. The telephone in the radio room would ring all the time. Once in a while, if the first broadcast ended a minute or two early, some people would be indignant because they are locked outside the factory and marked as late. They would then want to "pay us back" face to face. When meal times were approaching and the broadcast had not yet sounded, there would be a lot of noise in the canteen, and people would knock on the dishes and chopsticks, [shouting] "OK, start the broadcast!" Therefore, everything else could be a mess, but broadcasting could never be taken lightly. Thirty or forty years later, when we ran into a child from that time, he mentioned how the song "We Workers have Power" was still unforgettably familiar. As soon as it rang out, mom and dad could come home from work.[32]

Thoroughly integrated into everyday life, the wired broadcast systems were thus uniquely positioned to mediate a social collapse. The stations both received orders from above through the radio receiver and organized the individuals below to carry out policies. Someone sat in an office and controlled exactly what passed between "the Center" (Beijing and whoever constituted the party leadership at that moment) and "the People," manifesting Sun Yat-sen's vision of individuals regularly connected to the nation. In their overt acts of mediation lay great power. Beyond the influence they promised, the stations' simple form—they often consisted of not much more than a radio receiver, some records, and a microphone in an office—amplified the temptation to seize control. Carrying out a local coup without a means of coordinating action and administration is an intimidating task; one has to intimately understand the levers of power already. But with the ability to rally and instruct the great majority of individuals directly, this becomes a much more reasonable proposition. One need not know much beforehand. Once again, then, the dialectical nature of the technopolitical process becomes apparent. High politics reshaped the role of communications technologies in society, giving them new contexts (as with the wired stations) and sometimes a new importance (as with big-character posters). At the same time, the form of these technologies also influenced the expression of the revolution, enabling small groups to envision and enact a takeover.

One local cadre from Sanmenxia in Henan Province recalled how this unfolded in his district along the Yellow River. "The seizure of power," he writes, "started from the seizure of the broadcasting station. At that time communications conditions were not that developed, and Sanmenxia didn't have a newspaper. But every single household had installed a wired broadcast speaker, including in my own house, which had a large speaker from which I could hear the voice of the Center." The station, therefore, was the "vital point for controlling public opinion." In early 1967, following news of the worker uprisings in Shanghai, a local textile factory employee led a group to the broadcasting station and demanded its surrender. The county cadres were understandably reluctant: "As soon as we turned over the broadcasting station we would have no position in public opinion. . . . The influence on society would be immense . . . [and] it would become chaos." After an overnight siege, the broadcasting station, the node of county communications, fell. One by one every governmental, economic, and educational institution followed.[33] In many places like Yulin and Sanmenxia, a dizzying, multisided power struggle played out for months over the wires of the district.[34]

These encounters occurred in large cities, too, but often at the level of the factory, school, or neighborhood broadcast station. As elsewhere, their comprehensive power and susceptibility to seizure made these institutions scenes of intimate violence. In Wuhan, for instance, one of the city's fifty-odd Red Guards factions controlled a small wired broadcast station that reached a machine tool factory. Over thirty or forty hours in early January 1966, revolutionary workers repeatedly attempted to seize control over this critical local infrastructure. At first, "some workers were chatting and laughing, and some were even fighting, many of whom did not know what they were doing," wrote the Red Guard account published a few days later. Shortly, though, some workers began to denounce the broadcast station as counterrevolutionary propaganda. "You all are Hubei Provincial Committee Royalists!" they cried, before threatening to "crush our megaphones and destroy our broadcasting station."[35] Later, two or three hundred workers surged into the building that housed the broadcast studio. They smashed the window of the office next door, where three female students were sleeping, and cut the phone line to the now-barricaded broadcasters. Seizing some Red Guards, the workers tore off their hats and scarves, as the former college students shouted, "We want literary combat, not military combat!" The hardened workers ignored this feeble plea but could not gain entry to the studio. Instead, they began kidnapping whichever Red Rebel Guards they could get their hands on.

The violent contest over the means of communication in this small corner of Wuhan continued into the next day. The workers brought in three propaganda trucks with massive loudspeakers to pace back and forth in front of the broadcast building, calling the people inside bandits, criminals, and robbers. "We support the workers' revolutionary action! Whoever opposes the workers will have his dog-like head smashed." More organized now, the crowd of workers in front of the building made another attempt. At the sound of a whistle, two or three hundred people rushed up to the second-floor studio, now defended by six students and two workers. To avoid provocation, they finally stopped broadcasting yet still refused to surrender their equipment—the key to local power. The attackers then used a bench as a battering ram and completely destroyed the office door. At the same time, three workers broke the internal glass window and stomped on the broadcasters inside. A large number of Worker Federation members rushed in, throwing the statue of Chairman Mao aside, squeezing the female students into a corner, and tossing the loudspeaker from the window to the ground below. When one female broadcaster tried to stop them, a worker held her wrist and beat her hand until it was bloody and broken. They shoved another guard out of the window, but luckily he was caught by his feet and pulled back inside.[36]

The fighting elsewhere escalated even further. In Kaifeng the next month, room-to-room fighting occurred in the broadcast station offices with knives, chisels, bricks, tiles, pottery, and wooden clubs. Briefly reaching the transmission room, one faction began calling "hoarsely and furiously" to the portion of the city that could hear them for aid. "We have taken back the broadcasting station, and we ask the rebels to support us quickly!"[37] In Xichang, Sichuan, simple weapons were insufficient. There, the two major factions faced off along different sides of the same street, each with their own broadcast station. "Big character posters and slogan banners were everywhere, and the sound of radio loudspeakers rang out across the town," remember local residents. "Every kind of big news and little gossip flew out in all directions."[38] Each system of loudspeakers continually denounced the other and tried to rally residents to their cause. After days of being labeled "bastards" and "murderers," one of the factions decided to take out the heart of the enemy's power—its broadcast studio. They brought in experienced artillerymen to do the calculation and mapping work, but the weapon they fired into town landed at the wrong location. The shell exploded among residents queuing to buy wine from a local shop, killing seven.[39] This incident in a small rural district illustrates an important and unexpected aspect of the technopolitical process. In Xichang, news was everywhere, insistent and unavoidable, suffusing everything with politics. The newsscape can become intensely developed, but this does not necessarily bring the orderly unity that Sun Yat-sen and so many others envisioned. The technopolitical process and political stability are independent variables. Information can proliferate while things fall apart.

The Meanings of Technology

Throughout the ten years of the Cultural Revolution, as through the fifty years we have examined, the technologies and practices of news shaped the form of China's social unrest and determined the possibilities of its politics. Most obviously, communications infrastructures helped pace events across disparate geographies. Developments in Beijing and Shanghai were transmitted to tens of thousands of local broadcasting stations, which received and relayed the information. Self-styled rebel organizations took new ideological positions, new slogans, and, most important, new methods of action from the news. Red Guard rebellions among students at elite Beijing schools became the ten thousand Red Guards of poor, rural Yulin County. The worker uprising and commune of Shanghai became the seizure of power by tanners and miners in northern Shaanxi. The universal replication of information and behavior across class and space reified the formerly conceptual mass society.

But the significance of the development of communications infrastructure for the Cultural Revolution was not limited to its role in *national* synchronization. For when *local* groups attempted to seize power, the means of control (that is, the levers of political organization) were to a large degree coextensive with the technologies, behaviors, and geographies of communication. The form of the newsscape shaped the political possibilities of each small-scale revolution, of every individual's experience of the unrest.

The big-character posters, for instance, enabled large numbers of people to read a single message. In so doing, they helped to mobilize, radicalize, unite, or splinter individuals. They abetted the ramification of society, giving small groups the ability to form or the incentive to break off. Displayed in prominent places like schools and town squares, relaying information from the broadcasts, meetings, travelers, or individuals, these posters (like the wired networks they complemented) shifted the newsscape's power structure toward the local and even the personal.

In other countries, the dominance of larger-circulation newspapers meant that things like big-character posters, though technologically possible, were deemed socially irrelevant. Where people already had an abundance of news, why broadcast a message in such a way? Why write a newspaper with an animal-hair brush and lampblack? The desire for news was already being met in more efficient ways. People crafting their own posts in big letters would have resulted in less information, not more. Only when technological changes allowed average people to compose their own news in a way that resulted in an increase in the totality of information could a system resembling the big-character posters flourish.

Until then, the larger size of the media systems in the West meant fewer opportunities for sociopolitical splits and less power for small factions. Take the city of Wuhan, for instance, with its fifty independently operating Red Guard groups. Would this have been possible without the small wired broadcast networks? In China, technologies erected to enforce cohesion helped a society break apart in ways difficult to envision in other places. France, the United States, Czechoslovakia, Mexico, and Japan (to name just a few examples) all experienced profound social unrest in the 1966–68 period. In these locations, extensive local wired broadcasting systems like China's did not exist. Yes, a coup or protest could take control of the few large radio broadcast stations, but these seizures could be easily reversed. It was not something that a small group of university students could accomplish for lengthy periods. Furthermore, even if rebels did control a radio station in France or Germany or the United States, they would only reach those who chose to tune in on their radio receivers. No network of loudspeakers would make their views unavoidable to the average listener. Where there was no choice in media reception, as in China, those who controlled the system held more power at a more local level. This element of unavoidability heightened the stakes in controlling the broadcast stations and increased the level of competition over their control.

The particular infrastructures and practices of news shape the expression, duration, and depth of sociopolitical turmoil, just as they do the constructive processes we call state-building or economic development. In fact, an increasingly complex and thickly developed newsscape can lead to either of these phenomena—construction or destruction—depending on the choices made. In China, the politics of the Cultural Revolution changed how communications technologies operated in society, turning a totalized newsscape and a totalized polity against itself. Through this period of hyperabundant information and social breakdown, we can comprehend, finally, that the technopolitical process does not proceed in linear correlation with order. Rather, it lacks any innate moral value, whether for good or ill. Process is not progress.

CONCLUSION

Desire and the Transformation of the Newsscape

Change is constant, and history has no end date. The period after high Maoism continued to see rapid changes in the newsscape, the political and the technological again interwoven in lockstep. The advent of cheap transistor radios and cassette tapes were both particularly important developments. Beginning in the early 1970s, transistor radios allowed increasing numbers of people to listen to "enemy" broadcasts behind closed doors.[1] The circulation of nonrevolutionary music through these technologies quickly transformed into a culture touchstone. The swaying strains of Theresa Teng's love songs on illicit cassettes and transistor radios became shorthand for the hope of the Reform and Opening after 1976.[2] By the early 1980s, cassettes and stereo players became common enough to destabilize China's "sound governance."[3] During this period, news sources also diversified (even while remaining heavily controlled), as the number of radio stations and the range of their content increased.[4] This was not a clear-cut case of technology challenging authority, however, as government-run factories in Beijing and Shanghai made many of the radio sets.[5] Scarce resources were spent developing and then manufacturing millions of personal receivers, whose price dropped until they became ubiquitous. Politics constructed the technology as much as the other way around.

The same causal ambiguity applies to the most recent newsscape revolutions, those of the internet and social media. Histories of these phenomena in China privilege the place of government policy and regulation.[6] This has the effect of dissolving the complex interaction of technology, politics, social relations, and individual experience into a false oppression-resistance dichotomy.[7] This oversimplification has historical precedent. When China had vast wired broadcast networks, Western propaganda falsely assumed that the model brainwashed a passive population, juxtaposing this oppression with the supposed freedom of individual receivers. This misunderstood the complicated role of information in the lives of most people, for whom increased access to *any* kind of information was welcome. Such misinterpretation continues today. Though China developed a fortified, curated intranet that blocks or slows interaction with the outside world, most people did not experience its arrival as a loss. The average Chinese person did not mourn a unified, borderless internet. Instead they embraced the staggering increase in timely information it offered, as most people would. If it were not welcome, it would not have had the impact it did. As the history of the newsscape shows us, people must *want* communications technology for it to have enduring significance. And they do. In fact, one of the organizing principles of the technopolitical process is that people will always seek more information.

Seeking News, Driving Change

In the late summer of 1937, a reporter traveling through rural Hebei observed a moral sickness spreading among the farmers there, a depression of the spirit: "They wanted to know good news from the bloody battlefront. They wanted to know about all the brutal crimes of the enemy. They wanted to know. . . . They wanted to know. . . ." Maddeningly, however, they could not know. The news had ceased to flow. This blockage was unnatural, painful even—especially since this area had already come to regularly consume the news that emanated from neighboring cities' papers. In the absence of news—"the spiritual food" of the people, the reporter called it—there was malaise. Even the fragments of information that the lucky ones snatched from the air while huddled around a receiver "could not satisfy their common desire." While fragments were better than nothing, they were still, in a way, starving.[8]

News is a fundamental human need. Metaphors comparing it to food or water cross languages, cultures, and epochs. We hunger and thirst for it. It *flows* like a liquid through channels in our landscape and society, capable of being diverted or blocked or polluted. Like a drought, its absence gnaws away, creating a sense of discomfort and, finally, danger. Without the regular exchange of fresh information, most people, like the Hebei farmers, suffer a slow-building anxiety and experience a drive to slake the thirst. (Indeed, by the time of the reporter's visit, individuals were already reorganizing society to increase the flow of news through listening groups.) How long can the average person resist the urge to seek news of some form? A few days? A few weeks? Many cultures endow the recluse with spiritual powers precisely because their resilience in denying themselves these exchanges seems otherworldly.

Of the thousands of documents I consulted while researching this book, none produced an example of individuals who objected to an increased flow of news. Often people objected to the *content* of the information but only within a steady tide of alternatives. The people of Hebei, for instance, would choose to have Chinese-produced information, if it were possible, in addition to the Japanese news they received. But they would not choose to be without *any* information. Factory workers might be bored by the broadcast of production figures, but they did not want broadcasts to cease entirely. They wanted more variety, not less information. Of course, absence of evidence is not an evidence of absence. The archive could be overwhelmingly biased to record instances of enthusiasm if only because those recording information are inclined to believe that information is inherently valuable.

But still, the silence speaks volumes. Only in a situation of abundance can one be picky enough to object to certain stories, lucky enough to demand more entertainment, or privileged enough to create the category of "misinformation." In a situation of absence, people will prefer an increase in the volume or frequency of information almost without regard to the content. In fact, we crave the act of exchange perhaps as much as the information itself. Think about the evening news or endless scrolling through social media applications. Why do we watch? Why do we read or listen? We do not *need* information about car crashes and moral panics. National politics only affects our daily lives in the most abstract sense. Nor can most people do much to influence these occurrences; indeed, most people possess little motivation to improve anything that does not touch their lives directly. These news practices exist for their own sake. As social beings, we are hardwired to find pleasure in the act of exchanging information. Thus, when new technologies present the opportunity to increase the volume and frequency of information, people reorganize their quotidian behaviors and indeed their whole societies to feed that desire.

This human tendency to seek news has had immense implications for politics and history, even if for most people the political consequences are secondary. Indeed, the willingness to change to accommodate the desire has meant that the gulf between past and present is often vast. We live in such an information-saturated society that we forget that for most of the world's history, people struggled to get news, an adversity that made it precious across time and culture.[9] Receiving more than your neighbor's gossip, or salutations from a traveling stranger, or mangled rumor on a market day represents a fundamental break with the lived experience of our ancestors. Thus, a thickly developed newsscape is not part of a struggle with an external, tangible modernity but rather integral to the practices that form that phenomenon.

Understanding modernity as an information practice can lead to new insights about the convergent evolution of mass societies across ethnically, geographically, and temporally diverse cultures. Why have they emerged almost everywhere in the world? One answer is that people hunger for news. They eagerly embrace techniques and technology that allow for more information. Because people seek them out, information-rich methods of organizing the newsscape become a default, the groove toward which everything slides. Since these methods of receiving more information depend on people implicating themselves in highly organized, repetitive, and addictive behaviors (gathering, building, listening, sharing), mass societies can coalesce around them with minimal friction. The enduring appeal of news can overcome the stubborn will to independence. The advent and triumph of mass societies should not, therefore, be interpreted simply as a violent imposition from the fiscal military state on the putatively separate society below. Nor was this triumph simply a Darwinian process, where fit mass societies survived while other, frail forms of social organization failed. Though competition motivated governments to sponsor the technological development of the newsscape, this action only had significance and permanence because it met a human desire. People wanted a mass society because it was better at delivering large volumes of information. It was better at producing food, shelter, and material comfort, too, of course; but for most people in China (and perhaps elsewhere), these aspects of a mass society came after the arrival of regular news. As the example of collectivizing Chinese villages shows, the reorganization of the newsscape preceded economic restructuring—in contrast to Marx's argument about the primacy of industrialization over communications in *Das Kapital*.[10] Here the desire for news, not the need to transport industrially made materials, encouraged people to reorganize their behavior and thereby integrate themselves into a wider technopolitical process.

Likewise, it was not the appeal of an abstraction (the nation, the government, the party) that restructured society along mass lines. As we have seen, this process occurred regardless of whichever ethnicity, ideology, or regime supervised. Mass society is not nationalism. Instead, the attraction, indeed the pleasure, of the act of connection drew people in and only later allowed things like nationalism, intense governance, and economic integration to emerge. Still, it is important to recognize that speaking in terms of voluntarism and desire does not deny the violence of this process. Force is often the means to satisfy human needs. Examining these processes from below, as we have done, also does not deny a role for the state. Though the rise of a mass society can be understood as an externality derived from the technological ability to satisfy an innate desire for news, these tendencies were intentionally fostered by elite decision-making groups (corporations, parties, governments) to expand the possibilities of their own power and survival. The state built technology just as technology built the state. But these revolutions in power structures were also substantially consensual and rooted in everyday behavior. They happened because people desired to drink at an evergreen trough of information. To do so, they reconstructed the newsscape and reoriented their daily rituals. They read the morning paper, stopped by the news walls, and gathered under loudspeakers at the appointed hour.[11] They bought radios and built wired broadcast networks. These habits and actions were contingent and learned, but they responded to a fundamental part of the human condition. State ambitions, party ideology, and corporate greed certainly play a role in the *consequences* of mass societies, but they are not the sole cause.

The rhetorical displacement of the state and institutions would have been difficult for this book to accomplish if the society under examination possessed many preexisting, well-developed infrastructures like parties, national governments, large newspaper corporations, widespread railroads, and reliable telegraphy. These stories would have dominated and obscured the experiences of news. But the China at the beginning of our story lacked these to a substantial degree. This absence has allowed a more diffuse perspective on the information networks that tie society together to shine through. China's underdevelopment at the beginning of the twentieth century encourages a bottom-up analysis of the growth of mass society. Furthermore, the choice to look past institutions permits us to venture some critiques of classic perspectives on this phenomenon. For instance, the experiential history of information contradicts an important assumption of James C. Scott's *Seeing like a State*, which understands schemes of legibility and administration as essentially part of the state's intention, an imposition on the popular will.[12] In contrast, the history of news allows us to see the rise of a mass society not just as something constructed by the state but as a reorganization from within society—an account that most nearly resembles a Latourian mass action, the cumulative decisions of innumerable actors driven by their desires.[13] Becoming legible to the state, and being manipulated by it, was for some the price to pay for connection.

Given these themes, can we call the systems that arose in this period a public sphere? Yes and no. On the one hand, asking whether China had a public sphere or civil society evokes Max Weber's doubt whether "oriental" societies possessed cities.[14] Of course they had cities and the entire panoply of social formations that complex societies throw up, including a so-called public sphere. On the other hand, the spaces radio helped build were mutually constituted by the state (the part of society serving in administrative roles with coercive power) and the public (those outside those bodies). They cannot be understood as a civil society outside the state. But as many scholars have shown, government-sponsored systems were central to the rise of mass information in the West too.[15] To contend that there was an informational space free of the state in the West is to misunderstand that history. Thus, Chinese people organized and disciplined themselves to receive or contribute to news in ways fundamentally similar to Western people. Both constructed social and physical spaces to accelerate the speed and volume of information, even if the form of these spaces differed. Both systems reshaped the political world, creating the possibility of a mass society. But this mass society had a large spectrum of possible expressions, one of which was the rural communism of Mao Zedong. Ignoring this diversity, Habermas overstates the teleology of forms in the technopolitical process. Preserving Marxist and Hegelian tendencies toward progressive, stadial history (indeed, this is at the heart of his program), he argues in essence that a public sphere leads to certain determined consequences for politics and culture (i.e., bourgeois liberalism). Thinking in terms of class and space, he underappreciates the negotiation, chance, and decision-making that affects the technopolitical process.

I reject this reified "public sphere" because of the fixed framework it implies. The technopolitical process may help develop a mass society, but its resulting form varies. For instance, the creation of an active, participatory mass society in China led to an embrace of some of the values that also emerged from eighteenth-century coffee houses (egalitarianism, rationalism) and a rejection of others (rule of law, privately led development). Still other values, like freedom of speech, had an ambiguous, mutable status in the mass society that emerged in China. Accepting this complexity, this book has not tried to analyze the history of the newsscape and its political ramifications in terms of a Habermasian public. An Andersonian "imagined listening community" also plays little role.[16] Rather than these abstract nouns, we have focused on processes and behaviors. We have centered things people actually *do* rather than imagine. The newsscape analysis I have presented reveals not the idea of a nation but the act of one, not a discursive project of modernity but one organized in and of the material world.

Utopia Delayed

In the fall of 1958, images of a radiant socialist future reached the eyes of young wireless enthusiasts across China on the cover of the monthly *Radio* magazine. Old men still remember the joy these illustrations brought. Broadcast towers swaddle the world in radio waves. Sputnik races across a field of stars. A halo shines from Tiananmen to a distant row of factory smokestacks. A cornucopia of flowers and vacuum tubes, radios and televisions blanket the ground, as a striding worker offers electromagnetic waves to heaven with upturned palms.[17] These images conveyed the hope and expectation of a prosperous future. Despite their forward-looking orientation, these visions were rooted in history. Over the previous decades, people like the engineer Fang Ziwei, magistrate Zuo Xunyou, broadcaster Ding Yilan, and millions of ordinary Chinese people had dreamed that the radio would help bring about harmony, safety, satiety, and strength. As with the railroad before it and the internet after, political dreams accompanied technological innovation.[18] Technopolitical reconstruction would bring entertainment and information. It would deliver good governance. It would liberate them from the oppression of nature, foreigners, and each other.

Should we dismiss these dreams as utopian fantasies? Perhaps. Yet much of what was dreamed came to pass. Living conditions today far exceed the prosperity of Maoist propaganda posters. Radio really did what Sun Yat-sen said it would when he predicted it would allow the central government to administer distant areas. Just because a perfect world did not result does not mean that the consequences were not profound. In fact, we make these kinds of fallacious assumptions all the time—thinking that if something is not achieved absolutely, if its results are not universal or obvious, then the ambition was foolhardy. We see this today in a body of assumptions about the internet's impact on Chinese society.

Like the radio, neither the internet nor social media resulted in a peaceful democratization of knowledge. As we all know, liberal capitalism has not yet triumphed at the end of history. Instead, the most recent newsscape revolution has caused political polarization, social division, and even violence in societies as diverse as India and the United States. But the Chinese exception proves these patterns are not universal. Here, the internet and social media apparatuses are popularly understood as a stabilizing force or, at least, as one substantially less destabilizing than countries with other systems. Once again, the technopolitical process appears to have expressed itself differently in China. But is this ostensible stasis true? What can history teach us?

First, technology does not determine politics; it only determines the bounds of its possibilities. Nothing says that the increased politicization of American life has to be true of China, too, just because they have experienced similar technological shifts. The new social media–inflected newsscape enables a broad spectrum of outcomes for societies. Though the fundamental informational structures of newsscapes may be similar, they can be taken to different extremes. Second, the superficial stability of Chinese politics over the last several decades obscures deep changes in the newsscape and, consequently, in sociopolitical organization. As this book has suggested, we should look at quotidian, nonelite experience to understand the fundamental nature of these shifts in communications. Technological development has, for example, allowed Chinese government administration to reach a depth and breadth never before experienced in human history—to a point where every single person's movements, actions, and bodily functions are monitored and controlled for months on end. The effects of all this on an individual could be greatly mitigated if only they would forgo their cellphones and computers. But the desire for information is too great. The number of people willing to give up news for freedom is small indeed.

In reality, the vast majority of Chinese have not experienced the internet or social media or information technologies like smartphones as tools of oppression. Instead, they experience them as tools of liberation, just like the radio before, so long as they fulfill their news distribution function and do not *appear* to lack content. Delivering vast quantities of time-sensitive information, the internet responds to nearly insatiable demand. State-centered histories that focus on systems of information control thus overlook the experience of the Chinese internet as pleasurable and interesting. These studies ignore how contemporary technology has freed people to connect with their communities and avoid bargaining with vegetable sellers. (Like so many other daily interactions transformed by technology, the disappearance of haggling over the last twenty years will be a tempting topic for future historians.) Though there are many externalities to participating in the new newsscape, for most people the positives far outweigh the negatives.

A third and final historical lesson is perhaps the most important; even though change is always happening somewhere, moments of stasis occur, too. News decays. Interest mutates to boredom. People adapt, and disjunctures rebalance. The effects of a newsscape revolution play out, and people become less mesmerized by new forms of connection. Utopia and dystopia become an unremarkable everyday; a new future is imagined and then delayed. Given this return to the median, we should neither confide our hopes in technologies nor despair that any breakage they have caused will last forever.

Our story, then, lacks a resolution or a new utopia. Instead, this book has sought to convey the actual experience of fundamental social change. It has interpreted the long-term development of a mass-political society through the lens of technology, as seen through the techniques, behaviors, and processes of news. Understanding this development as a unitary technopolitical process, I have articulated a heuristic methodology—newsscape analysis—to study it over time. In a sense, such an interpretation is hardly new. Scholars have pointed out the close interrelation of communication and political construction since at least Harold Innis in the 1950s, though the intuition of their connection goes back millennia. What I hope I have brought out, however, are the ways in which the holistic analysis of news, technology, and political behaviors can, like the concept of coproduction described by historians of science, "offer new ways of thinking about power, highlighting the often invisible role of knowledges, expertise, technical practices and material objects."[19] Indeed, newsscape analysis helps reveal the often hidden connections between remote places, disparate classes, and political opponents within the physical and social world. It allows us to comprehend shifts in information orders and, therefore, in power structures far away from internationally connected cities, intellectual classes, and central governments.

Applying this analytic to the flow of information from the 1920s to the 1960s, this book has explored, as far as the sources have allowed, how engineers, politicians, and everyday interactions constructed newsscapes and how nonelite people perceived this change over time. It has paid particular attention to the most revolutionary element of the newsscape in the period: the bundles of technologies and behaviors we call radio. The Nationalist government's program to introduce radio newspapers in smaller counties allowed more regular and rapid news to arrive in rural areas beginning in the 1930s. This system increased in importance during the War of Resistance against Japan and the civil war that followed, as continual emergencies highlighted the demand for information. In the 1950s, after securing power, the Communist Party carried out a popular campaign to install wired broadcasting stations at increasingly local levels. By the time of the late Maoist period, many communes, farms, and factories had their own wired broadcast stations. Though radio existed only within the context of other media, it lay at the heart of the twentieth-century newsscape revolution.

That revolution resulted in many people hearing the news—local, national, and international—multiple times a day, even in rural areas and even among the poorest classes. During the late 1960s, the system became nearly comprehensive. Thus, the fifty years of the technopolitical process (1919–68) that this book has taken as its subject appear to offer a clear arc toward an ultimate and total politicization. But such an interpretation is too convenient. Indeed, the Cultural Revolution shows that the technopolitical process is not a teleology. It possesses neither a beginning nor an end. I have chosen these dates, 1919 and 1968, because they have social and cultural meaning, not because the technopolitical process began on a spring day in Beijing and reached a final apogee with the hourly connection of the people to the polity in a rural factory. Why, then, stop our story in the first years of the Cultural Revolution? Precisely because these were years of collapse and civil war, street battles, and internecine strife. They are years that show that the technopolitical process is not necessarily a progressive path to regularized power and a structured state, though this is possible. In fact, the technopolitical process can proceed and even deepen in times of disorder and violence. A proliferation of news can accompany social collapse. We must conclude, therefore, that people can direct the technopolitical process toward many distinct outcomes. Inherent in the political aspect of the dialectic are *choices*. We, as a collective humanity and as individuals, retain the ability to determine the ways in which the newsscape evolves. We decide how the technopolitical process shapes our society. In that, there is hope.

Notes

Introduction

1. "Guangbo ye ceng Dayuejin," in *Dangdai Beijing guangbo shihua* (Beijing: Dangdai Zhongguo Chubanshe, 2013); "Shoudu 300 wan ren weijiao maque," in *Tu wen gongheguo nianlun 1949–1959*, ed. Zhang Shujun (Shijiazhuang: Hebei Renmin Chubanshe, 2009); "Beijing shi maque weijiao zhan," in *1958 nian de gushi*, ed. Wang Xinghua (Beijing: Zhongguo Shaonian Ertong Chubanshe, 2001); "Fangbing baoliang, yifengyisu: shoudu chu sihai jiang weisheng tujizhou kaishi," *Renmin Ribao*, Dec. 23, 1957; *Spiegel Politik*, "Der Große Sprung nach vorn," video documentary, August 4, 2008, www.spiegel.de/video/mao-der-rote-kaiser-2-der-grosse-sprung-nach-vorn-video-33816.html.
2. "Shoudu 300 wan ren weijiao maque," 550.
3. "Bukanhuishou mie maque," in *Jinan Xiaoqinghe lishi wenhua congshu* ed. Zhang Jiping (Jinan: Jinan Chubanshe, 2008).
4. "Shanxi sheng zhengqu weisheng yundong da yue jin," *Renmin Ribao*, Jan. 5, 1958; Gu Ye, "Shanxi chu si hai jiang weisheng yundong xiang shen guang fazhan," *Renmin Ribao*, Jan. 14, 1958.
5. Foreign Broadcast Information Service (hereafter FBIS), "Daily Urges All-Out Sanitation Drive," Jan. 24, 1958.
6. Lingling xian aiguo weisheng yundong weiyuanhui, "Lingling xian weijian maque de chu bu zongjie," July 1, 1958, in Hunan sheng aiguo weisheng yundong weiyuanhui bangongshi, *Hunan sheng chu si hai jiang weisheng yundong ziliao huibian* (1958).

7. For contemporary coverage of the event in the West, see *Time* magazine, "Death to Sparrows," May 5, 1958; and *New York Times*, "Peiping vs. Pests," Feb. 15, 1959.

8. Dayton Lekner, "Echolocating the Social: Silence, Voice, and Affect in China's Hundred Flowers and Anti-Rightist Campaigns, 1956–58," *Journal of Asian Studies* 80, no. 4 (2021); Elizabeth Perry, "Moving the Masses: Emotion Work in the Chinese Revolution," *Mobilization* 7, no. 2 (2002).

9. For a discussion of the militarization of society, see Frank Dikotter, *Mao's Great Famine* (New York: Bloomsbury, 2017), 45–51.

10. China fell into civil strife after the collapse of the Qing dynasty in 1911, but problems with infrastructural collapse, internal disorder, and international depredation can be traced back further, certainly to the First Sino-Japanese War of 1895. See Benjamin A. Elman, "Naval Warfare and the Refraction of China's Self-Strengthening Reforms into Scientific and Technological Failure, 1865–1895," *Modern Asian Studies* 38, no. 2 (2004).

11. Benedict Anderson, *Imagined Communities: Reflections on the Origin and Spread of Nationalism* (London: Verso, 1983).

12. Stephen Kotkin, *Stalin*, vol. 1, *Paradoxes of Power, 1878–1928* (New York: Penguin, 2014), 4–5.

13. Michel Foucault, *The Order of Things: An Archaeology of the Human Sciences* (New York: Routledge, 2002), xxii.

14. As Michael Stamm shows, however, broadcast news and newspapers complemented one another in the first sixty years of radio. The replacement model that journalists feared would not come to pass until another communications revolution at the end of the century. See Michael Stamm, "Broadcasting News in the Interwar Period," in *Making News: The Political Economy of Journalism in Britain and America from the Glorious Revolution to the Internet*, ed. Richard John and Jonathan Silberstein-Loeb (Oxford: Oxford University Press, 2015), 133–63.

15. The dual revolutions were the (initially British) industrial revolution and the (initially French) political revolution. Eric Hobsbawm, *The Age of Revolution: 1789–1848* (New York: Vintage, 1996).

16. Sheila Jasanoff, "The Idiom of Co-production," in *States of Knowledge: The Co-production of Science and Social Order*, ed. Sheila Jasanoff (London: Routledge, 2004), 1–12, 2.

17. See Richard R. John, *Spreading the News: The American Postal System from Franklin to Morse* (Cambridge, MA: Harvard University Press, 1995); and Daniel Walker Howe, *What Hath God Wrought: The Transformation of America, 1815–1848* (Oxford: Oxford University Press, 2007).

18. Robert Darnton, *The Literary Underground of the Old Regime* (Cambridge, MA: Harvard University Press, 1982); Robert Darnton, *Poetry and the Police: Com-*

munications Networks in Eighteenth-Century Paris (Cambridge, MA: Harvard University Press, 2010).

19. Edward Anthony Wrigley, *Energy and the English Industrial Revolution* (Cambridge: Cambridge University Press, 2010); Victor Seow, *Carbon Technocracy: Energy Regimes in Modern East Asia* (Chicago: University of Chicago Press, 2021); Shellen Xiao Wu, *Empires of Coal: Fueling China's Entry into the Modern World Order, 1860–1920* (Stanford, CA: Stanford University Press, 2015).

20. Francesca Bray, *Technology and Gender: Fabrics of Power in Late Imperial China* (Berkeley: University of California Press, 1997); Jordan Sand, *House and Home in Modern Japan: Architecture, Domestic Space, and Bourgeois Culture, 1880–1930* (Cambridge, MA: Harvard University Press, 2003).

21. Anne Wilson, Victoria Parker, and Matthew Feinberg, "Polarization in the Contemporary Political and Media Landscape," *Current Opinion in Behavioral Sciences* 34 (2020).

22. Ann Blair, Paul Duguid, Anja-Silvia Goeing, and Anthony Grafton, eds., *Information: A Historical Companion* (Princeton, NJ: Princeton University Press, 2021), 272.

23. Blair et al., 501.

24. Blair et al., 130.

25. James Carey, *Communication as Culture: Essays on Media and Society* (Abingdon, UK: Routledge, 2009), 32.

26. Carey, 33.

27. Emily Thompson, *The Soundscape of Modernity: Architectural Acoustics and the Culture of Listening in America, 1900–1933* (Cambridge: Cambridge University Press, 2002); Alain Corbin, *Village Bells: Sound and Meaning in the 19th-Century French Countryside* (New York: Columbia University Press, 1998); R. Murray Schafer, *The Soundscape: Our Sonic Environment and the Tuning of the World* (Rochester, VT: Destiny, 1977).

28. This echoes a long tradition in the field. See Prasenjit Duara, *Rescuing History from the Nation: Questioning Narratives of Modern China* (Chicago: University of Chicago Press, 1995).

29. Reviel Netz, *Barbed Wire: An Ecology of Modernity* (Middletown, CT: Wesleyan University Press, 2004).

30. Elizabeth Köll, *Railroads and the Transformation of China* (Cambridge, MA: Harvard University Press, 2019).

31. For a history of shipping in China, see Anne Reinhardt, *Navigating Semicolonialism: Shipping, Sovereignty, and Nation-Building in China, 1860–1937* (Cambridge, MA: Harvard University Press, 2018).

32. Lydia Liu, "Scripts in Motion: Writing as Imperial Technology," *Past and Present* 140, no. 2 (March 2015).

33. Hal Langfur, *Adrift on an Inland Sea: Misinformation and the Limits of Empire in the Brazilian Backlands* (Stanford, CA: Stanford University Press, 2023); Christopher Bayly, *Empire and Information: Intelligence Gathering and Social Communication in India, 1780–1870* (Cambridge: Cambridge University Press, 1996); Harold Innis, *Empire and Communications* (Oxford: Clarendon, 1950); Daniel Headrick, *Tools of Empire: Technology and European Imperialism in the Nineteenth Century* (Oxford: Oxford University Press, 1981); Daniel Headrick, *The Tentacles of Progress: Technology Transfer in the Age of Imperialism, 1850–1940* (Oxford: Oxford University Press, 1988); Daqing Yang, *Technology of Empire: Telecommunications and Japanese Expansion in Asia, 1883–1945* (Cambridge, MA: Harvard University Press, 2010).

34. Heidi Tworek, *News from Germany: The Competition to Control World Communications, 1900–1945* (Cambridge, MA: Harvard University Press, 2019); Arthur Asseraf, *Electric News in Colonial Algeria* (Oxford: Oxford University Press, 2019); Julia Guarneri, *Newsprint Metropolis: City Papers and the Making of Modern Americans* (Chicago: University of Chicago Press, 2017).

35. Sei-Jeong Chin, "Print Capitalism, War, and the Remaking of the Mass Media in 1930s China," *Modern China* 30, no. 4 (2014).

36. Richard R. John and Jonathan Silberstein-Loeb, eds., *Making News: The Political Economy of Journalism in Britain and America from the Glorious Revolution to the Internet* (Oxford: Oxford University Press, 2015); Jürgen Habermas, *The Structural Transformation of the Public Sphere: An Inquiry into a Category of Bourgeois Society* (Cambridge, MA: MIT Press, 1989).

37. Michael Stamm, *Dead Tree Media: Manufacturing the Newspaper in Twentieth-Century North America* (Baltimore: Johns Hopkins University Press, 2018).

38. Emily Mokros, *The Peking Gazette in Late Imperial China: State News and Political Authority* (Seattle: University of Washington Press, 2021).

39. Eugenia Lean, *Public Passions: The Trial of Shi Jianqiao and the Rise of Popular Sympathy in Republican China* (Berkeley: University of California Press, 2007).

40. China had a thriving, unified capitalistic print culture from the mid-Ming dynasty (1368–1644) throughout the Qing dynasty (1644–1911) and into the Republican period (1911–49). Woodblock-printed classics, primers, novels, travelogues, and medicinal books circulated widely in a comprehensive ecosystem of print works. Cynthia Brokaw and Christopher Reed demonstrate that China had indigenous and adapted forms of print capitalism by the last years of the Qing dynasty. See Cynthia Brokaw, *Commerce in Culture: The Sibao Book Trade in the Qing and Republican Periods* (Cambridge, MA: Harvard University Asia Center, 2007); and Christopher Reed, *Gutenberg in Shanghai: Chinese Print Capitalism, 1876–1937* (Honolulu: University of Hawai'i Press, 2004). On the history of Chinese printing, see Joseph Needham and Tsien Tsuen-Hsuin, *Science and Civilization in China*, vol. 5 (Cambridge: Cambridge University Press. 1987).

41. After the lapse of a government monopoly in 1820, these gazettes were printed by private publishing houses, sometimes using wooden movable type. These gazettes differed, however, from modern newspapers in important respects. Their content remained the condensed reports of imperial court proceedings, edited to appeal to public interest. They contained no reports from correspondents, no advertisements, no editorials, and nothing that had not been filtered through Beijing. Emily Mokros, "Spies and Postmen: Communications Liaisons and the Evolution of the Qing Bureaucracy," *Frontiers of History in China* 14, no. 1 (2019).

Newspapers like *Shenbao* and *Shibao* (1904) represented an important departure from the court gazette format since they included news from foreign lands, letters to the editor, and reports from civilians across the country. Joan Judge, *Print and Politics: Shibao and the Culture of Reform in Late Qing China* (Stanford, CA: Stanford University Press, 1996); Weiping Tsai, *Reading Shenbao: Nationalism, Consumerism, and Individuality in China, 1919–37* (London: Palgrave Macmillan, 2010); Barbara Mittler, *A Newspaper for China? Power, Identity, and Change in Shanghai's News Media, 1872–1912* (Cambridge, MA: Harvard University Asia Center, 2004).

42. See Judge, *Print and Politics*, 79–140.

43. In 1910 the magazine *Eastern Miscellany* (*Dongfang Zazhi*) reached a circulation of fifteen thousand copies, making it the largest publication in China (Reed, *Gutenberg in Shanghai*, 215). In Japan in 1894, fifteen years earlier, one magazine had sold three hundred thousand copies of a single issue. Japan continued to outclass China in circulation figures for decades. Another Japanese magazine reached one million copies per issue in January 1927. Nathan Shockey, *The Typographic Imagination: Reading and Writing in Japan's Age of Modern Print Media* (New York: Columbia University Press, 2019), 29, 52.

44. Henrietta Harrison, "Newspapers and Nationalism in Rural China 1890–1929," *Past & Present*, no. 166 (Feb. 2000).

45. Shockey, *The Typographic Imagination*, 25–58.

46. Logographic scripts famously complicated and slowed the advent of typewriters. Handwriting was the most economical and timely method of written communication until the 1930s and 1940s, as seen by the predominance of handwritten memos in internal government documents. Thomas Mullaney, *The Chinese Typewriter: A History* (Cambridge, MA: MIT Press, 2017).

47. Erik Baark, *Lightning Wires: The Telegraph and China's Technological Modernization, 1860–1890* (Westport, CT: Greenwood Press, 1997).

48. For an insightful analysis of the persistent deficiencies of Chinese telegraphy, see Wook Yoon, "Dashed Expectations: Limitations of the Telegraphic Service in the Late Qing," *Modern Asian Studies* 49, no. 3 (2015).

49. C. Martin Wilbur, *The Nationalist Revolution in China, 1923–1928* (Cambridge: Cambridge University Press, 1984), 2.

50. For example, the norm-shattering strikes and protests in Canton and Hong Kong surrounding the May 30th movement, which protested the killing of Chinese civilians by police officers in the Shanghai international concession, claimed only one hundred thousand participants. Hans van de Ven, *From Friend to Comrade: The Founding of the Chinese Communist Party, 1920–1927* (Berkeley: University of California Press, 1991), 155.

51. Van de Ven, *From Friend to Comrade*; Lloyd Eastman, *The Abortive Revolution: China under Nationalist Rule, 1927–1937* (Cambridge, MA: Harvard University Press, 1974).

52. Charles Tilly, "Coercion, Capital, and European States, AD 990–1990," in *Collective Violence, Contentious Politics, and Social Change: A Charles Tilly Reader*, ed. Ernesto Castañeda and Cathy Lisa Schneider (London: Routledge, 2017), 140–54.

53. Starting from Hans van de Ven, *War and Nationalism in China, 1925–1945* (London: Routledge, 2003).

54. Rana Mitter, "Classifying Citizens in Nationalist China during World War II, 1937–1941." *Modern Asian Studies* 45, no. 2 (2011).

55. Rana Mitter, *Forgotten Ally: China's World War II, 1937–1945* (Boston: Houghton Mifflin Harcourt, 2013).

56. Anderson, *Imagined Communities*, 37–46.

57. See Cynthia Brokaw, *Commerce in Culture*; and Christopher Reed, *Gutenberg in Shanghai*.

58. Harrison, "Newspapers and Nationalism."

59. See Marshall McLuhan, *Understanding Media: The Extensions of Man* (Cambridge, MA: MIT Press, 1994).

60. Rebecca Scales, *Radio and the Politics of Sound in Interwar France, 1921–1939* (Cambridge: Cambridge University Press, 2016).

61. Isabel Huacuja Alonso, *Radio for the Millions: Hindi-Urdu Broadcasting across Borders* (New York: Columbia University Press, 2023); Stephen Lovell, *Russia in the Microphone Age: A History of Soviet Radio, 1919–1970* (Oxford: Oxford University Press, 2015).

62. Jie Li, "Revolutionary Echoes: Radios and Loudspeakers in the Mao Era," *Twentieth-Century China* 45, no. 1 (2020).

63. Lei also includes an empirically valuable chapter on radio news across the Maoist and post-Maoist period, focusing on a single broadcast channel. See Wei Lei, "Radio News and the Articulation of One Voice: Continuity and Transformation of China National Radio's Channel One," in *Radio and the Social Transformation of China* (Oxford: Routledge, 2019), 57–91.

64. Lekner, "Echolocating the Social: Silence"; Paulina Hartono, "'A Good Communist Style': Sounding Like a Communist in Twentieth-Century China," *Representations* 151, no. 1 (2020).

65. Several important works examine the social, political, and administrative histories of radio broadcasting in the Republican period. Carleton Benson, in an article and dissertation, examined the development and politicization of a radio storytelling genre *tanci* in Shanghai. See Carlton Benson, "From Teahouse to Radio: Storytelling and the Commercialization of Culture in 1930s Shanghai" (PhD diss., University of California, Berkeley, 1996); see also Michael Krysko, *American Radio in China: International Encounters with Technology and Communications, 1919–41* (New York: Palgrave Macmillan, 2011). Laura De Giorgi has written about the Nationalist's broadcasting policy and administration during the Nanjing decade, focusing on the centralization and "party-ification" *danghua* that occurred: "Communication Technology and Mass Propaganda in Republican China: The Nationalist Party's Radio Broadcasting Policy and Organisation during the Nanjing Decade (1927–1937)," *European Journal of East Asian Studies* 13, no. 2 (2014). Separately, she has written about Guomindang views on broadcast media and public education during the same period; see her "Media and Popular Education: Views and Policies on Radio Broadcasting in Republican China, 1920s–30s," *Interactions: Studies in Communication & Culture* 7, no. 3 (2016). Zhao Yuming has written many essays and collected volumes on the history of the Republican period administration of radio, as has Li Yu of the Communication University of China; see Zhao Yuming, *Zhongguo Xiandai Guangbo Jianshi* (Beijing: Zhongguo Guangbo Dianshi Chubanshe, 1987); Zhao Yuming, ed., *Xinxiu Difang Zhi Zaoqi Guangbo Shiliao*, 2 vols. (Beijing: Zhongguo Guangbo Yingshi Chubanshe, 2016); Li Yu, *Zhongguo Guangbo Xiandaixing Liubian* (Beijing: Zhongguo Chuanmei Daxue Chubanshe, 2016). Guo Zhenzhi has explored privately run Chinese radio stations in Shanghai during the 1930s, mirroring the strong focus on bourgeois Shanghainese radio to the exclusion of other locales and the working class; see his "A Chronicle of Private Radio in Shanghai," *Journal of Broadcasting & Electronic Media* 30, no. 4 (1986).

Chapter 1

1. Wang Peijun and Zhang Shanxiang, "Wusi Yundong zai Taizhou," in *Taizhou Wenshi Ziliao, Di Yi Juan*, ed. Wang Miaozeng (Zhejiang: n.p., 1998), 1–2; Li Jieren, "Wusi zhuiyi Wang Guangqi," *Chuan Xi Ribao*, May 4, 1950. Repr. in Zhong-Gong Sichuan sheng wei dang shi gongzuo weiyuanhui, *Wusi Yundong zai Sichuan* (Chengdu: Sichuan Daxue Chubanshe, 1989), 659–61; "Tianjin *Yi Shi Bao* baodao wusi yundong de baofa," May 5, 1919. Repr. in Tianjin Lishi Bowuguan, *Wusi Yundong zai Tianjin* (Tianjin: Tianjin Renmin Chubanshe, 1979), 2.

2. For an overview of the May Fourth Movement and its twentieth-century reverberations, see Rana Mitter, *A Bitter Revolution: China's Struggle with the Modern World* (Oxford: Oxford University Press, 2005).

3. See, e.g., Daniel Headrick, *The Invisible Weapon: Telecommunications and In-

ternational Politics, 1851–1945 (Oxford: Oxford University Press, 1991); and Erik Baark, *Lightning Wires: The Telegraph and China's Technological Modernization, 1860–1890* (Westport, CT: Greenwood Press, 1997).

4. For an introduction to the historiography of intermediality, see Juha Herkman, introduction to *Intermediality and Media Change*, ed. Juha Herkman, Taisto Hujanen, and Paavo Oinenan (Tampere, FI: Tampere University Press, 2012); Marshall McLuhan, *Understanding Media: The Extensions of Man* (Cambridge, MA: MIT Press, 1994); Elizabeth Eisenstein, *The Printing Revolution in Early Modern Europe* (Cambridge: Cambridge University Press, 1983); Benedict Anderson, *Imagined Communities: Reflections on the Origin and Spread of Nationalism* (New York: Verso, 1983); and Jürgen Habermas, *The Structural Transformation of the Public Sphere: An Inquiry into a Category of Bourgeois Society* (Cambridge, MA: MIT Press, 1989).

5. Ian Atherton, "The Itch Grown a Disease: Manuscript Transmission of News in the Seventeenth Century," *Prose Studies* 21, no. 2 (1998); Rachael Scarborough King, "The Manuscript Newsletter and the Rise of the Newspaper, 1665–1715," *Huntington Library Quarterly* 79, no. 3 (2016).

6. This lesson, I think, should be well appreciated by historians of China given the very late survival of handwritten communication (well into the Maoist era), for instance in business and internal government communications, in situations where in the West such documents would invariably be printed or typed.

7. Robert Darnton, "An Early Information Society: News and Media in Mid-Eighteenth Century Paris," *American Historical Review* 105, no. 1 (Feb. 2000).

8. Henrietta Harrison, "Newspapers and Nationalism in Rural China, 1890–1929," *Past & Present*, no. 166 (Feb. 2000).

9. Wang Dixian, "Wang Guangqi yu Wusi Shiqi de Sichuan Geming Douzheng," in Zhong-Gong Sichuan shengwei dangshi gongzuo weiyuanhui, *Wusi Yundong zai Sichuan* (Chengdu: Sichuan Daxue Chubanshe, 1989), 704–8.

10. James Carey, "Technology and Ideology: The Case of the Telegraph," in *Communication as Culture: Essays on Media and Society* (Abingdon, UK: Routledge, 2009), 155–76.

11. Li Jieren, "Wusi zhuiyi Wang Guangqi," 659.

12. Julean Arnold, *The Commercial Handbook of China* (Washington, DC: Government Printing Office, 1919), 311.

13. Thomas Mullaney, "Semiotic Sovereignty: The 1871 Chinese Telegraph Code in Historical Perspective," in *Science and Technology in Modern China, 1880s–1940s*, ed. Jing Tsu and Benjamin Elman (Boston: Brill, 2014), 153–83; Zhang Wenyang, "The Grammar of the Telegraph in the Late Qing: The Design and Application of Chinese Telegraphic Codebooks," *Journal of Modern Chinese History* 12, no. 2 (2018).

14. Michael Lindsay, *The Unknown War: North China 1937–1945* (London: Bergstrom & Boyle, 1975), n.p.

15. Correspondents from larger newspapers like *Shibao* did dispatch longer reports, up to three thousand characters. Thank you to Nataly Shahaf for bringing this to my attention. Bao Tianxiao, *Chuanyinglou huiyilu* (Hong Kong: Dahua chubanshe, 1971), 345–50.

16. Li Jieren, "Wusi zhuiyi Wang Guangqi," 659.

17. Li Jieren, 659.

18. "Waijiao jie zui jin zhi da heimu," *Guomin Gongbao*, May 8, 1919. Repr. in Zhong-Gong Sichuan sheng wei dang shi gongzuo weiyuanhui, *Wusi Yundong zai Sichuan* (Chengdu: Sichuan Daxue Chubanshe, 1989), 66.

19. "Sichuan ge Xiao quanti xuesheng zhi dian Beijing Xu Juren, Qian Ganchen," *Guomin Gongbao*, May 12, 1919. Repr. in Zhong-Gong Sichuan sheng wei dang shi gongzuo weiyuanhui, *Wusi Yundong zai Sichuan* (Chengdu: Sichuan Daxue Chubanshe, 1989), 70.

20. "Yao Dian Hui Zhi," *Guomin Gongbao*, May 9, 1919. Repr. in Zhong-Gong Sichuan sheng wei dang shi gongzuo weiyuanhui, *Wusi Yundong zai Sichuan* (Chengdu: Sichuan Daxue Chubanshe, 1989), 67.

21. "Beijing de xuesheng yundong," *Guomin Gongbao*, May 12, 1919. Repr. in Zhong-Gong Sichuan sheng wei dang shi gongzuo weiyuanhui, *Wusi Yundong zai Sichuan* (Chengdu: Sichuan Daxue Chubanshe, 1989), 68.

22. "Chuan Ji Xuesheng Bei Jin zhe," *Guomin Gongbao*, May 15, 1919. Repr. in Zhong-Gong Sichuan sheng wei dang shi gongzuo weiyuanhui, *Wusi Yundong zai Sichuan* (Chengdu: Sichuan Daxue Chubanshe, 1989), 68.

23. Li Jieren, "Wusi zhuiyi Wang Guangqi," 659–61.

24. Luo Dong, "Wu Yu zhi shou 'da kong jia dian,'" *Beijing Bao Digital*, www.bjnews.com.cn/feature/2019/04/17/568678.html.

25. Arnold, *Commercial Handbook of China*, 498.

26. Arnold, 499.

27. Compare Sichuan's 218,000 square miles to France's approximately 250,000. Sichuan's population is estimated from a 1910 government household census at seventy-eight million. Montague Bell, *The China Year Book, 1919–20* (New York: Routledge, 1919), 3: citing figures from the 1910 Ministry of the Interior census.

28. Bell, *The China Year Book*, "Appendix II: Mail Matter Received during the Year," 278–79.

29. Harrison, "Newspapers and Nationalism," 194.

30. The development of more reliable statistical measurement and modern census taking would have to wait until the Maoist period. Arunabh Ghosh, *Making It Count: Statistics and Statecraft in the Early People's Republic of China* (Princeton, NJ: Princeton University Press, 2020).

31. Don Patterson, "The Journalism of China," special issue, *University of Missouri Bulletin* 23, no. 34 (1922), Journalism Series, no. 26, ed. Robert S. Mann, 36.

32. Writing in 1922, an observer recorded that "the newspaper is passed from one hand to another, for a slightly smaller consideration each time, until it is ragged, and it is not uncommon to find such ragged copies in the less frequented sections of China such as Mongolia. A Chinese banker recently spoke of buying a month-old copy of the *Shun Pao* in Mongolia. Circulation of this type is assisted by the fact that the Chinese have not as yet been educated to demand spot news as have the newspaper readers of the United States and England, and consequently are as interested in a daily one week to one month old as is the American who receives his afternoon paper almost before the ink is dry. While the average American daily will claim from three to five readers to the copy, the Chinese publisher might without any breach of faith claim as a minimum twelve to fifteen." Patterson, "The Journalism of China," 58.

33. *N. W. Ayer & Son's American Newspaper Annual and Directory* (Philadelphia: N. W Ayer & Son, 1922), 11.

34. Melissa Brown and Damian Satterthwaite-Phillips find that less than 4 percent of women born before 1943 identify themselves as fully literate, defined as the ability to read a letter. A total of 9.7 percent identified themselves as half-literate, able to understand business signs. The remaining 86.8 percent could do neither. Melissa J. Brown and Damian Satterthwaite-Phillips, "Economic Correlates of Footbinding: Implications for the Importance of Chinese Daughters' Labor," *PloS One* 13, no. 9 (2018): 9.

35. If we count literacy as the ability to utilize a few dozen, or a few hundred, common characters in everyday life, then rates would be higher, perhaps substantially so. For more on the rich debate about what constitutes literacy (or "literacies") and its rates in nineteenth- and twentieth-century China, see Evelyn Rawski, *Education and Popular Literacy in Ch'ing China* (Ann Arbor: University of Michigan Press, 1979); Liu Yonghua, "Qing dai minzhong shizi wenti de zai renshi," *Zhongguo Shehui Kexue Pingjia*, no. 2 (2017); and Xu Yi [Bas van Leeuwen], "19 Shiji Zhongguo Dazhong shizi lü de zai gusuan," in *Qing Shi Lun Cong: 2013 nian hao* (Beijing: Zhongguo Guangbo Dianshi Chubanshe, 2013), 240–47.

36. Sun Fuyuan, "Huiyi Wusi Dang Nian," in Zhongguo Shehui Kexue Yuan Jindai Yanjiusuo, *Wusi Yundong Huiyi Lu (Shang)* (Beijing: Zhongguo Shehui Kexue Chubanshe, 1989), 254.

37. Tao Dun, "Wusi zai Shandong nongcun," in Quan guo zheng xie wenshi ziliao weiyuanhui, *Wusi Yundong Qin Li Ji* (Beijing: Zhongguo Wen Shi Chubanshe, 1999), 211.

38. Zhou Shizhao, "Xiang Jiang nuhou," in Quan guo zheng xie wenshi ziliao weiyuanhui, *Wusi Yundong Qin Li Ji* (Beijing: Zhongguo Wen Shi Chubanshe, 1999), 189–90.

39. Xu Dehang, "Huiyi wusi yundong," in Quan guo zheng xie wenshi ziliao

weiyuanhui, *Wusi Yundong Qin Li Ji* (Beijing: Zhongguo Wen Shi Chubanshe, 1999), 14–43, 20; Arnold, *Commercial Handbook of China*, 427.

40. The description of the postal system, rates, and train and shipping routes is taken from Arnold, *Commercial Handbook of China*. The description of the activities of the elder Xiang, headmaster of two local schools, and the local Linhai students is taken from Wang and Zhang, "Wusi Yundong zai Taizhou."

41. Wang and Zhang, "Wusi Yundong zai Taizhou," 3.

42. Wang and Zhang, 3.

43. Harrison, "Newspapers and Nationalism," 197.

44. Bian Donglei, "Wusi yundong zai xiangcun: chuanbo, dongyuan, yu minzu zhuyi," *Ershiyi Shiji* [Hong Kong] (April 2019).

45. Chen Jingqiu, "Wusi Yundong zai Xianju," in Xianju Wenshi Ziliao Weiyuanhui, *Xianju Wenshi Ziliao, Di Yi Juan* (n.p.: n.p., 1986), 23–25.

46. This is according to the paratextual information in the first reports of May 5, 1919, in *Yi Shi Bao*. Repr. in Tianjin Lishi Bowuguan, *Wusi Yundong zai Tianjin* (Tianjin: Tianjin Renmin Chubanshe, 1979), 2–3.

47. "Nankai xuexiao jiangyantuan fu jiao chu jiang yan," *Yi Shi Bao*, June 4, 1919. Repr. in Tianjin Lishi Bowuguan, *Wusi Yundong zai Tianjin* (Tianjin: Tianjin Renmin Chubanshe, 1979), 66.

48. "Nankai Xuexiao fu ge cun jiangyan," *Nankai Rikan*, June 3, 1919. Repr. in Tianjin Lishi Bowuguan, *Wusi Yundong zai Tianjin* (Tianjin: Tianjin Renmin Chubanshe, 1979), 52.

49. "Nankai Xuexiao fu ge cun jiangyan," *Nankai Rikan*, May 27, 1919. Repr. in Tianjin Lishi Bowuguan, *Wusi Yundong zai Tianjin* (Tianjin: Tianjin Renmin Chubanshe, 1979), 52.

50. "Beiyang Daxue jiangyantuan fu Tanggu ji jiaoqu jiangyan," *Yi Shi Bao*, June 2, 1919. Repr. in Tianjin Lishi Bowuguan, *Wusi Yundong zai Tianjin* (Tianjin: Tianjin Renmin Chubanshe, 1979), 64.

51. Zhang Tinghao, "Zai Shanghai canjia wusi yundong de huiyi," in *WuSi Yundong Qin Li Ji*, ed. Quan guo zhengxie wenshi ziliao weiyuanhui (Beijing: Zhongguo Wenshi Chubanshe, 1999), 174–77, 176.

52. Deng Yingchao, "Wusi yundong de huiyi." Repr. in Tianjin Lishi Bowuguan, *Wusi Yundong zai Tianjin* (Tianjin: Tianjin Renmin Chubanshe, 1979), 678–85.

53. Liu Qingyang, "Juexing le de Tianjin renmin." Repr. in Tianjin Lishi Bowuguan, *Wusi Yundong zai Tianjin* (Tianjin: Tianjin Renmin Chubanshe, 1979), 704.

54. Zhao Jingshen, "You guan wusi de yi dian huiyi," in *Wentan Huiyi*, ed. Zhao Jingshen (Chongqing: Chongqing Chubanshe, 1988), 12–14.

55. Zhou Gucheng, "Wusi Yundong yu Qingnian Xuesheng," in Zhongguo Shehui Kexue Yuan Jindai Yanjiusuo, *Wusi Yundong Huiyilu (Xia)* (Beijing: Zhongguo Shehui Kexue Chubanshe, 1989), 978.

56. Zhang, "Zai Shanghai canjia wusi yundong de huiyi," 174–75.

57. See Corbin, *Village Bells*.

Chapter 2

1. Parts of this chapter previously appeared in John Alekna, "Neither Nation nor Empire: Situating Shanghai Radio in a Global Technological Moment, 1922–25," *Technology and Culture* 63, no. 4 (2022): 1078–1105.

2. Erez Manela describes this as the postwar anticolonial "Wilsonian Moment" in *The Wilsonian Moment: Self-Determination and the International Origins of Anticolonial Nationalism* (Oxford: Oxford University Press, 2017).

3. Sun Yat-sen, *The International Development of China* (New York: Knickerbocker Press, 1922).

4. Sun Yat-sen, "Foundation for Rebuilding the Chinese Republic," originally published in *Xinwenbao* 30th anniversary commemorative edition (Shanghai, 1922). Translated and reprinted in Sun Yat-sen, *Fundamentals of National Reconstruction* (Taipei: China Cultural Service, 1953), 114–15.

5. Sun Zhongshan, *Sun Zhongshan Quanji, Di 7 Juan* (Beijing: Renmin Chubanshe, 2015), 48–51, 49.

6. Specifically, he called for the reduction of the total number of troops by half and the conversion of these demobilized men into a labor corps for the construction of roads, railways, and other works. "Manifesto by Sun Yat-sen," *China Press*, Jan. 26, 1923, 1–2.

7. Twice in January, on the eleventh and nineteenth, his most senior ministers had been in negotiations with the new British ambassador to China with the aim of improving relations between the Guomindang and the British government—and by implication winning backing for the Nationalists. See F. Gilbert Chan, "An Alternative to Kuomintang-Communist Collaboration: Sun Yat-sen and Hong Kong, January–June 1923," *Modern Asian Studies* 13, no. 1 (1979): 132. Bruce Elleman has also shown how Sun attempted to win favor with the US during this period to play off the Soviets; see Bruce Elleman, "Soviet Diplomacy and the First United Front in China," *Modern China* 21, no. 4 (1995): 465–66.

8. "Sun Zhongshan zuowan yan baojie jishi," *Shibao*, Jan. 26, 1923; "Sun Zhongshan fabiao heping tongyi xuanyan," *Xinwenbao*, Jan. 26, 1923; "Dr. Sun Host at Farewell Dinner Here, Tells Plans," *China Press*, Jan. 26, 1923.

9. "Sun Zhongshan zuowan yan baojie jishi," *Shibao*, Jan. 26, 1923.

10. "Dr. Sun Yat-sen Congratulates *China Press* Radio: Southern Leader Delighted with Innovation—Predicts Great Educational Value," *China Press*, Jan. 27, 1923. Zhang Yaojun speculates that Sun's friendship with the American owners of the *China Press* led him to work with the paper and Osborn's radio station as part of his publicity campaign. Zhang Yaojun, "Sun Zhongshan de heping tongyi xuanyan

jiujing heshi shouci gongbu?" Shanghai dang'an xinxi wang [Shanghai Archive News Web], accessed Feb. 8, 2020.

11. "Sun Zhongshan fabiao heping tongyi xuanyan," *Dagongbao*, Jan. 27, 1923.

12. "Russia Will Renounce All Czarist Exactions on China, Dr. Sun Told: Southern Leader and Joffe, Moscow Representative, Reach Understanding on All Disputed Points," *China Press*, Jan. 27, 1923.

13. First, the communiqué emphasized that both sides believed China unready for a fully communist system. Rather, the nation's most pressing problems were national unification and independence, aims that Russia was happy to help the Nationalists achieve. As a gesture of good faith, the Soviet ambassador confirmed that his country would renounce all rights and concessions in northeastern China, which Russia had gained possession of in the Czarist era. Sun, for his part, acknowledged the legitimacy of Russian presence in Outer Mongolia, so long as it did not lead to Mongolian independence. "Russia Will Renounce All Czarist Exactions on China."

14. Even after this generous Soviet agreement, though, Sun continued to treat with the British and included an official tour of Hong Kong and audiences with the colonial governor on his trip south. See Chan, "An Alternative to Kuomintang-Communist Collaboration."

15. In the summer of 1923, the CCP had only between 100–250 members. Hans van de Ven, *From Friend to Comrade: The Founding of the Chinese Communist Party, 1920–1927* (Berkeley: University of California Press, 1992), 100.

16. "Judgement Summonses," *Wellington Evening Post*, April 8, 1915; "The Wedding Day Arrest," *North China Herald*, March 10, 1923; "Forgery Charged: Remarkable Case in Police Court: Grave Allegations," *Singapore Free Press & Mercantile Advertiser*, May 4, 1929; "Trade Competitions Case: Another Arrest on Fraud Charge," *The Times* (London), Oct. 22, 1934; "N.Z. Chemist Gaoled," *Auckland Star*, June 9, 1943; N.Z. Press Association, "Gaol for N.Z.-Born Swindler–Paris," Dec. 7, 1953; "New Zealander Ernest George Osborn, 63, Arrested in London after Being Voluntarily Deported from France," *Courier-Mail* (NZ), March 31, 1954.

17. "News and Music Radio Broadcasting Programs for Shanghai Tuesday," *China Press*, Jan. 21, 1923.

18. For a critical overview of the development of the fact and idea of Shanghai as an international city, see Jeffrey Wasserstrom, *Global Shanghai, 1850–2010: A History in Fragments* (London: Routledge, 2008). For an introduction to implications of this internationality for the cultural and social life of the metropolis, see Wen-hsin Yeh, *Shanghai Splendor: A Cultural History, 1843–1945* (Berkeley: University of California Press, 2007). The theoretical foundations for analyzing China, its concession territories, and spheres of influence as "semicolonial" are explored by Jürgen Osterhammel in "Semi-colonialism and Informal Empire in Twentieth-Century China: Towards a Framework of Analysis," in *Imperialism and After: Continuities and Dis-*

continuities, ed. Wolfgang J. Mommsen and Jürgen Osterhammel (London: Allen & Unwin, 1986), 290–314.

19. Valuable analyses of the structures and consequences of the physical and legal apartheid of Shanghai can be found in Douglas Clark, *Gunboat Justice: British and American Law Courts in China and Japan, 1842–1943*, 3 vols. (Hong Kong: Earnshaw Books, 2015); and Frederick Wakeman Jr., *Policing Shanghai, 1927–1937* (Berkeley: University of California Press, 1995).

20. "S'Hai to Have Radio Phones Soon Says Expert," *China Press*, Dec. 19, 1922.

21. Details about this transmitter are few. Later in the year, we do have a description of the transmitter Osborn used in Hong Kong, which may give us some idea of its technical parameters. This second, two-hundred-watt transmitter was British-made, though assembled entirely in Shanghai, and utilized "eight of the latest type vacuum tubes." "Broadcasting in Hongkong [sic]," *South China Morning Post*, August 2, 1923. For the full program of music and events, see "News and Music Radio Broadcasting Programs for Shanghai Tuesday," *China Press*, Jan. 21, 1923: ". . . equipment has been en route from the United States, the center of the radiophone activities" "Program Starts at Eight O'Clock; News, Music, Entertainment," *China Press*, Jan. 23, 1923. Journalist Cao Zhongyuan gives us the figure of fifty watts but no information about the make and model of machine; see Cao Zhongyuan, "San nian lai Shanghai wuxiandian zhi shiqing," *Dongfang Zazhi* 24, no. 18 (Sept. 25, 1924): 49–66, 50.

22. Cao, "San nian lai Shanghai wuxiandian," 50.

23. "News and Music Radio Broadcasting Programs for Shanghai Tuesday," *China Press*, Jan. 21, 1923.

24. "Radio 'Interference' Mars Concert Here; Good Program Tonight," *China Press*, Jan. 25, 1923.

25. The front and rear panels of the box dropped down to reveal the innards of the machines, so that students could ogle the interrelation of power, transmitting, receiving, and amplifying units, the oscillator. When ready to move, Robertson locked the panels back in place, and the radio became "a strong trunk ready for the road." C. H. Robertson, "10,000 Miles of Radio Lectures in China," *Radio Broadcast Magazine* 3, no. 10 (Sept. 1923): 382–91.

26. "First Radio Program Broadcasted Last Night, Big Success," *China Press*, Jan. 24, 1923.

27. "Radio 'Interference' Mars Concert Here; Good Program Tonight," *China Press*, Jan. 25, 1923.

28. "China's President Wants to Hear Radio Concerts," *China Press*, Feb. 11, 1923.

29. The previous year Li Yuanhong had been photographed at an experimental receiver listening with his grandson and granddaughter. Robertson, "10,000 Miles of Radio Lectures."

30. See Hans van de Ven, *Breaking with the Past: The Maritime Customs Service and the Global Origins of Modernity in China* (New York: Columbia University Press, 2014).

31. These men were H. J. Brett (Britain), D. Charlot (France), J. E. Jacobs (US), B. Renborg (Sweden), A. W. Van der Star (Netherlands), C. Brun (Denmark), J. D'Hondt (Belgium), H. Yokotake (Japan), and O. Thoresen (Norway)

32. "Minutes of a Meeting Held at the Offices of the Superindendent [sic] of Customs at 2 pm on the 29th September, 1925, on the Subject of the Munitions of War," in *Copies of Minutes and Correspondence to and from the Doyen of the Consular Corps*, 1925, 186–93, Foreign Office (FO) 671/537, National Archive at Kew.

33. "Minutes."

34. Foreigners fully understood this, as such requests and restrictions were discussed in all the major English-language newspapers in China. See, e.g., "Untitled," *North China Herald*, Nov. 24, 1923. See also "Beiyang zhengfu dui guangbo shiye de guanli," in Shanghai Shi Danganguan et al., *Jiu Zhongguo de Shanghai Guangbo Shiye* (Beijing: Dang'an Chubanshe, 1985), 40–53.

35. Cao, "San nian lai Shanghai wuxiandian," 61.

36. "Romance and the Affidavit: A Shanghai Man Detained after a Wedding," *North China Herald*, March 3, 1923.

37. "Deaton and Chang Take Over Affairs of Radio Corporation," *China Press*, March 22, 1923. The Chinese engineer was one K. L. Zhang of Shanghai, about whom little is known.

38. Cao, "San nian lai Shanghai wuxiandian," 51.

39. "Osborn to Head Federal Radio Administration," *China Press*, March 23, 1923; "Wing On Tower Broadcasting to Start Thursday," *China Press*, May 29, 1923.

40. "Radio Sending from Wing On's Begins Thursday," *China Press*, May 30 1923.

41. See "Jiaotong bu dian zheng si guangyu yan xing qudi si she diantai deng ji shi" (February–October 1923), in Shanghai Shi Danganguan et al., *Jiu Zhongguo de Shanghai Guangbo Shiye* (Beijing: Dang'an Chubanshe, 1985), 40–41.

42. Cao, "San nian lai Shanghai wuxiandian," 50.

43. "A Radio Federation," *North China Herald*, April 28, 1923.

44. Cao, "San nian lai Shanghai wuxiandian," 61.

45. Cao, 51.

46. "Fang Ziwei jiang you Mei gui guo," *Shenbao*, May 14, 1924.

47. "Fang Ziwei zuo ri dao Hu," *Shenbao*, May 18, 1924.

48. "Who's Who in China: Mr. Geo. T. W. Fong," *China Weekly Review*, Oct. 22, 1927.

49. "Fang Ziwei jun zhi wuxian dianhua tan," *Shenbao*, June 14, 1924.

50. Fang Ziwei, "Suowen: Meiguo wuxian dianbao guangbo zhan zhi xin fazhan," *Kexue* [Science] 10, no. 7 (1925): 921.

51. Benjamin Elman, "Rethinking the Twentieth Century Denigration of Traditional Chinese Science and Medicine in the Twenty-First Century." Paper prepared for the Sixth International Conference on the New Significance of Chinese Culture in the Twenty-First Century, Prague, Czech Republic, Nov. 2003. www.princeton.edu/~elman/documents/Rethinking_the_Twentieth_Century_Denigration.pdf.

52. Cao Zhongyuan, "Shouyinji san ji zhenkongguan bing ji qi zhi fa," *Kexue* [Science] 10, no. 7 (1925): 898–903.

53. See, e.g., Yuan Fang, "Meiguo feizhiye wuxiandian zhi fada," *Dian you* [Telegraphists' Companion] 1, no. 1 (1925): 7.

54. Such homemade sets were capable of very good local reception. A student magazine contributor reported that his homemade radio was very easy to build and could receive broadcast within a range of twenty to thirty *li* of Shanghai, hearing all music and news very clearly. More than forty *li* away from the city it was not so clear. But radiogram call signs you can hear up to three hundred *li* away. K. C. San, "Jianyi wuxian dianhua shouyinji zhizao fa," *Xuesheng Zazhi* 12, no. 4 (1925): 2–4.

55. Xia Yan, "Jianbian wuxiandian jieshouji zhi zhizao fa," *Dian you* 1, no. 1 (1925): 2–5; Xia Yan, "Jianbian wuxiandian jieshouji zhi zhizao fa (Xu)" *Dian you* 1, no. 2 (1925): 2–3.

56. "Zhu Qiqing jun zhi wuxiandian hua tiaoli tan Jiaotong Bu kalu zhi zhenxiang," *Shenbao*, June 15, 1924.

57. "This is because the cost of English exports is too high, [and] French and German imports are too delayed and too infrequent," the businessman argued.

58. Cao, "San nian lai Shanghai wuxiandian," 56.

59. Zhu Qiqing, "Hu Shang guangbo wuxiandian shiye gailun," *Dian you* 1, no. 6 (1925): 4–6.

60. Fang Ziwei, "Guohuo shikuang shouyinji chuxian," *Kexue* 10, no. 7 (1925): 921.

61. "Tongzi Bu you tian she dianxue ban," *Dagongbao*, Feb. 10, 1923.

62. Cao, "San nian lai Shanghai wuxiandian," 52. See also "Jiaotongbu dianzhengsi qudi Shenbao guan yu Kailuo gongsi lianhe jing ying guangbo diantai you guan wenjian," in Shanghai Shi Danganguan et al., *Jiu Zhongguo de Shanghai Guangbo Shiye* (Beijing: Dang'an Chubanshe, 1985), 44–46. The newspaper stopped providing content while the station continued broadcasting: apparently the authorities in Beijing did realize the two institutions used the same facility.

63. Cao had been a founding member of the Radio Club of China (*Zhongguo Wuxiandian Julebu*) in 1919. In September 1924, he wrote a survey of the development of wireless in China over the previous three years, interviewing many of the leading lights of communications himself.

64. Cao, "San nian lai Shanghai wuxiandian," 59.

65. Cao, 61.

66. Cao, 63.

67. Cao, 60.

68. Cao, 60.

69. Cao, 55.

70. Cao, 54. The ambiguities rife in the world of radio made determining the number of sets nigh impossible. Most crystal receivers were handmade with parts purchased from vendors outside the specialty radio shops. Furthermore, the installation of radio receivers was illegal, and most individuals living outside the foreign concessions took measures to hide the fact they had done so. In 1924, Cao Zhongyuan went around Shanghai querying "all sorts of relevant people and long-time radio personnel," but he could not get a single answer. Some said there were four hundred radio sets in the city; some said six hundred.

71. Cao, 50; Robertson, "10,000 Miles of Radio Lectures."

72. Robertson, "10,000 Miles of Radio Lectures."

Chapter 3

1. I use the name *Mukden*, the old Manchu name for the capital of Manchuria, when it is translated into English this way by Chinese people themselves, as in this case. Otherwise, I use the Chinese name *Fengtian* for the capital. The city is called *Shenyang* today. Captain Spear to Military Attaché, Peking. May 26, 1925. *General Correspondence: China*, FO 371/10916, 37–42.

2. For an introduction to the various names and changing identities of this region, see Mark Elliott, "The Limits of Tartary: Manchuria in Imperial and National Geographies," *Journal of Asian Studies* 59, no. 3 (2000).

3. Heidi Tworek, *News from Germany: The Competition to Control World Communications, 1900–1945* (Cambridge, MA: Harvard University Press, 2019), 63–67, 213–23.

4. Daqing Yang, *Technology of Empire: Telecommunications and Japanese Expansion in Asia, 1883–1945* (Cambridge, MA: Harvard University Press, 2010).

5. Michael Krysko, *American Radio in China: International Encounters with Technology and Communications, 1919–41* (New York: Palgrave Macmillan, 2011).

6. See, e.g., Louise Young, *Japan's Total Empire: Manchuria and the Culture of Wartime Imperialism* (Berkeley: University of California Press, 1998); Prasenjit Duara, *Sovereignty and Authenticity: Manchukuo and the East Asian Modern* (Lanham, MD: Rowman & Littlefield, 2003); Rana Mitter, *The Manchurian Myth: Nationalism, Resistance, and Collaboration in Modern China* (Berkeley: University of California Press, 2000).

7. Victor Seow, *Carbon Technocracy: Energy Regimes in Modern East Asia* (Chicago: University of Chicago Press, 2022); Koji Hirata, "Made in Manchuria: The Transnational Origins of Socialist Industrialization in Maoist China," *American Historical Review* 126, no. 3 (2021).

8. See the special issue of *Twentieth Century China* reconsidering the contributions of so-called warlord governments to Chinese state-building: *Twentieth Century China* 47, no. 1 (2022).

9. Frederick Simpich, "Manchuria, Promised Land of Asia: Invaded by Railways and Millions of Settlers, This Vast Region Now Recalls Early Boom Days in the American West," *National Geographic*, Oct. 1929, 379–428.

10. On the complexities of early Chinese telegraphy, see my discussion in chapter 1 above (pp. 29–31).

11. Chen Ertai, *Zhongguo guangbo zhi fu: Liu Han zhuan* (Beijing: Zhongguo Guangbo Dianshi Chubanshe, 2006). This detailed account of Liu Han's life resulted from historical research and interviews by the author Chen Ertai with Liu's son, widow, and surviving colleagues in the 1980s.

12. Chen Ertai, *Zhongguo Guangbo shikao* (Beijing: Zhongguo Guangbo Dianshi Chubanshe, 2008), 5.

13. Chen, 6.

14. Chen, 10.

15. Geng Li, the first head of the Tianjin Broadcasting Station—founded in 1927—also graduated with Beijing Communications University's class of 1914. The first head of the Beijing Broadcasting Station, Shen Zonghan, graduated in 1914 and earned a masters in electrical engineering the following year. The man who succeeded Liu Han in 1928 as head of the Harbin Radio Station was a graduate, as was Ye Shaofan, the head of the Telegraph Administration (*Dianzhengsi*). Gao Jiyi, who as head of the Northeast Communications Committee (*Dongbei Jiaotong Weiyuanhui*) administered all broadcasting and telegraphy in three eastern provinces, was an alumnus and may have had a hand in recruiting Liu Han there.

16. "Fengtian sheli Sansheng wuxiandian zhi choubei," *Shenbao*, Nov. 10, 1922.

17. "The Ghost of Chang Tso-lin: A Fanciful Interview in Changchun Following Rumours Elsewhere: Tampering with Wireless Waves," *North China Herald*, Sept. 1, 1929.

18. Robertson and his lecture tour on radio technology are discussed in chapter 2.

19. C. H. Robertson, "10,000 Miles of Radio Lectures in China," *Radio Broadcast Magazine* 3, no. 10 (Sept. 1923): 382–91.

20. Account adapted from "Dr. Sun's Arrival in Peking," *North China Herald*, Jan. 1, 1925.

21. "Dongsansheng jun she wuxiandian," *Xin Wenhua*, Feb. 1923.

22. "Hu-Ha jian wuxiandian jijiang tongdian," *Shenbao*, Dec. 21, 1922.

23. "Fengtian sheli Sansheng wuxiandian zhi choubei," *Shenbao*, Nov. 10, 1922; "The Wireless World: News and Notes," *Otago Daily Times* (New Zealand), Oct. 12, 1923.

24. "Fengtian sheli Sansheng wuxiandian zhi choubei," *Shenbao*, Nov. 10, 1922;

"Ha-Bu wuxian diantai zhi shuaxin: Neirong jiao E-ren shi you wanbei," *Shenbao*, March 6, 1923.

25. The Japanese businessmen were, however, rebuffed by the Japanese government, which likely opposed Zhang possessing an independent communications network. "Fengtian sheli Sansheng wuxiandian zhi choubei," *Shenbao*, Nov. 10, 1922; "Foreign Office Denies Wireless Petition," *Japan Times*, Nov. 27, 1922.

26. "Ha-Bu wuxian diantai zhi shuaxin: Neirong jiao E-ren shi you wanbei," *Shenbao*, March 6, 1923.

27. "Ha-Bu wuxian diantai zhi shuaxin."

28. "Guonei zhuandian," *Shenbao*, May 12, 1923.

29. Several newspapers republished the article of March 6 quoted above. Logically, it appeared in the hometown paper, the Harbin-Heilongjiang Times *Bin-Jiang Shibao*, on March 17 under the headline "The Future Prospects of Wireless" (*wuxiandian zhi qiantu*). [Partially transcribed in Chen Ertai, *Zhongguo Guangbo Shikao* (Beijing: Zhongguo Guangbo Dianshi Chubanshe, 2008), 27.] Less logically, the *Louisville (KY) Courier-Journal* republished an abridged translation under the headline "Harbin, China Has Up-to-Date Radio: Former Russian Station Now Used in Modern Way," on May 6. Chinese radio historians Zhao Yuming and Chen Ertai have had a long-running debate over whether this institution can be labeled a broadcasting station. Zhao Yuming, "Zai tan Zhongguo xiandai guangbo shi yanjiu zhong de ruogan wenti," *Xiandai Chuanbo* 163, no. 2 (2010): 131–37.

30. "Harbin Press's Latest," *North China Herald*, April 12, 1924.

31. Fu Xingpei, "Di-er ci Zhi-Feng zhanzheng jishi," in *Wenshi Ziliao Jingxuan Di 4 Ce*, ed. Wenshi Ziliao Xuanji (Beijing: Zhongguo Wenshi Chubanshe, 1990), 93–101, 94.

32. Marie-Claire Bergère, *Sun Yat-sen*, trans. Janet Lloyd (Stanford, CA: Stanford University Press, 1998), 396–97.

33. "Marshal Chang Is Up-to-Date: He leads in Demonstrating Uses for Radio," *Japan Times*, Dec. 4, 1924.

34. "Guonei Zhuandian," *Shenbao*, Dec. 19, 1924.

35. "Guonei Zhuandian."

36. "Beijing zheng wenji," *Shenbao*, Dec. 30, 1924.

37. "Guonei Zhuandian," *Shenbao*, Jan. 15, 1925; "Guonei Zhuandian," *Shenbao*, Jan. 31, 1925.

38. Benton Byrd, "Canton Leads!" *South China Morning Post*, Sept. 7, 1923.

39. Byrd.

40. This would have been today's Zhongshan Park. It lies immediately adjacent to the Forbidden City.

41. "Beijing Zhongyang Gongyuan Shiyan Guangbu Wuxiandianji," *Dian you* 1, no. 2 (1925): 14.

42. "Government at Peking Plans Broadcasting," *China Press*, Feb. 12, 1925.

43. Charles Dailey, "Look to Radio to Help Unify Chinese People: Peking More Liberal toward New Science," *Chicago Daily Tribune*, April 5, 1925.

44. "Peking Hears a Radio Concert and Is Amazed," *Japan Times*, March 16, 1925.

45. Dailey, "Look to Radio."

46. "Dr. Sun Yat-sen Congratulates *China Press* Radio," *China Press*, Jan. 27, 1923.

47. "Sun Zhongshan shishi zhi aidao," *Shenbao*, March 14, 1925.

48. "Sun Zhongshan shishi zhi aidao."

49. "Loud-Speakers Installed for Sun Memorial Rites," *China Press*, April 10, 1925.

50. "Radio-Broadcasting in China," *China Press*, Jan. 9, 1925.

51. This was the so-called Anti-Fengtian War (*fan Feng zhanzheng*).

52. Radio dealers paid 10 percent of the price of all apparatuses on import, after having secured a license and paid a $1,000 deposit as guarantee of good behavior. At sale, all radios needed to be registered for a $2 fee. Finally, as in Britain, listeners paid an annual listening fee; the price was set at $12 for vacuum tube sets and $6 for crystal sets. See Chen, *Zhongguo Guangbo zhi Fu*, 109–20; and "Radio Catches On in Manchuria: Far Ahead of Rest of China," *North China Herald*, August 21, 1926. For an official translation of the regulations as issued, see *General Correspondence: China*, FO 371/12458, 190–92 (1927). The regulations were finally issued in October 1926; "Ben Guan Zhuandian," *Shenbao*, Oct. 9, 1926.

53. Julean Arnold, "Special Circular No. 566—Electrical Equipment Division: Radio Broadcasting in China," Department of Commerce, Washington, DC, March 23, 1928.

54. Fengtian Province Office Public Order No. 254, Oct. 27, 1926. Repr. in *Shenyang guangbo zhi ziliaoxing wengao*, ed. Liu Jiufu (Shenyang: Shenyang Renmin Guangbo Diantai, 1987), 20–25; British Foreign Office, *General Correspondence: China*, FO 371/12458, 67; "Dongbei wuxiandian mingchun tongbao," *Shenbao*, Dec. 12, 1926.

55. "Eastern Radio: Developments in China and Ceylon," *South China Morning Post*, April 20, 1927.

56. "Ben guan zhuandian," *Shenbao*, May 14, 1927.

57. "Guangbo wuxian diantai tian fang: Shangye guanggao," *Dagongbao*, Dec. 11, 1927.

58. It had a power of five hundred watts. "Huabei zhi guangbo wuxiandian shiye," *Dagongbao*, August 7, 1927; "Jing-Jin guangbo wuxiandian," *Dagongbao*, March 31, 1927.

59. "Eastern Radio: Developments in China and Ceylon," *South China Morning Post*, April 20, 1927.

60. "Huabei zhi guangbo wuxiandian shiye," *Dagongbao*, August 7, 1927.

61. "Untitled," *HuaBei Huakan* 49, no. 3 (1929); "Za dian yi shu," *Shenbao*, August 28, 1927.
62. "The Activities of Harbin: New Broadcasting Station," *North China Herald*, Jan. 28, 1928.
63. U.S. Department of Commerce, "Asiatic Radio Broadcasting Stations," May 22, 1929.
64. "The Activities of Harbin: New Broadcasting Station," *North China Herald*, Jan. 28, 1928; "Grand Opera on the Radio: Harbin Performances: A Little Misunderstanding," *North China Herald*, March 17, 1928.
65. "Manchuria Linked by Radio with All Eastern World," *China Press*, March 5, 1928.
66. "A Radio Confucius," *Los Angeles Times*, July 22, 1927.
67. "Radio to Prove Invaluable for Chinese People," *The Campus* (of Allegheny College), Oct. 6, 1927.
68. "Fengtian shengzhang gongshu wei Yi xian cheng gou shoutingqi si ju bing zhuangshe didian qing jian heshi," Liaoning Provincial Archive, JC010-01-008901 (Document 8901), Dec. 1928.
69. "Liaoning sheng zhengfu wei wuxiandian ye xuexiao jiaoyuan Chen Shiyi zhu '*dui Dong-Sheng dianye jihua shu*' yi ben shi," Liaoning Provincial Archive, JC010-01-003228 (Document 3228), 1930.
70. "China Goes on the Air," *Washington Post*, June 5, 1928.
71. Department of Commerce, "Radio Markets of the World, 1928–1929" (Washington, DC: Government Printing Office, 1928 [Archive Unbound]), 26.
72. "China Goes on the Air," *Washington Post*, June 5, 1928.
73. "China Goes on the Air."
74. "Dongsheng wuxiandian ye gaikuang: xian you wuxiandiantai shiyi chu: Fengtian diantai wei shijie di yi," *Shenbao*, Oct. 31, 1928.
75. "Dongsheng wuxiandian ye gaikuang."
76. Later, authorities claimed that the generals had in fact been tried by a special court and convicted on all the main charges: insubordination, gross corruption, treason. It seems unlikely that such formalities had been observed. Japanese reports discovered after the Second World War suggest Yang had indeed been the focus of efforts to create a more pliable Manchuria in the wake of the old marshal's assassination. Indeed, after the executions, the Japanese premier, Baron Tanaka, extolled Yang as "the most able statesman in Manchuria, [whose] death would probably result in a temporary check being given to the progress of Japan's railway" and economic interests in the region. "New Reports from Mukden Tell about Executions," *China Press*, Jan. 14, 1929.
77. "New Reports from Mukden."
78. Department of Commerce: Bureau of Foreign and Domestic Commerce,

"Asiatic Radio Broadcasting Stations," May 22, 1929, USNARA Archives Unbound, www.gale.com/intl/primary-sources/archives-unbound (subscription required); Department of Commerce: Bureau of Foreign and Domestic Commerce, "Asiatic Radio Broadcasting Stations," Oct. 1, 1929, USNARA Archives Unbound, www.gale.com/intl/primary-sources/archives-unbound (subscription required); "Radio Broadcast," *South China Morning Post*, Feb. 27, 1930.

79. Jiaotong Congbao, "Dongsansheng ge dianju quanti zhiyuan dian," no. 20 (1929). Letter dated Jan. 26, 1929.

80. See, e.g., "Ultimatum by Nanking," *North China Herald*, Dec. 30, 1930.

81. "Malicious Propaganda: Nanking and Mukden Working in Full Harmony," *South China Morning Post*, August 19, 1929.

82. "Manchurian 'War' Waged over Radio: Chinese Broadcast Charges, Each Side Blames Other," *Boston Globe*, August 21, 1929.

83. "Heavy Chinese Troop Concentration Indicated," *Los Angeles Times*, August 23, 1929. The intensity of this conflict caused a crisis in Harbin, where several people were arrested under suspicion of Communist tendencies. See "Reds in Harbin Jailed to Prevent Rioting," *New York Times*, Nov. 7, 1929.

84. "Sir Robert Ho Tung on Manchuria: Interesting Speech on Province Broadcast from Mukden," *North China Herald*, Sept. 7, 1929.

85. "Chinese Turning Big Arsenal into Peace Factory," *Christian Science Monitor*, April 5, 1929; Rana Mitter, "The Last Warlord," *History Today*, Feb. 1, 2004.

86. "Itinery [sic] Drives in North of Anti-Opium Association Are Given Popular Support," *China Press*, August 12, 1929.

87. Chen, *Zhongguo Guangbo zhi Fu*, 169–71. It was not until a couple days later, when the first refugees reached Beijing, that firsthand news of the events in Fengtian were broadcast to Chinese listeners.

88. "Japanese Occupy Manchuria," *North China Herald*, Sept. 22, 1931.

89. "Eye-Witness Tells of Japan Mukden Invasion: Artillery Opened Fire on Airdrome: Police and Soldiers Clash," *China Press*, Sept. 26, 1931.

90. "Japanese Occupy Manchuria."

91. "Japanese Occupy Manchuria."

92. "Japan's Grip on Manchuria," *Manchester Guardian*, Jan. 9, 1932.

93. "Japan's Grip on Manchuria."

94. "Half of Globe's Broadcasting Stations in U.S.: Russia Exhibits Growing Interest in Radio," *Chicago Daily Tribune*, Sept. 13, 1931. At two thousand kilowatts, the Fengtian station was four times more powerful than the broadcasting station in Nanjing.

95. "Tungliao Junction of Ssupingkai-Taonan Railway Taken by Japanese," *China Press*, Oct. 6, 1931.

96. Yang, *Technology of Empire*, 78.

97. Chi Man Kwong, "Building a 'Total Mobilization State': Thinking about War and Society in 1920s Manchuria," *American Journal of Chinese Studies* 26, no. 1 (April 2019).

Chapter 4

1. "Ge di minzhong fenkai ji'ang," *Dagongbao*, July 12, 1937.
2. Meng Yiqi, "Cong zhongxue xiaoyuan toushen kangzhan de huiyi," *Jianghuai Wenshi* 4 (1999).
3. Zhang Yan and Huang Wenxuan, "Kangzhan shouyin yiwen," *Hongyan Chunqiu* 1 (2006).
4. Zhang and Huang.
5. Marshall McLuhan, *The Gutenberg Galaxy: The Making of Typographic Man* (Toronto: University of Toronto Press, 1962); Marshall McLuhan, *Understanding Media: The Extensions of Man* (Cambridge, MA: MIT Press, 1994).
6. China did not manufacture its own vacuum tubes until after 1949.
7. A best-selling American tabletop model with five tubes cost between seventy-five and one hundred Chinese yuan, though prices ranged as high as several hundred yuan. Anne Allison, "Review of Radio Listening in China, 1937–1945," Office of War Information, June 6, 1945, US National Archives and Records Administration (hereafter USNARA) RG 208, entry 370, box 385.
8. "Gansheng wancheng guangbo shouyinwang," *Dagongbao*, May 15, 1937.
9. "Gansheng wancheng guangbo shouyinwang." In 1939, Japan furnished 55 percent of the total number of radios sold in China (though this represented only 27 percent of the total value of the radio market). These affordable Japanese radios helped expand personal radio use among the petit-bourgeois. Still, even the lowest-priced Japanese set (Chinese $4.50) was out of reach of the working class. In 1936, the best-paying iron foundry in Peking paid a minimum monthly wage of Chinese $4.00 and a maximum of $15.00, while the average wage of an industrial worker in Shanghai in 1937 was Chinese $18.00. Several currencies circulated during the period, at least three of which were called Chinese dollars, or "$ yuan." The number and legal validity of these currencies changed over the course of the 1930s. The form of tender also varied according to one's location, north or south, whether in a concession or Chinese territory. I have expressed values here in terms of currencies and days of labor as described in the sources.
10. A crystal radio set is quoted as being around ten Chinese yuan, including installation cost, in 1927. See "Shuodao guangbo wuxiandian," *Dagongbao*, Sept. 16, 1927. Prices only declined over the course of the subsequent decade. Edgar Snow cites the daily wage of a laborer at 0.2 yuan per day, or 6 yuan a month in 1933. Edgar Snow, "Awakening the Masses," *New York Herald Tribune*, Dec. 17, 1933.
11. Carlton Benson describes aspects of this phenomenon through the growth of

radio *tanci*, a kind of instrumental storytelling ballad. Carlton Benson, "From Teahouse to Radio: Storytelling and the Commercialization of Culture in 1930s Shanghai" (PhD diss., University of California, Berkeley, 1996).

12. Ming Xin, "Tushuguan li yingdang kaifang wuxiandian ma?" *Dagongbao*, August 19, 1936.

13. Meng Tang, "Qiwang ben shi dangju qudi naoshi boyin," *Dagongbao*, Feb. 15, 1935.

14. Benson, "From Teahouse to Radio," 114.

15. Meng, "Qiwang ben shi dangju qudi naoshi boyin."

16. Benson, "From Teahouse to Radio," 141–211.

17. "Radio in China: Loudspeakers Placed in Squares and Parks," *Times of India*, July 10, 1934.

18. Ming, "Tushuguan li yingdang kai wuxiandian ma?"

19. This was, in reality, the source of endless headaches for broadcasters and listeners. The number of stations crowded onto a limited number of wavelengths made the problem of interference severe. Given the complicated legal-administrative status of Shanghai, with three different responsible authorities—Chinese, International, and French—the issue was nigh impossible to resolve. Indeed, it festered until the final Japanese takeover in 1941. See "Broadcasting in Shanghai," *North China Herald*, June 3, 1936.

20. Guo Zhenzhi, "A Chronicle of Private Radio in Shanghai," *Journal of Broadcasting & Electronic Media* 30, no. 4 (1986).

21. "Government Buys Station XGHE," *North China Herald*, June 30, 1937.

22. Chen Guofu, "Zhongyang Guangbo Diantai Chuangban Jingguo," in *Zhonghua Minguo Xinwen Nianjian* (Taipei: Taibei shi Xinwen Jizhe Gonghui, 1961), 46–47.

23. For more on the establishment of the Guomindang Broadcasting System, see Laura De Giorgi, "Communication Technology and Mass Propaganda in Republican China: The Nationalist Party's Radio Broadcasting Policy and Organisation during the Nanjing Decade (1927–1937)," *European Journal of East Asian Studies* 13 (2014).

24. Wang Xueqi, *Di si zhanxian: Guomindang zhongyang guangbo diantai duoshi* (Beijing: Zhongguo Wen Shi Chubanshe, 1988), 12–13. Citing *Guangbo Zhoubao*, Oct. 20, 1934.

25. Wang Xueqi, *Di si zhanxian*, 15.

26. "Loudspeakers to Tell of Developments: Nanking to Install News System in Streets to Keep People Informed," *China Press*, Feb. 22, 1933.

27. "Fenyang shouyin maizuo," *Dagongbao*, Sept. 15, 1936.

28. "Nanling xian chengni zhuangshe duanbo wuxiandian," *Anhui Jianshe*, no. 18 (1930): 214–15.

29. "Nanling xian chengni zhuangshe duanbo wuxiandian," 215.
30. "Huangbang Ling zhuangshe wuxiandian shouyinji," *Banghu Yuekan*, no. 12 (1930): 94.
31. "Yishui Minjiaoguan Wuxiandian yi kaishi zhuangshe," *Shandong Minzhong Jiaoyu Yuekan* 3 no. 5 (1932): 169.
32. "Yi xiang shi tichang kexue zhuangshe shouyinji," *Anhui Jiaoyu Zhoukan*, no. 66 (1934): 4.
33. Allison, "Review of Radio Listening in China," 3.
34. Allison, 3.
35. Tian Yili, "Ba xian zhi xing," *Dagongbao*, Feb. 13, 1936.
36. Wu Daoyi, *Zhongguang Sishi Nian* (Taipei: Zhongguo Guangbo Gongsi, 1968), 17.
37. Wu, 17.
38. Wu, 17.
39. "Zhongyang Guangbo Diantai guanlichu shouyinyuan xunlianban gaikuan," *Wuxiandian* 1, no. 1 (1934).
40. "Shouyinyuan lianxi yi biye," *Zhejiang Sheng Jianshe Yuekan* 4, no. 4 (1930): 7–8.
41. "Zhejiang sheng ge xian shouyinji tongji biao," *Zhejiang Sheng Zhengfu Xingzheng Baogao*, no.4 (April 1931): 49–52; "Shouyinyuan lianxi yi biye," *Zhejiang sheng Jianshe Yuekan*.
42. Ai Peijun, "Tianquan di yi tai shouyinji," in Zhengxie Tianquan Xian Weiyuanhui, *Tianquan Wenshi Ziliao, Di-si Ji* (n.p.: n.p., 1986), 19.
43. "Wuchow Matters: Radio Connection for Leading Cities," *South China Morning Post*, Dec. 30, 1932; "Kwangsi to Have Radio Broadcasting," *China Press*, July 23, 1933.
44. "Listeners' Club: Nanning's Schedule," *South China Morning Post*, Feb. 8, 1934.
45. Huang Xiaoying, "Mingguo shiqi guangyin jiaoyu de lishi huigu," *Dianhua jiaoyu yanjiu* 6 (2011): 110–20; Nanjing Second Historical Archive [hereafter NJSHA], Doc. 5.12100, *Quanguo zhongdeng xuexiao yu minjiaoguan zhuangshe wuxiandian shouyinji banfa dawang, ge sheng shi jiaoyu dingju fenqi jie jiao wuxiandian shouyinji jiakuan banfa ji youguan wenshu*, 1935.7–1939.7; "Chuan ge xian shezhi shouyinji tongji," *Sichuan Yuebao* 9, no. 6 (1936): 178.
46. NJSHA Doc. 5.12184, *Jiaoyubu shehui jiaoyu sibian "1935 niandu yiqian ge sheng shi ge jiaoyuguan zizhuang shouyinji yilan biao" ji youguan wenshu*, 1936.11–1944.07.
47. "Ge sheng shi xiaoguan zhuangshe shouyinji tongji," *Shenbao*, Jan. 20, 1937.
48. "Chuan ge xian shezhi shouyinji tongji," 178.
49. "Gan sheng wancheng guangbo shouyinwang qi yu zuiduan shijian nei meiqu you yi shouyinji," *Dagongbao*, May 15, 1937.

50. Madame Chiang Kai-shek, "China's 'New Life' Movement," originally published in *China Press*; republished in *China's Leaders and Their Policies: Messages to the Chinese People*, by Wang Ching-Wei and General Chiang Kai-Shek (Shanghai: China United Press, 1935).

51. Hubei Provincial Archive [hereafter HBPA], Doc. LS31-1-886-001, *Hubei Sheng Zhengfu guanyu shezhi shouyinji xianqi wancheng de xunling*, 1936.

52. Hu Baoren, "Manhua Shang-Xian de diyi bu shouyinji," in Shaanxi Sheng Shang-Xian Weiyuanhui Wenshi Ziliao Yanjiu Weiyuanhui, *Shang-Xian Wenshi Ziliao, Di-si Ji* (1987), 186–87.

53. Li Qing, "Laiyuan de di yi tai shouyinji," in Zhengxie Laiyuan Xian Weiyuanhui, *Laiyuan Xian wenshi ziliao: Di-yi Ji* (1996), 261.

54. Ai, "Tianquan di yi tai shouyinji," 19.

55. Ai, 19.

56. "Radio in China: Loudspeakers Placed in Squares and Parks," *Times of India*, July 10, 1934.

57. "Modern China: Lecture to Real Estate Institute," *Construction and Real Estate Journal*, March 11, 1936.

58. HBPA, Doc. LS31-1-886-001, *Hubei Sheng Zhengfu guanyu shezhi shouyinji xianqi wancheng de xunling ji xingzhengyuan de xunling*, Oct. 30, 1936.

59. Julia Strauss, *Strong Institutions in Weak Polities: State Building in Republican China, 1927–1940* (Oxford: Oxford University Press, 1998); Rana Mitter, *Forgotten Ally: China's World War II, 1937–1945* (Boston: Houghton Mifflin Harcourt, 2013).

60. "Zhuangshe wuxiandian shouyinji," *Tongzhong* 1, no. 8/9 (1935): 50.

61. "Yan Huiqing qing junsuo daibiao yao ling Shanghai pao sheng: zheng yu Rui dangju jieqia boyin shiyi," *Shenbao*, Feb. 13, 1932.

62. "Hangchow and the Crisis: Sensation Caused by News from Shanghai," *North China Herald*, Feb. 16, 1932.

63. Sei-Jeong Chin, "Print Capitalism, War, and the Remaking of the Mass Media in 1930s China," *Modern China* 30, no. 4 (2014); "Canton Keeps Close Watch: Close Attention Paid to Shanghai Events," *North China Herald*, Feb. 16, 1932.

64. "Much Excitement in Hangchow: News by Loud-Speakers and Bulletins," *North China Herald*, March 1, 1932.

65. "Inland China Getting Thrills from Radio, Salesmen Report Here: Natural Curiosity of Chinese Displayed by Growth of Interest in Broadcasts; Radio Called Force for Unity in this Country; Recent Fighting Caused Greater Interest," *China Press*, July 17, 1932.

66. "Inland China Getting Thrills."

67. "A New Danger," *North China Herald*, March 22, 1932.

68. "Xi'an Guangbo Diantai Gaikuang," *Guangbo Zhoubao*, Sept. 19, 1936. Repr. in *Shaanxi guangbo dianshi zhi* (Xi'an: 1992), 469–70.

69. Wang Xueqi, *Di si zhanxian*, 61.
70. Xibei Daxue Lishixi Zhongguo xiandai shi yanjiushi, *Xi'an shibian ziliao xuanji* (1978), 113–18.
71. John Service, *The John S. and Caroline Service Oral History Project*, vol. 1 (Berkeley: Regional Oral History Office, Bancroft Library, 1978), 132.
72. Wu, *Zhongguang Sishi Nian*, 19.
73. "Shan bian hou zhi Zhengzhou," *Dagongbao*, Dec. 23, 1936.
74. Wang Xueqi. *Di si zhanxian*, 60.
75. Zhang and Huang, "Kangzhan shouyin yiwen."
76. Wu Shui, "Guofang yu wuxiandian," *Dagongbao*, March 2, 1937.
77. "China Rejoices at Safe Return of Gen. Chiang: Gen. Chang 'Awaits Punishment' at Nanking," *China Weekly Review*, Jan. 2, 1937.
78. Wu, *Zhongguang Sishi Nian*, 17.

Chapter 5

1. Daqing Yang, *Technology of Empire: Telecommunications and Japanese Expansion in Asia, 1883–1945* (Cambridge, MA: Harvard University Press, 2010), 2–3.
2. Yang, 87–121; Aaron Stephen Moore, *Constructing East Asia: Technology, Ideology, and Empire in Japan's Wartime Era, 1931–1945* (Palo Alto, CA: Stanford University Press, 2013); Koji Hirata, "Steel Metropolis: Industrial Manchuria and the Making of Chinese Socialism," *Enterprise & Society* 21, no. 4 (2020); Victor Seow, *Carbon Technocracy: Energy Regimes in Modern East Asia* (Chicago: University of Chicago Press, 2022).
3. Mitter demonstrates that the war witnessed an increase in the measurement and classification of society. Greater legibility allowed a growth in the state provision of social welfare. See Rana Mitter, "Classifying Citizens in Nationalist China during World War II, 1937–1941," *Modern Asian Studies* 45, no. 2 (2011).
4. Timothy Brook, *Collaboration: Japanese Agents and Local Elites in Wartime China* (Cambridge, MA: Harvard University Press, 2005); Yun Xia, *Down with the Traitors: Justice and Nationalism in Wartime China* (Seattle: University of Washington Press, 2017); Rana Mitter, *The Manchurian Myth: Nationalism, Resistance, and Collaboration in Modern China* (Berkeley: University of California Press, 2000).
5. "3,000 Japanese Troops Take Control of Peiping," *Baltimore Sun*, August 9, 1937.
6. "Beifang ge xianju you zhanshi," *Dagongbao*, August 23, 1937.
7. Zhu Jingxin, "Gu du xianluo hou zhi xingxing sese," *Shenbao*, Nov. 4, 1937.
8. "Japan Reports China Getting Russian Planes: Japanese Advance Continues," *New York Herald Tribune*, Sept. 3, 1937.
9. "Huilei hua Jin," *Dagongbao*, Sept. 22, 1937.
10. "Beifang Qianxian: Neidi xiaoxi xuyao goutong," *Dagongbao*, Sept. 26, 1937.

11. "Jingu Za chou," *Dagongbao*, June 6, 1938.
12. "Ri zai Huabei xuanchuan gongzuo," *Dagongbao*, April 21, 1939.
13. Zhu Jingxin, "Gu du xianluo hou zhi xingxing sese," *Shenbao*, Nov. 4, 1937.
14. "Huilei hua Jin," *Dagongbao*, Sept. 22, 1937.
15. Shanghai Municipal Police Records, case-file D.8126. See also Carlton Benson, "The Resurgence of Commercial Radio in Gudao Shanghai," in Henriot and Yeh, eds. In the Shadow of the Rising Sun (Cambridge: Cambridge UP, 2004).
16. "Jin ri jiu-shiba nian shimin juxing kangdi xuanshi," *Dagongbao*, Sept. 18, 1937.
17. Shanghai Municipal Police Records, case-file D.8167.
18. Shanghai Municipal Police Records, case-file D.8122.
19. Shanghai Municipal Police Records, case-file D.8090.
20. Shanghai Municipal Police Records, case-file D.8178.
21. Shanghai Municipal Police Records, case-file D.8167.
22. Wireless telegraph operators did stay behind to maintain communications with partisan fighters in the fallen city. They were not removed from the radio station at Sassoon House until early January 1938. See Wu Daoyi, *Zhongguang Sishi Nian* (Taipei: Zhongguo Guangbo Gongsi, 1968), 78. Unsubstantiated reports in the foreign press claimed some of these operators were killed live on-air while trying to transmit one last radio message—a plea for help—the day of the Japanese takeover.
23. Wu, *Zhongguang Sishi Nian*, 78–79.
24. "New Central Radio Station to Broadcast Daily in Chungking," *China Press*, March 12, 1938.
25. Wu, *Zhongguang Sishi Nian*, 64–66.
26. Wu, 64–66.
27. NJSHA, Doc. 5.12097, "Jiaoyubu guanyu zengshe houfang ge xian shi shouyinji zhixing fang'an yu Neizhengbu ji Zhonyang Guangbo Shiye Guanli Chu deng lai wang wenshu," June 6, 1938.
28. NJSHA, Doc. 5.12196, "Sichuan sheng minjiaoguan guanyu pei gou dian jiao qicai deng shixiang wenshu. (Nei you '26 nian du Sichuan sheng xiaoxue ji minzhong xuexiao zhuangshe shouyinji diaochabiao.')," 1937–38, 119–55.
29. NJSHA, Doc. 5.12196, 29–30.
30. HBPA, Doc. LS3-1-417, "Ge sheng pu she shouyinji ji yunyong banfa, shouyinyuan xunlian banfa dawang, xian shi shouyinshi banshi tongce, shouyinji diaochabiao," May 8, 1940. See also NJSHA, Doc. 719.189, "Zhongyang guangbo shiye guanlichu suoshu ge tai zhudi biao, sheli guangbo shouyinji wang jihua cao'an, Zeng Ranxi chengbao jieshou Hankou, Zhongshan, Minguo wei baojing guo ji ge sheng pu she shouyinji ji yingyong banfa deng wenjian," Sept. 1941, 44.
31. NJSHA, Doc. 719.189, 44.
32. Guoshiguan, "Zhongyang wuxiandian qicaichang qing guanbo shiye guanli-

chu fajiao wuxiandian jijian xianlu chengshi," correspondence dated Jan. 29, 1942, to Feb. 18, 1942.

33. NJSHA, Doc. 5.12203, "Shaanxi Sheng qing pei gou shouyinji dianchi deng dian jiao qicai de wenshu," 1939–41.

34. HBPA, Doc. LS3-1-417, 8.

35. NJSHA, Doc. 5.12196, 5.

36. Anne Allison, "Review of Radio Listening in China, 1937–1945," Office of War Information, June 6, 1945, USNARA RG 208/370/385, US National Archives, College Park, MD.

37. The wartime capitals of Hunan and Guangdong, respectively. Office of War Information, "Excerpt from a Field Report on Hunan-Kwangtung, May 1944," USNARA RG 208, NC-148, entries 523, 524, 525, box 3021.

38. Matthew D. Johnson, "Propaganda and Sovereignty in Wartime China: Morale Operations and Psychological Warfare under the Office of War Information," *Modern Asian Studies* 45, no. 2 (2011): 340.

39. "Dui diren xuanchan zhou," *Dagongbao*, Sept. 16, 1942.

40. Zhengzhi Bu Wenhua Gongzuo Weiyuanhui, "Riben guanbing dui wofang dui di xuanchuan yingxiang zuotanhui," *Diqing Cankao Ziliao*, no. 27, interrogation dated Dec. 27, 1943.

41. Zhengzhi Bu Wenhua Gongzuo Weiyuanhui.

42. His few writings appeared in the "Students World" (xueshengjie) section of the newspaper. Judging from his later writings and the paper he was printed in, he was left-leaning. His description of the town of Guangning appears in "Liuwang yi ye," *Dagongbao*, Sept. 14, 1939.

43. "Liuwang yi ye."

44. Chen Yih, "Radio Chungking Calling," *New York Times*, August 29, 1943.

45. "Ri zai Huabei xuanchuan gongzuo," *Dagongbao*, April 21, 1939.

46. "Ri zai Huabei xuanchuan gongzuo."

47. Sun Yude, "My Older Brother's First Radio," in *China in Family Photographs*, ed. Ed Krebs and Hanchao Lu (Los Angeles: Bridge21 Publications, 2018), 313–17.

48. Yude, 313–17.

49. Meng Yiqi, "Cong zhongxuexiaoyuan toushen kangzhan de huiyi," *Jianghuai Wenshi* 4 (1999): 33.

50. Meng, 35.

51. "Liu Tsan-Ch'i's Report on Trips to Kunming, Kweilin, Hengyang, & Kukong," Jan. 12, 1944, USNARA RG 208, entry 370, box 385.

52. "Radio is now the only source from which over 500 newspapers and thousands of mobile newspapers receive their news and make it available to the people," reported one Chongqing official in 1943. Quoted in Chen Yih, "Radio Chungking Calling," *New York Times*, August 29, 1943.

53. "Memorandum: Conditions at Shanghai," Sept. 28, 1943; "Excerpt from Interview with Chinese refugee from Pieping [sic]," Nov. 20, 1943, USNARA RG 208, NC-148, entries 523, 524, 525, box 3021.

54. Regrettably, the Republican period archives of China Central Broadcasting are not open to research, but they would presumably contain similar evidence focused on Chongqing's broadcasting.

55. See, e.g., Theodore Herman, Field Intelligence Officer, "Attitudes in Shanghai: Interview with a Chinese Businessman Who Left Shanghai at the Beginning of February 1945," June 8, 1945, USNARA RG 208, entry 370, box 376.

56. Ye Xianglong to Bob Hanson, Miss Yung Tseng Chang, and Mr. Lee Tung, May 21, 1943, USNARA RG 208, NC-148, entries 523, 524, 525, box 3021. Letters by Ye Xianglong cited below can all be found in the same USNARA box.

57. Ye Xianglong to Mr. Li Ming-tung, Miss Chang Yung Tseng, and Sirs, Oct. 17, 1943.

58. Ye Xianglong to Mr. Li Ming-tung, Miss Chang Yung Tseng, and Sirs.

59. Naval Intelligence to OWI, "Conditions in Shanghai," August 1944, USNARA RG 208, entry 370, box 379.

60. Naval Intelligence to OWI.

61. Ye Xianglong to Mr. Li Ming-tung, Miss Chang Yung Tseng, and Sirs, Oct. 17, 1943.

62. "Conditions in Shanghai," Sept. 28, 1943, USNARA RG 208, NC-148, entries 523, 524, 525, box 3021.

63. "Attitudes in Shanghai," June 7, 1945, USNARA RG 208, entry 370, box 376.

64. Beijing Municipal Archive (hereafter BJMA), Doc. J181-026-13913, "Beijing tebie shi jingchaju zhenjidui guanyu song Liu Yu tou wuxiandian shouyinji deng yi an de cheng," Jan. 27, 1940.

65. See BJMA, Doc. J181-022-09983, "Beijing tebie shi jingchaju guangyu yi ming nanzi yong guashi zhipiao goumai Riben ren chumai de yi jia shouyinji tongling chaji de pi," ca. June 26, 1940; see also BJMA Doc. J181-026-05932, "Beijing tebie shi jingchaju zhenjidui guanyu Zhang Zhenqing pian shouyinji de cheng," Oct. 1944.

66. Shanghai listener Cheng Wei to Business Manager [of KWID San Francisco], Feb. 2, 1944, USNARA RG 208, NC-148, entries 523, 524, 525, box 3021.

67. Ye wrote in Mandarin; the odd grammar and syntax is the work of translators at the OWI office. USNARA RG 208, NC-148, entries 523, 524, 525, box 3021.

68. US Embassy, Chungking, to Secretary of State, "Conditions in Shanghai Prior to December 14, 1943," Feb. 19, 1944, USNARA Archives Unbound, www.gale.com/intl/primary-sources/archives-unbound (subscription required).

69. "Excerpt from Interview with Chinese Refugee from Peiping," Nov. 20, 1943, USNARA RG 208, NC-148, entries 523, 524, 525, box 3021.

70. Ye Xianglong to Miss Yung Tseng Chang, Ming Tung Lee, and Mr. Bob Hanson, June 7, 1943.

71. Jin Yi, "Yi Ha-er-bin," *Dagongbao*, March 8, 1940.

72. The decline can be attributed primarily to Japanese harassment and suppression, but the cost of running such an operation had also increased along with the price of paper.

73. "A University Student Describes Peiping," June 8, 1945, USNARA RG 208, entry 370, box 376.

74. "A University Student Describes Peiping."

75. US Embassy, Chungking to Secretary of State, "Conditions in Shanghai Prior to December 14, 1943."

76. Ye Xianglong to Bob Hanson, Miss Yung Tseng Chang, and Mr. Lee Tung, May 21, 1944.

77. Ye Xianglong to Miss Chang Yun-tseng and Sirs, Sept. 4, 1943.

78. Ye Xianglong to Bob Hanson, Miss Yung Tseng Chang, and Mr. Lee Tung, June 7, 1943.

79. "Excerpt from the TA KUNG PAO (CHUNGKING) June 1, 1944," USNARA RG 208, NC-148, entries 523, 524, 525, box 3021.

80. Ye Xianglong to Bob Hanson, Miss Yung Tseng Chang, and Mr. Lee Tung, May 21, 1944.

81. Ye Xianglong to Bob Hanson, Miss Yung Tseng Chang, and Mr. Lee Tung, June 7, 1943.

82. F. McCracken Fisher to John K. Fairbank, Office of War Information, "Description of Conditions in Shanghai, Spring 1944," August 15, 1944, USNARA RG 208, entry 370, box 375.

83. Ye Xianglong to KGEI San Francisco, May 21, 1943.

84. "Mei you wuxiandian jiu shi mei you guojia," in BJMA, Doc. file J070.003.00023.

85. Zhao Yuming, ed., *Riben Qin Hua Guangbo Shiliao Xuanbian* (Beijing; Zhongguo Guangbo Dianshi Chubanshe, 2015), 157.

86. Richard Mitchell, *Censorship in Imperial Japan* (Princeton, NJ: Princeton University Press, 1983), 211.

87. NHK, *The History of Broadcasting in Japan* (Tokyo: History Compilation Room, Radio & TV Culture Research Institute, Nippon Hoso Kyokai, 1967), 127.

88. NHK, 113.

89. Janis Mimura, *Planning for Empire: Reform Bureaucrats and the Japanese Wartime State* (Ithaca, NY: Cornell University Press, 2011), 81.

90. Yang, *Technology of Empire*, 117.

91. NHK, *History of Broadcasting in Japan*, 114; Zhao Yuming, "Beijing Shi," in

Xinxiu Difang Zhi, ed. Zhao Yuming (Beijing, Zhongguo Guangbo Yingshi Chubanshe, 2016), 32; Christopher Rand, OWI, to American Embassy Chungking, "Notes on Japanese Dominated Medium-Wave Stations," Dec. 22, 1943, USNARA RG 208, entry 370, box 376.

92. Zhao, *Xinxiu Difang Zhi*, 709.
93. NHK, *History of Broadcasting in Japan*, 114.
94. NHK, 103.
95. Yang, *Technology of Empire*, 179.
96. NHK, *History of Broadcasting in Japan*, 102.
97. Christopher Rand, OWI, to American Embassy Chungking, "Notes on Japanese Dominated Medium-Wave Stations," Dec. 22, 1943.
98. *The Manchoukuo Yearbook, 1942* (Hsinking, Manchoukuo: Manchoukuo Year Book Co., 1942), 627.
99. Christopher Rand, OWI, to American Embassy Chungking, "Notes on Japanese Dominated Medium-Wave Stations," Dec. 22, 1943.
100. Office of War Information, News and Intelligence Bureau, "Radio Stations in China," March 23, 1945, 3, USNARA RG 208, NC-148, entry 522, box 3020.
101. Office of War Information, News and Intelligence Bureau, 9.
102. Office of War Information, News and Intelligence Bureau, 3.
103. Office of War Information, News and Intelligence Bureau, 10.
104. Office of War Information, News and Intelligence Bureau, 6.
105. Associated Press, "News Late, China Exults at Pact," *Minneapolis Morning Tribune*, Dec. 3, 1943.
106. John C. Caldwell, "General Report on Fukien Province," Feb. 26, 1944, 4, USNARA Archives Unbound, www.gale.com/intl/primary-sources/archives-unbound (subscription required).
107. "Excerpt from China Newspaper, November 1944: Four Days in Ishan," USNARA RG 208, NC-148, entries 523, 524, 525, box 3021.
108. "Liu Tsan-Ch'i's Report on Trips to Kunming, Kweilin, Hengyang, & Kukong," Jan. 12, 1944, 7.
109. NHK, *History of Broadcasting in Japan*, 128. After 1943, the number of radios declined owing to wartime bombing damage and a drop in new production.
110. NHK, 128.
111. NHK, 136.
112. NHK, 128.
113. Sheldon Garon, *Molding Japanese Minds: The State in Everyday Life* (Princeton, NJ: Princeton University Press, 1998).
114. BJMA, Doc. J183-002-27631, "Beiping shi jingchaju guanyu shouting guangbo shixiang de xunling," May 1942.
115. BJMA, Doc. J183-002-27631.

116. BJMA, Doc. J183-002-27694, "Beiping shi jingchaju guanyu puji shouyinji banfa de xunling," Nov. 3, 1943.

117. The population of Beijing was estimated at two million in 1945. See Office of War Information, News and Intelligence Bureau, "Radio Stations in China," March 23, 1945, 9.

118. BJMA, Doc. J183-002-27694.

119. Order from the Executive Yuan to Shanghai Municipal Government of September 22, 1942, "Wang wei Xingzhengyuan deng guanyu puji guangbo xuanchuan ling chi ge jiguan gouzhi Ri zhi shouyinji de xunling," in Shanghai Shi Danganguan et al., *Jiu Zhongguo de Shanghai Guangbo Shiye* (Beijing: Dang'an Chubanshe, 1985), 430–31.

120. Foreign Broadcast Information Service (hereafter FBIS), "Permits required for use of Radios," March 5, 1945. FBIS, "Radio Repair Service Open to Public," May 15, 1945.

121. FBIS, "Care of Radio Receivers Explained," August 14, 1944.

122. Zhao Yuming, "Shanxi Sheng," in *Xinxiu Difang Zhi*, 163.

123. Zhao, *Xinxiu Difang Zhi*, 753.

124. BJMA, Doc. J183-002-36284, "Beijing shi jingchaju wu fen ju guanyu jie nei wuxiandian shouyinji diaochabiao." In the first radio census under the returned Chinese government in 1946, the strong majority of sets were Japanese. The radios are almost universally listed as only being able to receive domestic broadcasts.

125. Barak Kushner, *The Thought War: Japanese Imperial Propaganda* (Honolulu: University of Hawai'i Press, 2007).

126. Rana Mitter, *Forgotten Ally: China's World War II, 1937–1945* (Boston: Houghton Mifflin Harcourt, 2013).

127. NJSHA, Doc. 5.12098, "Jiaoyu Bu guanyu yan kong qixiang guangbo zeng she fen tai yu qudi zhuangshe shouyinji guice, shezhi jiaoyu boyi deng shixiang yu junshi weiyuanhui bangongting, Nanjing guangbo diantai deng laiwang wenshu," Dec. 1938, 16. Zhejiang Provincial Department of Education memo submitted to Ministry of Education: "Wei yan xing tongzhi guangbo xuanchuan jiaqiang guangbo diantai li yi fangzhi diren fan xuanchuan qing."

128. F. McCracken Fisher to John K. Fairbank, Office of War Information, "Description of Conditions in Shanghai, Spring 1944," August 15, 1944.

129. Ye Xianglong to Mr. Li Ming-tung, Miss Chang Yung Tseng, and Sirs, Oct. 17, 1944.

130. The most vociferous anti-Soviet propaganda programs were outsourced to the useful puppets in Vichy Saigon, and included (accurate) reports of Soviet atrocities committed against the Poles and Ukrainians. See Office of War Information, "Differential Japanese Radio Treatment of Russia," July 15, 1943, USNARA RG 208, entry 370, box 405.

131. "Ri zai Huabei xuanchuan gongzuo," *Dagongbao*, April 21, 1939.

132. Office of War Information. "Japanese Propaganda, and Chinese Beliefs and Attitudes in Occupied China," August 27, 1945, USNARA RG 208, entry 370, box 375: "However, the Japanese propaganda has never gone strong on openly claiming that Yenan's anti-Japanese effort is superior to that [of] Chungking's, although this would certainly be useful to widen the split between the two Chinese groups."

133. *Dagongbao*, "Ri zai Huabei xuanchuan gongzuo," April 21, 1939.

134. Ye Xianglong to Gentlemen of the San Francisco Short-wave Station, Miss Yung-Tsen Chang (Elsie Eng), Mr. Bob Hanson, April 18, 1943.

135. Ye Xianglong to Bob Hanson, Miss Yung Tseng Chang, and Mr. Lee Tung, May 21, 1943; see also Ye's letter of June 7, 1943.

136. US Embassy, Chungking, to Secretary of State, "Conditions in Shanghai Prior to December 14, 1943," Feb. 19, 1944. The observer continues: "This favorable impression seems to derive from three factors: 1. He has made no money personally (they believe); 2. His past history and importance; and 3. The excellent Chinese he writes."

137. Office of War Information, News and Intelligence Bureau. "Radio Stations in China," March 23, 1945.

138. "Ri zai Huabei xuanchuan gongzuo," *Dagongbao*, April 21, 1939.

139. Theodore Herman, Field Intelligence Officer, "Attitudes in Shanghai: Interview with a Chinese Businessman Who Left Shanghai at the Beginning of February 1945," June 8, 1945, USNARA RG 208, entry 370, box 376.

140. "Broadcast by Mr. Wang Hsiao-lai, Former Chairman of the Shanghai Chamber of Commerce on March 15," USNARA RG 208, entry 370, box 375.

141. Jin Mai, "Zai Chongqing guangbo Riben touxiang de xiaoxi," *Shiji* 5 (1997): 57.

142. San Shui, "Guangbo Riben touxiang xiaoxi jishi," *Shi Ting Jie*, Jan. 1, 1988, 50.

143. Wang Jiaqi, "Kongbu shijie," in *Xuelei de huiyi: Lao yi bei yu dangdai shaonian tan "Jiu-Yi-Ba,"* ed. Guan Jiahe (Shenyang: Liaoning shaonian ertong chubanshe, 1991).

144. Mei Yi, "Woguo renmin guangbo shiye gaikuang," *Renmin Ribao*, April 25, 1950.

145. Chiang Kai-shek Diaries, August 15, 1945, Hoover Institute, Stanford University; transcribed in Ye Yonglie, "Zai Meiguo kan Jiang Jieshi Riji," *Tongzhou Gongjin*, no. 2 (2008).

Chapter 6

1. Cheng Mei, Chen Daofu, and Xue Xiayuan, eds. *Ding Yilan Zhuan* (Beijing: Zhongguo guoji guangbo chubanshe, 2011), 35–37.

2. Timothy Cheek, *Propaganda and Culture in Mao's China: Deng Tuo and the*

Intelligentsia (Oxford: Oxford University Press, 1997). See, esp., chap. 2: "The Revolutionary Propagandist: Life in the Jin Cha Ji Border Region 1937–1945."

3. Wang Xijian, "Huiyi Kang-Zhan shiqi Bei Jun 112 shi dixiadang de gongzuo," in *Zhong-Gong Dongbei Jun dixia dang gongzuo huiyi*, ed. Zhong-Gong Dongbei Jun dang shi zu (Beijing: Zhong-Gong Dang shi chubanshe, 1995), 413–18.

4. Wang, 415.

5. Historian Dagfinn Gatu, citing General Peng Dehuai, reports that in 1940 there were 184 underground newspapers in North China. According to Japanese reports, there were 60 newspapers in Communist-held southern Shanxi alone. Dagfinn Gatu, *Village China at War: The Impact of Resistance to Japan, 1937–1945* (Copenhagen: NIAS Press, 2007), 84–86.

6. FBIS, "Yenan Reports Newspapers Established," Nov. 17, 1944.

7. Li Mai, "Ji-Zhong diqu de xinwen gongzuo," in *Xinwen yanjiu ziliao congkan 1981 nian di 2 ji*, ed. Zhongguo shehui kexue yuan xinwen yanjiusuo (Beijing: Xinhua Chubanshe, 1981), 143–55.

8. FBIS, "Yenan Reports Newspapers Established," Nov. 17, 1944.

9. FBIS.

10. Li, "Ji-Zhong diqu de xinwen gongzuo."

11. Cheng Gang, "Liuliang Meng," in *Qinli Kang-Zhan: Beijing jiaoyujie lao tongzhi Kang-Zhan huiyilu*, ed. Zhong-Gong Beijing shi wei jiaoyu gongzuo weiyuanhui (Beijing: n.p., 2005), 461–65.

12. Lu Tianhong, "Kang-Zhan shi wo zhuban de youji xiaobao," in *Wuhan Wenshi Ziliao*, no. 4 (2012).

13. Lu, 31.

14. Chen Yunhao, "Xie hua he duan xinwen—huiyi Yanfu Dazhong bao," *Xinwen Zhanxian*, no. 6 (1979).

15. Lu Kaitai, "Huiyi Fangcheng Kang-Zhan Ribao," in *Fangcheng Wenshi Ziliao, Di 8 Ji*, ed. Fangcheng Xian weiyuanhui wenshi ziliao yanjiu weiyuanhui. (n.p.: n.p., 1991), 20–22.

16. Liang Yi, "Kang-zhan chuqi, Wuyang Xian dixiadang dui di douzheng de pianduan huiyi," in Luohe Wenshi Ziliao, Di 6 Ji: Kang-Ri Zhanzheng Zhuanji, ed. Zhongguo Renmin Zhengzhi xieshang huiyi Luohe Shi weiyuanhui wenshi ziliao weiyuanhui (n.p.: Ma Dian Yinshuachang, 1995), 196–205, 202.

17. Liang, "Kang-zhan chuqi," 203–4.

18. Lu, "Kang-Zhan shi wo zhuban de youji xiaobao."

19. Liang, "Kang-zhan chuqi," 200.

20. Lu Tianhong, "Kang-Zhan shi wo zhuban de youji xiaobao."

21. Cheek, *Propaganda and Culture in Mao's China*, 75.

22. Hsiao Li Lindsay, *Bold Plum: With the Guerillas in China's War against Japan* (Morrisville, NC: Lulu Press, 2007), 190–91.

23. He Zhiping, "Muping gongzuo de huiyi," in Zhong-Gong Shandong sheng-wei dang-shi ziliao zhengji yanjiu weiyuanhui, ed., *Shandong dang-shi ziliao*, no. 3 (1985): 12.

24. Hsiao Li Lindsay began writing her memoir, *Bold Plum*, in 1947, soon after leaving Yan'an, though it was not published for more than fifty years. Michael Lindsay's memoirs were written somewhat later, to accompany a book of wartime photos he published. Unfortunately, the pages of his book are unnumbered. Michael Lindsay, *The Unknown War: North China, 1937–1945* (London: Bergstrom & Boyle, 1975).

25. Hsiao Li Lindsay, *Bold Plum*, 87.

26. Hsiao Li Lindsay, 88; Michael Lindsay, *The Unknown War*, n.p.

27. Hsiao Li Lindsay, *Bold Plum*, 93.

28. Hsiao Li Lindsay, 99–100.

29. Hsiao Li Lindsay, 114–17.

30. Hsiao Li Lindsay, 152.

31. Michael Lindsay, *The Unknown War*, introduction.

32. Hsiao Li Lindsay, *Bold Plum*, 130.

33. Michael Lindsay, *The Unknown War*.

34. Visiting the People's Republic in 1973, Lindsay found that many of the top positions in telecommunications administration were graduates of his class.

35. Hsiao Li Lindsay, *Bold Plum*, 148, 177.

36. Hsiao Li Lindsay, 190–91.

37. Tian Jianping and Zhang Jinfeng, eds. *Jinchaji kang Ri genju di shubao chuanbo shilue, 1935–1945* (Baoding: Hebei Daxue Chubanshe, 2010), 49, 66, 169.

38. Ding Yilan, "Hutuo hepan ding xin meng," *Shehui kexue zhanxian*, no. 2 (1986).

39. Deng Tuo would later become famous as the editor of the *People's Daily*. Ding was the station chief at Beijing's primary radio station. Deng Tuo is the subject of Timothy Cheek's *Propaganda and Culture in Mao's China*. A full biography of Ding Yilan, the aforementioned *Ding Yilan Zhuan*, containing long excerpts from oral histories given by her, was published in 2011.

40. In 1940 Zhou Enlai visited the Soviet Union and brought back with him the radio transmitter that enabled CCP headquarters to begin small-scale voice broadcasts to the neighboring regions of North China.

41. "Hua-Bei zhi guangbo wuxiandian shiye," *Dagongbao*, August 7, 1927.

42. Alan P. L. Liu, *Radio Broadcasting in Communist China* (Cambridge, MA: MIT Center for International Studies, 1964), 37.

43. David Strand, *An Unfinished Republic: Leading by Word and Deed in Modern China* (Berkeley: University of California Press, 2011).

44. Mary Beard, "The Public Voice of Women," *London Review of Books* 36, no. 6 (March 2014).

45. Tina Tallon, "A Century of 'Shrill': How Bias in Technology Has Hurt Women's Voices," *New Yorker*, Sept. 3, 2019.

46. Michele Hilmes, *Radio Voices: American Broadcasting, 1922–1952* (Minneapolis: University of Minnesota Press, 1997). See, esp., chap. 5, "The Disembodied Woman," 130–50.

47. Hu Fang, "Nanjing zhi ying—minguo boyinyuan Liu Junying," *Xinwen Yanjiu Dao Kan* 7, no. 24 (2016): 77, 81.

48. Strangely, they went through extensive physical exams by local doctors as part of the interview process; there was more concern about their organs functioning than their voices. See Wen Long, "Yu sheng fu guangbo diantai nü baogaoyuan jianyan tige quwen," *Fu-er-mou-si* (Famous), Dec. 23, 1934.

49. Yu Wang has recently pointed out that even CCP propaganda officials argued that higher-pitched, female voices carried better under conditions of interference and jamming. Yu Wang, "Listening to the State: Radio and the Technopolitics of Sound in Mao's China" (PhD diss., University of Toronto, 2019), 87, citing a 1947 letter of Lu Dingyi, chief of the CCP Propaganda Department. See also Wang Yu, "Geming de fudiao: Boyinyuan yu shehuizhuiyi boyin fengge de zaidihua." in *Ting Xiandai Zhongguo*, ed. Tang Xiaobing (Shanghai: Fudan Daxue Chubanshe), 58–69; and Wang Yu, "Nuxing de pinlu: Xingbie, guangbo yu 20 shiji 40 niandai zhongguo geming de tingjue wenhua," *Funu Yanjiu Luncong: Journal of Chinese Women's Studies* 166, no. 4 (July 2021).

50. Report of Monitor 'I' Kunming, July 6–July 19, 1945, USNARA RG 208, NC-148, entry 522, box 3020.

51. Marjorie K. M. Chan, "Gender Differences in the Chinese Language: A Preliminary Report," in *Proceedings of the Ninth North American Conference on Chinese Linguistics (NACCL9)*, ed. Hua Lin, 2:35–52 (Los Angeles: GSIL Publications, University of California, 1997). Differences in speech patterns between men and women are common across cultures and languages.

52. Chan, 36–41.

53. Gina Tam, *Dialect and Nationalism in China, 1860–1960* (Cambridge: Cambridge University Press, 2020); Janet Chen, *The Sounds of Mandarin* (New York: Columbia University Press, 2023).

54. Xu Ruizhang, "50 nian qian de hong se dianbo—yi Yan'an Xinhua Guangbo Diantai," *Dangshi Zongheng*, no. 4 (1991).

55. Wang Ying, "Kang-Zhan shiqi de Yan'an Xinhua Guangbo Diantai," *Zhonghua Nuzi Xueyuan Xuebao*, no. 5 (2015).

56. Xu, "50 nian qian de hongse dianbo," 14–16.

57. Xu Ruizhang (Mai Feng) and Yao Wen transferred from Yan'an University at that time, Xiao Yan (Chang Lihua) and Sun Xi from Yan'an women's University,

and the wife of Cheng Mingsheng, president of the communication school of the Third Bureau of the Central Military Commission, a Japanese woman.

58. Wang, "Kang-Zhan shiqi de Yan'an Xinhua Guangbo Diantai."

59. Xu, "50 nian qian de hongse dianbo."

60. Wang, "Kang-Zhan shiqi de Yan'an Xinhua Guangbo Diantai."

61. Ding Yilan, "Wushi nian qian de kuanghuan zhi ye," *Xinwen Aihaozhe*, no. 9, (1995).

62. Ding, 20.

63. Ding, 20.

64. Ding, 20.

65. Hsiao Li Lindsay, *Bold Plum*, 327–28.

66. Alain Corbin, *Village Bells: Sound and Meaning in the 19th-Century French Countryside* (New York: Columbia University Press, 1998).

67. Ding, "Wushi nian qian de kuanghuan zhi ye."

68. FBIS, "Chu Teh Rejects Chiang's Instructions," August 14, 1945; FBIS, "Chu Teh Issues Surrender Instructions," August 11, 1945.

69. Tian and Zhang, *Jinchaji kang Ri genju di shubao chuanbo shilue*, 43.

70. "Military Intelligence Service: Weekly Propaganda Summary: August 26, 1945," USNARA RG 208, entry 370, box 385.

71. Ding, "Wushi nian qian de kuanghuan zhi ye."

72. Zhang Fan, "Deng Tuo furen Ding Yilan: Kai Guo da dian guangboyuan," *Wen Shi*, no. 5 (2015): 28–29.

73. Zhang. From the Fuping County base, Ding Yilan began to train new announcers. She taught them in the valley during the day and at night climbed up the mountain, where the broadcasting studio was located. She worked day and night, sometimes broadcasting for eight hours in a row.

74. Wang Yizhi, "Jieguan Changchun Shi guangbo diantai de di yi ge huihe," in *Chengshi jieguan qinli ji* (Beijing: Zhongguo Wen Shi Chubanshe, 1999), 93–95.

75. Charity Interview 713, April 26, 1954, USNARA RG 59, entry 56_D_454, box 4.

76. Kang Minzhuang, "Xuexi zhangwo he yunyong guangbo gongju: Huiyi Dalian diantai jian tai chuqi qingkuang," in *Dalian guangbo huiyilu*, ed. Zhu Qingyan (n.p.: n.p., 1986), 3–20, 9.

77. Bai Quanwu, "Zhengqu Ri-qiao hezuo chuangjian renmin Guangbo" (Striving to work with overseas Japanese to found People's Broadcasting), in *Dalian guangbo huiyilu*, ed. Zhu Qingyan (n.p.: n.p., 1986), 21–31, 22.

78. Kang, "Xuexi zhangwo he yunyong guangbo gongju," 8–9.

79. Bai, "Zhengqu Ri-qiao hezuo chuangjian renmin Guangbo," 24.

80. Bai, 28.

81. For more on the movement of Japanese from mainland Asia in the postwar

period, see Lori Watt, *When Empire Comes Home: Repatriation and Reintegration in Postwar Japan* (Cambridge, MA: Harvard University Press, 2009).

82. Bai, "Zhengqu Ri-qiao hezuo chuangjian renmin Guangbo."

83. Kang, "Xuexi zhangwo he yunyong guangbo gongju," 14.

84. Kang, 14.

85. BJMA, Doc. J001-001-01188, "Hua-Bei jiaofei zongbu qudi minzhong shouting Gongchandang guangbo bing banli shouyinji dengji de dai dian," Nov. 25, 1948.

86. Chongqing Municipal Archive (hereafter CQMA), Doc. 0053.0023.00040.0000.067.000, "Guanyu yanjia guanzhi shouyinji jinzhi shouting de han," May 11, 1949.

87. HBPA, Doc. LS31-2-684. "Hubei sheng jianshe ting yu Shishou, Guangji, Yunmeng deng xian youguan jinzhi shouting jianfei guangbo banfa shixiang zhi ling dai dian," Feb. 1, 1949, 1.

88. Guoshiguan, Doc. 001-070000-00047-099, "Jiaoyu Bu cheng Guomin zhengfu wei fangzhi Fudan Daxue jian wei fenzi shouting Yan'an huangmiu guangbo yi an," Nov. 30, 1945.

89. Guoshiguan, Doc. 003-010303-0856, "Kongjun zongzi lingbu han su ziyuan weiyuanhui chanpin xiang mubiao deng an," Nov. 2, 1948, 24–25; Yao Hsin-nung, "Memorandum re Radio Broadcasting in China," August 13, 1948, British Archives Kew FO 953/321.

90. Ma Sanli, "Xin jiu shehui liang zhong tian," in *Chengshi jieguan qinli ji* (Beijing: Zhongguo Wen Shi Chubanshe, 1999), 226–28.

91. "Shouting Xinhua Guangbo Diantai guangbo jishi," in *Xinxiu Difang Zhi*, ed. Zhao Yuming, 945–50.

92. FBIS, "Kuomintang Troops Urged to Surrender," May 27, 1947.

93. "Shouting Xinhua Guangbo Diantai guangbo jishi."

94. Hu Guanzhong, with Fang Zhenhuo, eds., "Wo suo qinli de Xiamen jiefang de san jian shi," in *Koushu lishi: qinli Xiamen jiefang*, ed. Zhong-Gong Xiamen shi wei xuanchuan bu. (Xiamen: Xiamen Daxue Chubanshe, 2009), 137–45.

95. Hu, with Fang.

96. Cheng, Chen, and Xue, *Ding Yilan Zhuan*, 129–30; Hu, with Fang, "Wo suo qinli de Xiamen jiefang de san jian shi," 145.

Chapter 7

1. "Beijing Xinhua Guangbo Diantai jintian quanbu zhuanbo Zhongyang Renmin Zhengfu chengli qingzhu dahui shikuang," *Renmin Ribao*, Oct. 1, 1949. Huang Bo, "Yong bu xiaoshi de shengyin—Qi Yue," *Shiji Fengcai*, Nov. 2019. For a short biography of Ding Yilan, see Zhang Fan, "Deng Tuo furen Ding Yilan: Kai guo da dian guangboyuan," *Gonghui Xinxi*, no. 5 (2015).

2. Andrew F. Jones, *Circuit Listening: Chinese Popular Music in the Global 1960s*

(Minneapolis: University of Minnesota Press, 2020); Stephen Lovell, *Russia in the Microphone Age: A History of Soviet Radio, 1919–1970* (Oxford: Oxford University Press, 2015).

3. Theodor Adorno, "Schema of Mass Culture," in *The Culture Industry: Selected Essays on Mass Culture*, ed. J. M. Bernstein (London: Routledge, 2001), 61–97; Theodor Adorno, "The Culture Industry: Enlightenment as Mass Deception," in Theodor Adorno and Max Horkheimer, *Dialectic of Enlightenment* (Stanford, CA: Stanford University Press, 2002), 94–136.

4. Adorno, "The Culture Industry," 129.

5. Theodor Adorno, "On the Fetish Character of Music and the Regression of Listening," in *The Culture Industry: Selected Essays on Mass Culture*, ed. J. M. Bernstein (London: Routledge, 2001), 29–60, 46.

6. Theodor Adorno, "Radio Physiognomics," in *Current of Music: Elements of a Radio Theory*, ed. Robert Hullot-Kentor (Cambridge: Polity, 2009), 41–132, 70.

7. Kate Lacey, *Listening Publics: The Politics and Experience of Listening in the Media Age* (Cambridge: Polity, 2013); Carolyn Birdsall, *Nazi Soundscapes: Sound, Technology, and Urban Space in Germany, 1933–1945* (Amsterdam: Amsterdam University Press, 2012).

8. Frederick Teiwes and Warren Sun, *China's Road to Disaster: Mao, Central Politicians and Provincial Leaders in the Great Leap Forward, 1955–59* (New York: Routledge, 1999).

9. Yang Jisheng, *Tombstone: The Great Chinese Famine, 1958–1962* (New York: Farrar, Straus and Giroux, 2008), 8–9.

10. Sun Yat-sen, *Fundamentals of National Reconstruction* (Taipei: China Cultural Service, 1953), 115.

11. Jennifer Altehenger's recent study of the dissemination of laws in post-1949 China will be a classic of the genre, demonstrating as it does the combinations of propaganda strategies and party hierarchy that brought information to the grassroots in coordinated campaigns. She is right to point out, however, the deficiencies in the system of published propaganda printing and distributions (see esp. her chap. 2: "Paper Trails"). Even when massive print-runs did occur, as with the 1954 draft constitution, one cannot help feeling that such dry material must have fallen flat at the level of reception (see her introduction). Radio was more dynamic, more reliable, and ultimately cheaper than mass production of booklets, magazines, and illustrates volumes. Jennifer Altehenger, *Legal Lessons: Popularizing Laws in the People's Republic of China: 1949–1989* (Cambridge, MA: Harvard University Asia Center, 2018).

12. For a discussion of the militarization of society, see Frank Dikotter, *Mao's Great Famine* (New York: Bloomsbury, 2017), 45–51. The profound influence of China's long twentieth-century wars has been explored in a number of recent vol-

umes, most prominently by Rana Mitter in her *Forgotten Ally: China's World War II, 1937–1945* (Boston: Houghton Mifflin Harcourt, 2013); and Hans van de Ven in *China at War: Triumph and Tragedy in the Emergence of the New China* (Cambridge, MA: Harvard University Press, 2018).

13. Jeremy Brown and Matthew Johnson, eds., *Maoism at the Grassroots* (Cambridge, MA: Harvard University Press, 2015), 6.

14. According to the Greek Maître d'Hotel, "Project Charity, Interview 719," April 30, 1954, USNARA RG 59, entry 56 D 454, box 4.

15. The station had been forfeited by the German government after 1945. See "Project Charity, Interview 726," June 9, 1954, USNARA RG 59, entry 56 D 454, box 5.

16. Broadcast from the Shanghai People's Station in Mandarin, Foreign Broadcast Information Service (hereafter FBIS), "Shanghai Underground Radio," June 9, 1950.

17. Shanghai Municipal Archive (hereafter SHMA), Doc. A22.2.9.6, "Miao Lichen guanyu qudi duanbo shouyinji ji siying diantai chuli wenti de baogao," ca. Nov. 1950.

18. Gu Zhenzhi, "A Chronicle of Private Radio in Shanghai," *Journal of Broadcasting & Electronic Media* 30, no. 4 (1986).

19. "Dismantling of United Kingdom Government Radio Equipment in China," British Foreign Office Record (FO), *General China Correspondence*, Doc. FO 371/8365. The US consulate at Mukden also had a long running standoff with Communist authorities after liberation in that city, where they refused to surrender their radio equipment. The consul was eventually placed into custody.

20. SHMA, Doc. A22-2-9-3, "Shanghai shi renmin zhengfu xinwen chuban chu guanyu qudi Hu shi duanbo shouyinji wenti de yijian," 1950.

21. These men tended to quickly become "unlucky" under the new regime—though one radio shop owner in Chongqing, a former KMT soldier, lasted in the business until March 1951, when he was arrested in a mass purge. "Project Charity Interrogation No. 630," April 1, 1954, USNARA RG 59, lot 56 D 454, box 4.

22. SHMA, Doc. A22.2.9.3, "Shanghai shi renmin zhengfu xinwen chuban chu guanyu qudi Hu shi duanbo shouyinji wenti de yijian," 1950.

23. SHMA, Doc. A22.2.9.3.

24. US Information Agency, "Examples of VOA Effectiveness," Feb. 1, 1953, USNARA RG 306, lot P.292, box 5.

25. SHMA, Doc. A22.2.9.6, "Miao Lichen guanyu qudi duanbo shouyinji ji siying diantai chuli wenti de baogao," ca. Nov. 1950.

26. "Uplift Interview 38," Oct. 20, 1953, USNARA RG 59, lot 56 D 454, box 7.

27. "'Voice of America' Listening in Communist China," September 5, 1962, USNARA RG 0306, entry P 142, box 10.

28. "Voice of America Handbook: A Summary of Available Data on VOA Effectiveness," August 1, 1950, USNARA RG 59, entry P 311, box 6.

29. FBIS, "Workers' Pledge Bans Voice of America," Nov. 29, 1950.

30. "'Voice of America' Listening in Communist China," Sept. 5, 1962, USNARA RG 306, entry P 142, box 10.

31. FBIS, "Voice of America," Dec. 13, 1950.

32. "Charity Interview 634," Jan. 26, 1954, USNARA RG 59, lot 56 D 454, box 4.

33. FBIS, "Kwangtung Radio Owners," April 6, 1951.

34. "Uplift Interview 119," March 30, 1954, USNARA RG 59, lot 56 D 454, box 7.

35. "Charity Interview 649," Feb. 1, 1954, USNARA RG 59, lot 56 D 454, box 4.

36. SHMA, Doc. A22.2.9.3, "Shanghai shi renmin zhengfu xinwen chuban chu guanyu qudi Hu shi duanbo shouyinji wenti de yijian," 1950.

37. SHMA, Doc. A22.2.9.6, "Miao Lichen guanyu qudi duanbo shouyinji ji siying diantai chuli wenti de baogao," ca. Nov. 1950.

38. "Charity Interview 713," April 26, 1954, USNARA RG 59, lot 56 D 454, box 4.

39. A witness "learned the story during his visits to the police when he overheard the fellow complaining of his misfortune to the police and loudly decrying the quality of this particular Czech product." "Interrogation of Mr. Lawrence Dignanese of Shanghai," Feb. 23, 1954, USNARA RG 84, entry UD 2689, box 02.

40. Hong Kong consulate to Secretary of State, Nov. 6, 1950, USNARA RG 469, entry UD 237, box 18.

41. Correspondence regarding re-export of radio tubes and other equipment from Hong Kong to Mainland, Jan. 1950 to Jan. 1951, USNARA RG 469, entry UD 237, box 18.

42. FBIS, "Radio Parts," Dec. 24, 1952.

43. In 1953 the government announced it would require sixty thousand loudspeakers for the whole of China. Shanghai factories raised money, bought up aluminum sheets, and overfilled the quota, with the result that the informant's company was overstocked and near bankruptcy, unable to meet current salaries. "Charity Interview 754," June 14, 1954, USNARA RG 59, lot 56 D 454, box 5.

44. Lin Ta-kuang, "Broadcasting for the People," *China Reconstructs* 4, no. 8 (August 1955).

45. See, e.g., FBIS, "Instructions on Production," April 2, 1951.

46. For Sunday's broadcast schedules, see "Renmin Guangbo," *Jiefang Ribao*, April 22, 1951; "Xinan, Chongqing Renmin Guangbo Diantai jintian zhongyao guangbo jiemu," *Xinhua Ribao* (Chongqing), April 22, 1951; and "Ben shi renmin diantai jin ri yao mu," *Tianjin Ribao*, April 22, 1951. The "So, Mama Has Had a Little Brother" segment aired Saturday: "Renmin Guangbo." *Jiefang Ribao*, April 22, 1951.

47. In Chongqing, Russian-language lessons were given on the public street broadcasting system three times a day: 7:00–7:30 a.m., 12:30–1:00 p.m., and 9:00–

9:30 p.m., according to one report from 1954: "Charity Interview 675," March 30, 1954, USNARA RG 59, lot 56 D 454, box 4. On Sunday April 22, 1951, only the Guangzhou station listed a Russian-language program, which lasted a half hour, from 7 to 7:30 p.m. See "Guangzhou Renmin Guangbo Diantai jin ri bo yin yao mu," *Nanfang Ribao*, April 22, 1951.

48. "Charity Interview 822," Oct. 4, 1954, USNARA RG 59, lot 56 D 454, box 5.

49. At a meeting in March 1951, only twenty thousand people physically attended: "Zuo huanying zhiyuan jun daibiao: diantai zuo wan juxing huangying guangbo hui," *Jiefang Ribao*, March 16, 1951. At Broadcast meetings a few weeks later, 2.8 million people participated: *Jiefang Ribao*, April 30, 1951 (photo caption).

50. CQMA, Doc. 0304.0001.02802.0000.045.000, Chongqing shi gejie renmin daibiao huiyi xieshang weiyuanhui [Chongqing City People's Consultative Conference Committee], "Guanyu jieshao shouting Chongqing renmin guangbo diantai youguan fante douzheng baogao yu zhenya fan geming xinlun deng de tongzhi." No date is given, but other documents about this same broadcast listening session date it to August 9, 1951.

51. "Kai hao guangbo dahui," *Jiefang Ribao*, April 20, 1951.

52. CQMA, Doc. 0295.0001.02071.0000.033.000, "Guanyu juxing huanying Zhongguo renmin zhiyuanjun gui guo daibiao guangbo wan hui bing guiding zuzhi shouting banfa de tongzhi," May 6, 1951.

53. CQMA, Doc. 0263.0005.00004.0000.059.000, Chongqing shi Yaoye Tongyehui Choubei Weiyuanhui [Pharmacist's Association Committee], "Guanyu shouting Zhongguo renmin zhiyuanjun gui guo daibiao guangbo wanhui de tongzhi," May 7, 1951.

54. "Huadong renmin qingzhu wu-yi-jie: guangbo dahui ding ershiwu ri juxing," *Jiefang Ribao*, April 20, 1951.

55. "Zenyang zuzhi shouting guangbo dahui," *Jiefang Ribao*, April 22, 1951.

56. "Guangbo dahui jin ri kaishi," *Jiefang Ribao*, April 25, 1951.

57. "Huadong Renmin Guangbo dahui jiemu: Rao zhuxi, Chen Yi tongzhi jianghua," *Jiefang Ribao*, April 26, 1951.

58. "Gejie renmin relie shouting guangbo," *Jiefang Ribao*, April 26, 1951.

59. FBIS, "Chekiang Rally Hits Rearming of Japan," April 23 1951.

60. Cf. Stephen Kotkin's study of socialist speech patterns ("speaking like a Bolshevik") in his study of Soviet culture. Stephen Kotkin, *Magnetic Mountain: Stalinism as a Civilization* (Berkeley: University of California Press, 1995).

61. CQMA, Doc. 0317.0001.00206.0000.066.000, Jianye Yinhang Chongqing Fenhang [Jianye Bank Chongqing Branch], "Jianye Yinhang Chongqing fenhang guanyu baosong ben hang shouting huanying Zhongguo zhiyuanjun gui guo daibiao guangbo wan hui zongjie zhi Chongqing shi jinrongye tongye gonghui choubei weiyuanhui de han," May 10, 1951.

62. "Guangbo dahui jifa aiguo rechao: ge di tingzhong fenfen jianju feite," *Jiefang Ribao*, April 27, 1951.

63. "Huadong renmin guangbo dahui shengli bimu: Gedi renmin xianqi aiguo rechao: Tingzhong biaoshi jue xie zhu zhengfu suqing feite," *Jiefang Ribao*, April 28, 1951.

64. FBIS, "Large Bandit Ring Arrested in Shanghai," May 1, 1951. For a short account of the arrests in the context of the campaign to suppress counterrevolutionaries, see Nara Dillion, "New Democracy and the Demise of Private Charity in Shanghai," in *Dilemmas of Victory: The Early Years of the People's Republic of China*, ed. Jeremy Brown and Paul Pickowicz (Cambridge, MA: Harvard University Press, 2007), 80–102, 88.

65. "Renmin de Shengyin: Erbai duo wan ren shouting shi gejie daibiao kuo dahui guangbo," *Jiefang Ribao*, April 30, 1951.

66. "Renmin de Shengyin."

67. It is unclear whether the sound of the executions were carried live over the air, but executions in front of large crowds were common in the period, so it is not implausible.

68. These accounts of listener behavior are found in "Renmin de Shengyin: Er bai duo wan ren shouting shi gejie daibiao kuo dahui guangbo," *Jiefang Ribao*, April 30, 1951. The figure for the total number of listeners to the Sunday trial broadcast is taken from a picture caption in *Jiefang Ribao*, April 30, 1951.

69. FBIS, "Eight Years," March 24, 1953.

70. A broadcast meeting in Chongqing, during the August 1951 Movement against Enemy Agents and for the Suppression of Counter-Revolutionary Speech (titles got longer, too), lasted a total of eight hours, from 9 a.m. to 12 p.m. and from 1 p.m. to 6 p.m. There was an hour break for lunch. Businesses, institutions, and schools were presumed to be listening the whole time. Specially trained cadres led the listening masses in discussion, encouraging people to express their opinions and accusations. Audience reactions could be instantaneously phoned-in to either the meeting place or the radio station so that organizers could gauge opinion and make adjustments accordingly. CQMA, Doc. 0304.0001.02802.0000.045.000 (see note 51 above).

71. For instance, later in 1951 during the promulgation of the new marriage law, crowds were ordered to listen to public denunciations ("trials") of violators of the statute. CQMA, Doc. 0263.0005.00004.0000.264.000, "Guanyu juxing wei fan hunyin fa gongpan guangbo dahui qing zuzhi shouting de tongzhi," Dec. 5, 1951. During this campaign, the banking association and pharmacists of Chongqing dutifully submitted their reactions to the "feudal" and "toxic" behaviors of polygamists and wife-sellers. CQMA, Doc. 0300.0001.01623.0000.023.000, "Guanyu gaozhi Chongqing shi jinrong ye tongye gonghui choubei weiyuanhui huiyuan yinhang

zuzhi renyuan shouting wei fan hunyin fa gongpan guangbo bing jian song shouting yijian deng de han," Dec. 22, 1952. Some vestiges of the ever-present broadcast system remain. On some Chinese university campuses, for instance, wired loudspeaker broadcast at certain times of day.

72. FBIS, "Resist-U.S. Movement to Be Reviewed," May 1, 1951.

73. Zhao Hanzhou, "Yiyuan Xian Shouyinzhan," in Zhengxie Yiyuan Xian Weiyuanhui, *Yiyuan Xian Wenshi Ziliao, Di Liu Ji* (Shandong: n.p., 1997), 254–57.

74. Jiang Shiying, "Jiefang chuqi Chuxiong zhuanqu shouyinzhan de xuanchuan gongzuo," in Zhongguo Renmin Zhengzhi Xieshang Huiyi Yunnan Sheng Chuxiong Yizu Zizhi Zhou Weiyuanhui Jiao Ke Wen Wei Wenshi Ziliao Weiyuanhui, *Chuxiong Zhou Wenshi Ziliao Xuanji, Di 21 Ji: Wenha tiyu shi qiu wenji* (Yunnan: n.p., 2004), 162–67.

75. HBPA, Doc. SZ118-02-0018-006, "Guanyu chao zhuan Zhongnan Junweihui guanyu jinali guangbo shouyinzhan gongzuo tongzhi," June 26, 1950.

76. For instance, in Guizhou, see Fu Changsheng, "Yi jiefang chuqi shouyinzhan gongzuo," in *Zui shi nanwang de yi ye huiyi xin de xuan*, ed. Hu Shihui and Zhang Xiangqi (Guizhou: Guizhou Renmin Chubanshe, 1999), 84–89. A local history from Hunan recounts how *shouyinyuan* had been active in their county since 1938, but the two radio sets that had been left behind by the GMD government in 1949 were broken beyond repair: "Di-er Zhang: Guangbo," in Rucheng Xianzhi Bianzuan Weiyuanhui, *Rucheng xian zhi* (Hunan: Hunan Renmin Chubanshe, 1997), 681.

77. Numerous local histories and memoirs (particularly in the state-sponsored *Wenshi Ziliao*) preserve the story of early CCP broadcasting in rural counties and villages. For this chapter, I consulted more than two dozen radio-operator narratives from every region of China proper, citing representative examples.

78. Jiefang Ribao, "Bixu zhongshi guangbo," April 27, 1951.

79. Clear statistics of the gender breakdown of *shouyinyuan* as a body are lacking. Anecdotally, however, there were many offices with all-female staffs, including the leadership. See Wang Fulai, "Taizhou Di Yi Jia Guangbozhan," in *Taizhou Wenshi Ziliao Di 5 Ji*, ed. Wang Miaozeng (Zhejiang: n.p., 1998), 193–96.

80. Fu, "Yi jiefang chuqi shouyinzhan gongzuo."

81. Zhao, "Yiyuan Xian Shouyinzhan."

82. Zhao.

83. HBPA, Doc. SZ118-02-0018-006.

84. "Di-er Zhang: Guangbo," in *Rucheng Xianzhi*, 681. See also Jiang, "Jiefang chuqi Chuxiong zhuanqu shouyinzhan de xuanchuan gongzuo."

85. "Suilu Xian Di Yi Bu Shouyinji," in Zhengxie Fusui Xian Weiyuanhui Wenshi Ziliao Bianji Weiyuanhui, *Fusui Wenshi Ziliao, Di 5 Ji* (Guangxi: n.p., 1998), 20–24.

86. For instance, in Yunnan: Jiang, "Jiefang chuqi Chuxiong zhuanqu shouyinzhan de xuanchuan gongzuo."
87. Fu, "Yi jiefang chuqi shouyinzhan gongzuo."
88. A short-lived province located mainly in southern Henan.
89. FBIS, "What a Great Impetus Radio Broadcast Publicity Work in Rural Villages Has Given to the Resist-America and Aid-Korea Campaign Since May 1," July 30, 1951.
90. Huang Qixuan, "Shouyinzhan zai Dong xiang," in Zhengxie Sanjiang Dongzu Zizhi Xian Weiyuanhui, *Sanjiang Wenshi Ziliao, Di 6 Ji* (Nanning: Guangxi Renmin Chubanshe, 2002), 118–20.
91. Fu, "Yi jiefang chuqi shouyinzhan gongzuo."
92. Wang Helin, "Yi Nanping guangbo shouyinzhan chengli hou de ji jianzhi," in Fujian Sheng Nanping Shi Weiyuanhui Wenshi Ziliao Weiyuanhui, *Nanping Wenshi Ziliao, Di 12 Ji* (Fujian: n.p., 1991), 10–16.
93. Geng Wenbin, "Taixing shouyinzhan shimo," in Zhengxie Taizhou Shi Xuexi Wenshi Weiyuanhui, *Taizhou Wenshi Ziliao, 1949–1952* (Jiangsu: n.p., 2008), 202–206.
94. Bu Zifang, "Huigu Yao'an Xian shouyinzhan de jianli," in Zhongguo Renmin Zhengzhi Xieshang Huiyi Yunnan Sheng Chuxiong Yizu Zizhi Zhou Weiyuanhui Jiao Ke Wen Wei Wenshi Ziliao Weiyuanhui, *Chuxiong Zhou Wenshi Ziliao Xuanji, Di 21 Ji: Wenha tiyu shi qiu wenji* (Chuxiong: n.p., 2004), 162–167.
95. Zhao, "Yiyuan Xian Shouyinzhan."
96. Huang Qixuan, "Shouyinzhan zai Dong xiang."
97. Bu Zifang, "Huigu Yao'an Xian shouyinzhan de jianli."
98. Bu.
99. "Suilu Xian Di Yi Bu Shouyinji."
100. Zhao, "Yiyuan Xian Shouyinzhan."
101. Lin, "Broadcasting for the People."
102. In late 1955, the central government began hinting at a shift in rural broadcasting policy. Various memoranda urged rural development and proclaimed that every village and cooperative would be able to hear wired radio as part of the project of socialist modernization. Mao Zedong, "Nongye hezuohua de yi chang ban lun he dangqian de jieji douzheng," Oct. 11, 1955. Repr. in Dangdai Zhongguo de Guangbo Dianshi Bianji Bu, *Zhongguo de Youxian Guangbo* (Beijing: Beijing Chubanshe, 1988), 1; Mao Zedong, "Zhengxun dui nongye shi qi tiao de yijian," Dec. 21, 1955. Repr. in *Zhongguo de Youxian Guangbo*.
103. Note that the Chinese terminology for local wired broadcast via post, *guangbozhan*, is distinct from the term for wireless broadcast station, *guangbo diantai*. The words for "wired" and "wireless" themselves are rarely used except in technical or

highly official documents. Wired broadcasting posts were sometimes called "rediffusion networks" or "rediffusion posts" in English.

104. For instance if a "broadcast" meeting comes through the speaker of a telephone handset rather than a loudspeaker, is it a telephone call or radio broadcast?

105. Lovell, *Russia in the Microphone Age*.

106. Alan Liu, *Radio Broadcasting in Communist China* (Cambridge, MA: Center for International Studies at MIT, 1964), 14; citing Alex Inkeles, *Public Opinion in Soviet Russia: A Study in Mass Persuasion* (Cambridge, MA: Harvard University Press, 1958), 244.

107. Wired broadcasting was not confined to totalitarian regimes, though it suited their needs remarkably well. British Hong Kong in the 1950s constructed a similar radio regime. That government also desired to block out enemy broadcasts, though in this case the enemy was Communist China.

108. Wang, "Taizhou Di Yi Jia Guangbozhan."

109. Fu, "Yi jiefang chuqi shouyinzhan gongzuo."

110. Che Jixin, ed., *Qi Lu Wenhua Da Cidian* (Jinan, Shandong: Shandong Jiaoyu Chubanshe, 1988), 353.

111. Hanyang Xian Guangbozhan, "Guanyu wo zhan wu ge yue lai de gongzuo yu wenti baogao," June 4, 1957, in HBPA, Doc. SZ7-2-46, "Ge xian guangbozhan guanyu youxian guangbo gongzuo de zongjie, baogao," 1957.

112. Gucheng xian nongcun youxian guangbozhan, "1957 shang ban nian gongzuo zongjie," August 1957, in HBPA, Doc. SZ7-2-46.

113. Che, *Qi Lu Wenhua Da Cidian*.

114. Liu, *Radio Broadcasting in Communist China*, 17–18, citing (1) *Communist China Digest*, no. 20 (July 26, 1960); and (2) Ching-kwe Chu. "The Development of Broadcasting for the National Minorities," *Hsin-Wen Chan-Hsien* (News Front), no. 11 (1959), 25.

115. Yin Huazhen, "Jiyi zhong de Jun Xian renmin guangbo zhan," in Danjiangkou shi Zhengxie Wenshi Ziliao Weiyuanhui, *Danjiangkou Wenshi Ziliao, Di 9 Ji* (Danjiangkou: n.p., 2008), 298–301.

116. Yin, 300.

117. Yin, 298.

118. Macheng Xian Guangbozhan, "Macheng xian 1956 nian jian zhan gongzuo zongjie," in HBPA, Doc. SZ7-2-46, "Ge xian guangbozhan guanyu youxian guangbo gongzuo de zongjie, baogao," 1957. Macheng County later became a model county, exalted by the *People's Daily* and visited by half a million cadres. See Dikotter, *Mao's Great Famine*, 38.

119. Hanyang Xian Guangbozhan, "Guanyu wo zhan wu ge yue lai de gongzuo yu wenti baogao," June 4, 1957, in HBPA, Doc. SZ7-2-46.

120. Hanyang Xian Guangbozhan.

121. Gucheng Xian Nongcun Youxian Guangbozhan, "1957 shang ban nian gongzuo zongjie," August 1957, in HBPA, Doc. SZ7-2-46.

122. Gucheng Xian Nongcun Youxian Guangbozhan.

123. Daye Xian Guangbozhan, "1957 nian shang nian du xuanchuan gongzuo zongjie," Sept. 4, 1957, in HBPA, Doc. SZ7-2-46.

124. Gucheng Xian Nongcun Youxian Guangbozhan, "1957 shang ban nian gongzuo zongjie," August 1957.

125. Wang, "Taizhou Di Yi Jia Guangbozhan."

126. "Di Qi Zhang: Guangbo Dianshi," in Yunnan Sheng Xundian Huizu Yizu Zizhui Xian Zhi Bianzuan Weiyuanhui, *Xundian Huizu Yizu Zizhi Xian Zhi* (Yunnan: Yunnan Renmin Chuban She, 1999), 795–97.

127. Yin, "Jiyi zhong de Jun Xian renmin guangbo zhan."

128. Beginning in 1957, the "Local News" program at Shiquan County, Shaanxi Broadcasting Post, asked advanced "front line" individuals to broadcast speeches to their compatriots. In 1959, the broadcasting post transmitted stories from 178 such individuals, interspersed with phrases like "Communism is heaven, and the People's Communes are the stairway there." See "Di-er Zhan: Guangbo Dianshi," in Shiquan Xian difangzhi bianzuan weiyuanhui, *Shiquan Xian Zhi* (1991), 630–31.

129. Wang, "Taizhou Di Yi Jia Guangbozhan." Wang attributed the large number of daily contributions to competition between districts, who sought to outdo one-another.

130. Wang.

131. Jiayu Xian Guangbozhan, "Baogao," August 17, 1957, in HBPA, Doc. SZ7-2-46.

132. Gail Hershatter, *The Gender of Memory: Rural Women in China's Collective Past* (Berkeley: University of California Press, 2011), 323n10.

133. Dikotter, *Mao's Great Famine*, 27–29.

134. Huanggang Xian Guangbozhan, "Women ju ban le yi ci 'shuili gaochao guangbo dahui,'" Dec. 10, 1957, in HBPA, Doc. SZ7-2-46.

135. Daye Xian Guangbozhan, "1957 nian shang nian du xuanchuan gongzuo zongjie."

136. Daye Xian Guangbozhan.

137. Dikotter, *Mao's Great Famine*, 37–39.

138. Elihu Katz, "Disintermediating the Parents: What Else Is New?" in *The Wired Homestead*, ed. Joseph Turow and Andrea L. Kavanaugh (Cambridge, MA: MIT Press, 2003), 46.

139. Wufeng Xian Guangbozhan, "Guanyu zai kaizhan xiang nongcun renkou zhong jinxing shehui zhuyi jiaoyu xuanchuan de qingkuang he gongzuo de baogao." Oct. 12, 1957, in HBPA, Doc. SZ7-2-46.

140. Five reports from the Xishui County wired broadcasting station are preserved in the Hubei Provincial Archive. HBPA, Doc. SZ7-2-46.

141. Yang, *Tombstone*, 8.

142. Xishui Xian Guangbozhan, "Women de qunzhong gongzuo," Nov. 29, 1957, in HBPA, Doc. SZ7-2-46.

143. Xishui Xian Guangbozhan, "Xuanchuan yewu gongzuo baogao," ca. 1957, in HBPA, Doc. SZ7-2-46.

144. Xishui Xian Guangbozhan, "Women de qunzhong gongzuo."

145. Xishui Xian Guangbozhan, "Xuanchuan yewu gongzuo baogao."

146. Xishui Xian Guangbozhan.

147. Xishui Xian Guangbozhan, "Women shi zenyang genju guangbo tedian kaizhan xuanchuan gongzuo de," Nov. 30, 1957, in HBPA, Doc. SZ7-2-46.

148. Xishui Xian Guangbozhan, "Women de qunzhong gongzuo."

149. Xishui Xian Guangbozhan.

150. Xishui Xian Guangbozhan, "Women de shehui zhuyi xuanchuan de qingkuang, wenti ji jinhou de dasuan," Nov. 8, 1957, in HBPA, Doc. SZ7-2-46.

151. Tongcheng Xian Guangbozhan, "Wo zhan dui nongcun jinxing shehui zhuyi jiaoyu de qingkuang baogao," Oct. 15, 1957, in HBPA, Doc. SZ7-2-46.

152. Xishui Xian Guangbozhan, "Women shi zenyang genju guangbo tedian kaizhan xuanchuan gongzuo de."

153. Xishui Xian Guangbozhan, "Chongfen fahui guangbo xuanchuan zuoyong geng hao de wei zhongxin gongzuo fuwu," Dec. 2, 1957, in HBPA, Doc. SZ7-2-46.

154. Xishui Xian Guangbozhan, "Women shi zenyang genju guangbo tedian kaizhan xuanchuan gongzuo de."

155. Xishui Xian Guangbozhan.

156. Yang, *Tombstone*, 3–10.

Chapter 8

1. Li Dai, *Chunfeng huakai: Wo suo jingli de 1976–1985 nian* (Hangzhou: Zhejiang Gong Shang Daxue Chubanshe, 2018), 3–7.

2. "Wo guo nongcun youxian guangbo shiye pengbo fazhan," *Renmin Ribao*, Sept. 15, 1975. Alan P. Liu estimated that as early as 1959, 350 million people regularly listened to wired broadcasting in China. Alan P. Liu, *Radio Broadcasting in Communist China* (Cambridge, MA: Research Program on Problems of International Communication and Security, MIT, 1964).

3. See, e.g., "Meishan Xian Wei juban lilun xuexi guangbo jiangzuo: Tuidong nongcun pi Deng douzheng bubu shenru," *Renmin Ribao*, August 13, 1976; "Jia li tian jian dou keyi tingdao guangbo," *Renmin Ribao*, Nov. 12, 1975.

4. "Woguo nongcun youxian guangbo shiye pengbo fazhan," *Renmin Ribao*, Sept. 15, 1975.

5. "Woguo shi zhi chu ba zhong bandaoti shouyinji," *Renmin Ribao*, Sept. 10, 1963; "Jianchi duo kuai hao sheng di jianshe shehui zhuyi zongluxian woguo shouyinji chanliang zengjia zhiliang tigao shouyinji de xiaoshou jiage zhunian you suo jiangdi," *Renmin Ribao*, May 14, 1972; "Zengchan shouyinji wei guangbo shiye zuochu gongxian," *Renmin Ribao*, May 24, 1976.

6. Susan Ou and Heyu Xiong, "Mass Persuasion and the Ideological Origins of the Chinese Cultural Revolution," *Journal of Development Economics* 153 (2021). The authors of the study attempt to address the possibility that radio-reception quality was simply a function of state power by controlling for the number of Communist Party members in a given county in 1956. I find this unpersuasive. First, party membership rates and state power are not necessarily correlated. Second, while I sympathize with their overall conclusions, I have argued that the technological and political are difficult if not impossible to distinguish. In other words, a stronger radio network by definition meant a stronger more capable state, because that state had to build it or, at least, create the conditions for its construction. Thus the attempt to distinguish between state power and technological (propaganda) power is unnecessary. Indeed, their conclusion only serves to show that communication technologies are a central aspect of state power.

7. Student groups still use the intersection today to advertise clubs, classes, and activities.

8. Roderick MacFarquhar and Michael Schoenhals, *Mao's Last Revolution* (Cambridge, MA: Harvard University Press, 2006), 54–58.

9. FBIS, "Article Attacking Sung Shuo," June 6, 1966.

10. MacFarquhar and Schoenhals, *Mao's Last Revolution*, 58.

11. MacFarquhar and Schoenhals, 59.

12. Ye Xin, *Ye Xin Yanzhong de Shanghai* (Beijing: Xinhang Chubanshe, 2018), 216–19.

13. Ye, 217.

14. FBIS, "Article Attacking Sung Shuo," June 6, 1966.

15. Ye, 217.

16. "WenGe zhong de suiyue," in Liu Chuansheng, *Beijing Shifan Daxue dashi mingjia koushu shi* (Beijing: Guangming Ribao Chubanshe, 2012), 108–11.

17. Sidney Rittenberg, *The Man Who Stayed Behind* (Durham, NC: Duke University Press, 2001), 299.

18. Rittenberg, 301.

19. "Wenhua Dageming yu boluan fanzheng," in Yulin Shi zhi bianzuan weiyuanhui, *Yulin Shi Zhi* (Xi'an: Sanqing Chubanshe, 1996), 476–88. Hereafter, *Yulin Shi Zhi*.

20. FBIS, "Peking University Committee," June 3, 1966; FBIS, "People's Daily Editorial on CCP Decision," June 6, 1966.

21. *Yulin Shi Zhi*, 477.
22. See, e.g., FBIS, "Hunan Party Official's Counterattack Fails," July 1, 1966; and FBIS, "Kwangtung Radio Director Tien Wei Ousted," July 6, 1966.
23. FBIS, "Cultural Revolution Activists Meet in Wuhan," July 11, 1966; FBIS, "People's Daily Editor's Note," July 11, 1966.
24. *Yulin Shi Zhi*, 477.
25. Mao announced his decision to withdraw the work teams on July 24. The withdrawal of work teams from Beijing's schools was announced on July 28 and read to a large gathering of students and teachers on July 29. Recordings of this rally, including apologies from Deng Xiaoping and Liu Shaoqi, were heard throughout Beijing. For the high-level discussions over the proper role of the working groups, which could be withdrawn from late July to early August, see MacFarquhar and Schoenhals, *Mao's Last Revolution*, 82–85.
26. FBIS, "CCP Decision on Cultural Revolution Issued," August 9, 1966; FBIS, "Red Flag on Decision on Cultural Revolution," August 11, 1966.
27. *Yulin Shi Zhi*, 477.
28. MacFarquhar and Schoenhals's translation in *Mao's Last Revolution*, 93.
29. FBIS, "Peking Rally Celebrates Cultural Revolution," August 18, 1966.
30. FBIS, "Red Guards Attack Old Ways in Peking," August 23, 1966; FBIS "People's Daily Editorial," August 23, 1966.
31. *Yulin Shi Zhi*, 477–78.
32. Yan Boyuan, "Wo dang le qi nian yeyu guangboyuan," in *Women renmin chang: Jiangxi "Xiao san xian" 9333 chang shilu* (Shanghai: Shanghai Renmin Chubanshe, 2015), 730–32.
33. Wang Guoqing, "Zhuixun WenGe de zongying zhi er," in Sanmenxia shi weiyuanhui xuexi wenshi ziliao weiyuanhui, *Sanmenxia Wenshi Ziliao Di 17 Ji* (2007), 41–57.
34. *Yulin Shi Zhi*, 478.
35. Mao Zhuxi luxian hongweibing Hubei diqu geming zaofan silingbu, "Fennu kongsu Wuhan diqu 'Geming zhigong lianhehui' de taotian zuixing," Jan. 6, 1967. Author's personal collection.
36. Mao Zhuxi luxian hongweibing.
37. Kaifeng shi wei jiguan hongqi geming zaofan zongbu, "Er yue shiba ri Kaifeng shi guangbozhan shijian de zhenxiang," Feb. 28, 1967. Author's personal collection.
38. Ma Mingyi and Fang Chunli, "Jiang Botong, He Wenfang zhi si," in Xichang Wenshi Bianshen weiyuanhui, *Xichang Wenshi Di 16 Ji* (2001): 106–9.
39. Zhou Zuguo, "Xichang WenGe qin wen qin li," in Xichang Wenshi Bianshen weiyuanhui, *Xichang Wenshi Di 16 Ji* (2001): 96–105.

Conclusion

1. Andrew Jones, *Circuit Listening: Chinese Popular Music in the Global 1960s* (Minneapolis: University of Minnesota Press, 2020); Yu Wang, "Listening to the Enemy: Radio Consumption and Technological Culture in Maoist China, 1949–1965," *Twentieth-Century China* 47, no. 2 (2022).

2. Tao Dongfeng, "Teresa Teng and the Spread of Pop Songs in Mainland China in the Early Reform Era," *Inter-Asia Cultural Studies* 23, no. 2 (2022).

3. Xu, Chuan, "From Sonic Models to Sonic Hooligans: Magnetic Tape and the Unraveling of the Mao-Era Sound Regime, 1958–1983," *East Asian Science, Technology and Society* 13, no. 3 (2019).

4. Charles Hamm, "Music and Radio in the People's Republic of China," *Asian Music* 22, no. 2 (1991).

5. This point is also made in Laurence Coderre, *Newborn Socialist Things: Materiality in Maoist China* (Durham, NC: Duke University Press, 2021), 34, 45; and Masumi Sato, "Technological Development in China Viewed through the Electronics Industry: An Engineer's View," *Developing Economies* 9, no. 3 (1971).

6. Fang Xingdong and Chen Shuai, "Zhongguo hulianwang 25 nian," *Xiandai Chuanbo* 273, no. 4 (2019); Gianluigi Negro, *The Internet in China: From Infrastructure to Nascent Civil Society* (London: Palgrave Macmillan, 2017).

7. Angela Xiao Wu, "Historicizing Internet Use in China and the Problem of the User Figure," *IEEE Annals of the History of Computing* 37, no. 4 (2015).

8. "Beifang Qianxian: Neidi xiaoxi xuyao goutong," *Dagongbao*, Sept. 26, 1937.

9. Ann Blair, Paul Duguid, Anja-Silvia Going, and Anthony Grafton, *Information: A Historical Companion* (Princeton: Princeton University Press, 2021).

10. Karl Marx, *Capital*, vol. 1, trans. Ben Fowkes (Harmondsworth: Penguin, 1990), 505–6.

11. I am thinking here of the "ritual view of communication" theorized in James Carey, *Communication as Culture: Essays on Media and Society* (Abingdon, UK: Routledge, 2009), 11–18.

12. James C. Scott, *Seeing like a State: How Certain Schemes to Improve the Human Condition Have Failed* (New Haven, CT: Yale University Press, 1998).

13. Bruno Latour, *The Pasteurization of France*, trans. Alan Sheridan and John Law (Cambridge, MA: Harvard University Press, 1988).

14. For a discussion, see Vatro Murvar, "Some Tentative Modifications of Weber's Typology: Occidental versus Oriental City," *Social Forces* 44, no. 3 (1966).

15. Richard John and Jonathan Silberstein-Loeb, eds., *Making News: The Political Economy of Journalism in Britain and America from the Glorious Revolution to the Internet* (Oxford: Oxford University Press, 2015); Heidi Tworek, *News from Germany: The Competition to Control World Communications, 1900–1945* (Cambridge, MA:

Harvard University Press, 2019); Donald Read, *The Power of News: the History of Reuters* (Oxford : Oxford University Press, 1999).

16. Carolyn Birdsall, *Nazi Soundscapes: Sound, Technology, and Urban Space in Germany, 1933–1945* (Amsterdam: Amsterdam University Press, 2012), 109.

17. *Wuxiandian* (Wireless), Oct. and Nov. 1958.

18. Technocratic utopianism is a widely discussed theme in the history of technology. See Elizabeth van Meer, "The Transatlantic Pursuit of a World Engineering Federation: For the Profession, the Nation, and International Peace, 1918–48," *Technology and Culture* 53, no. 1 (2012); Frederick Turner, *From Counterculture to Cyberculture: Stewart Brand, the Whole Earth Network, and the Rise of Digital Utopianism* (Chicago: University of Chicago Press, 2010); and James Carey, with John J. Quirk, "The History of the Future," in James Carey, *Communication as Culture* (Abingdon, UK: Routledge, 2009), 133–54.

19. Sheila Jasanoff, "The Idiom of Co-production," in *States of Knowledge: The Co-production of Science and Social Order*, ed. Sheila Jasanoff (London: Routledge, 2004), 1–12, 4.

Bibliography

Archival Sources

FOREIGN
British Foreign Office Record, National Archive at Kew (FO)
Foreign Broadcast Information Service (FBIS)
Shanghai Municipal Police Files, 1929–1945
United States National Archives, College Park, MD (USNARA)

CHINESE
Academia Historica Taipei 國史館 (Guoshiguan)
Beijing Municipal Archive 北京市档案馆 (BJMA)
Chongqing Municipal Archive 重庆市档案馆 (CQMA)
Hubei Provincial Archive 湖北省档案馆 (HBPA)
Liaoning Provincial Archive 辽宁省档案馆 (LNPA)
Nanjing Second Historical Archive 第二历史档案馆 (NJSHA)
Shanghai Municipal Archive 上海市档案馆 (SHMA)

Principal Print Sources
China Press
Dagongbao 大公報 [Ta Kung Pao; l'Impartial]
Dian You 电友 [Telegraphists' Companion]
Dongfang Zazhi 東方雜誌 [Eastern Miscellany]

Jiefang Ribao 解放日报 [Liberation Daily]
Kexue 科學 [Science]
North China Herald
Renmin Ribao 人民日报 [People's Daily]
Shenbao 申報 [Shanghai News]
Wuxiandian 无线电 [Radio]
Xinhua Ribao 新华日报 [New China Daily] (Chongqing edition)

Published Sources
Adorno, Theodor. *The Culture Industry: Selected Essays on Mass Culture*. Edited by J. M. Bernstein. London: Routledge, 2001.
———. *Current of Music: Elements of a Radio Theory*. Edited by Robert Hullot-Kentor. Cambridge, UK: Polity, 2009.
Adorno, Theodor, and Max Horkheimer. *Dialectic of Enlightenment*. Stanford, CA: Stanford University Press, 2002.
Ai Peijun 艾培君. "Tianquan di yi tai shouyinji" 天全第一台收音机. In Zhengxie Tianquan Xian Weiyuanhui, *Tianquan Wenshi Ziliao, Di-si Ji* 天全文史资料, 第四辑. N.p.: n.p., 1986.
Alonso, Isabel Huacuja. *Radio for the Millions: Hindi-Urdu Broadcasting across Borders*. New York: Columbia University Press, 2023.
Altehenger, Jennifer. *Legal Lessons: Popularizing Laws in the People's Republic of China, 1949–1989*. Cambridge, MA: Harvard University Asia Center, 2018.
Anderson, Benedict. *Imagined Communities: Reflections on the Origin and Spread of Nationalism*. London: Verso, 1983.
Arnold, Julean. *The Commercial Handbook of China*. Washington, DC: Government Printing Office, 1919.
Asseraf, Arthur. *Electric News in Colonial Algeria*. Oxford: Oxford University Press, 2019.
Atherton, Ian. "The Itch Grown a Disease: Manuscript Transmission of News in the Seventeenth Century." *Prose Studies* 21, no. 2 (1998): 39–65.
Baark, Erik. *Lightning Wires: The Telegraph and China's Technological Modernization, 1860–1890* (Westport, CT: Greenwood Press, 1997).
Bai Quanwu 白全武. "Zhengqu Ri-qiao hezuo chuangjian renmin Guangbo" 争取日侨合作 创建人民广播. In *Dalian guangbo huiyilu* 大连广播回忆录, edited by Zhu Qingyan 朱青宴, 21–31. N.p. [Dalian?]: n.p., 1986.
Bao Tianxiao 包天笑. *Chuanyinglou huiyilu* 釧影樓回憶錄. Hong Kong: Dahua chubanshe, 1971.
Bayly, Christopher. *Empire and Information: Intelligence Gathering and Social Communication in India, 1780–1870*. Cambridge: Cambridge University Press, 1996.

Beard, Mary. "The Public Voice of Women." *London Review of Books* 36, no. 6 (March 2014): 11–14.

Bell, Montague. *The China Year Book, 1919–20.* New York: Routledge, 1919.

Benson, Carlton. "From Teahouse to Radio: Storytelling and the Commercialization of Culture in 1930s Shanghai." PhD diss., University of California, Berkley, 1996.

———. "The Resurgence of Commercial Radio in Gudao Shanghai." In *In the Shadow of the Rising Sun: Shanghai under Japanese Occupation*, edited by Christian Henriot and Wen-Hsin Yeh, 279–301. Cambridge: Cambridge University Press, 2004.

Bergère, Marie-Claire. *Sun Yat-sen.* Translated by Janet Lloyd. Stanford, CA: Stanford University Press, 1998.

Bian Donglei 卞冬磊. "Wusi yundong zai xiangcun: Chuanbo, dongyuan, yu minzu zhuyi" 五四運動在鄉村：傳播、動員與民族主義. *Ershiyi Shiji* 二十一世紀 [Hong Kong] (April 2019): 90–102.

Birdsall, Carolyn. *Nazi Soundscapes: Sound, Technology, and Urban Space in Germany, 1933–1945.* Amsterdam: Amsterdam University Press, 2012.

Blair, Ann, Paul Duguid, Anja-Silvia Goeing, and Anthony Grafton, eds. *Information: A Historical Companion.* Princeton, NJ: Princeton University Press, 2021.

Bray, Francesca. *Technology and Gender: Fabrics of Power in Late Imperial China.* Berkeley: University of California Press, 1997.

Brokaw, Cynthia. *Commerce in Culture: The Sibao Book Trade in the Qing and Republican Periods.* Cambridge, MA: Harvard University Asia Center, 2007.

Brook, Timothy. *Collaboration: Japanese Agents and Local Elites in Wartime China.* Cambridge, MA: Harvard University Press, 2005.

Brown, Jeremy, and Matthew Johnson, eds. *Maoism at the Grassroots.* Cambridge, MA: Harvard University Press, 2015.

Brown, Melissa J., and Damian Satterthwaite-Phillips. "Economic Correlates of Footbinding: Implications for the Importance of Chinese Daughters' Labor." *PloS One* 13, no. 9 (2018): https://journals.plos.org/plosone/article?id=10.1371/journal.pone.0201337.

Bu Zifang 布子芳. "Huigu Yao'an Xian shouyinzhan de jianli" 回顾姚安县收音站的建立. In Zhongguo Renmin Zhengzhi Xieshang Huiyi Yunnan Sheng Chuxiong Yizu Zizhi Zhou Weiyuanhui Jiao Ke Wen Wei Wenshi Ziliao Weiyuanhui, *Chuxiong Zhou Wenshi Ziliao Xuanji, Di 21 Ji: Wenhua tiyu shi qiu wenji* 楚雄州文史资料选辑，第21辑：文化体育史秋专辑, 162–67. Chuxiong: Published by committee, 2004.

Carey, James. *Communication as Culture: Essays on Media and Society.* Abingdon, UK: Routledge, 2009.

Chan, F. Gilbert. "An Alternative to Kuomintang-Communist Collaboration: Sun Yat-sen and Hong Kong, January–June 1923." *Modern Asian Studies* 13, no. 1 (1979): 127–39.

Chan, Marjorie K. M. "Gender Differences in the Chinese Language: A Preliminary Report." In *Proceedings of the Ninth North American Conference on Chinese Linguistics (NACCL9)*, edited by Hua Lin. Vol. 2, 35–52. Los Angeles: GSIL Publications, University of California, 1997.

Che Jixin 车吉心, ed. *Qi Lu Wenhua Da Cidian* 齐鲁文化大辞典. Jinan, Shandong: Shandong Jiaoyu Chubanshe, 1988.

Cheek, Timothy. *Propaganda and Culture in Mao's China: Deng Tuo and the Intelligentsia*. Oxford: Oxford University Press, 1997.

Chen Ertai 陈尔泰. *Zhongguo Guangbo shikao* 中国广播史考. Beijing: Zhongguo Guangbo Dianshi Chubanshe, 2008.

———. *Zhongguo Guangbo zhi fu: Liu Han zhuan* 中国广播之父: 刘瀚传. Beijing: Zhongguo Guangbo Dianshi Chubanshe, 2006.

Chen Guofu 陳果夫. "Zhongyang Guangbo Diantai Chuangban Jingguo" 中央廣播電臺創辦經過. In *Zhonghua Minguo Xinwen Nianjian* 中華民國新聞年鑒. Taipei: Taibei shi Xinwen Jizhe Gonghui, 1961.

Chen, Janet. *The Sounds of Mandarin*. New York, Columbia University Press, 2023.

Chen Jingqiu 陈静秋. "Wusi Yundong zai Xianju" 五四运动在仙居. In *Xianju Wenshi Ziliao Weiyuanhui, Xianju Wenshi Ziliao, Di Yi Juan Xi* 仙居文史资料第1辑, 23–25. N.p.: n.p., 1986.

Chen Yunhao 陈允豪. "Xie hua he duan xinwen—huiyi Yanfu Dazhong bao" 写话和短新闻——回忆《盐阜大众》报. *Xinwen Zhanxian* 新闻战线, no. 6 (1979): 43–44.

Cheng Gang 成刚. "Liuliang Meng" 吕梁梦. In *Qinli Kang-Zhan: Beijing jiaoyujie lao tongzhi Kang-Zhan huiyilu* 亲历抗战: 北京教育界老同志抗战回忆录, edited by Zhong-Gong Beijing shi wei jiaoyu gongzuo weiyuanhui, 461–65. Beijing: Zhongguo Guangbo Dianshi Chubanshe, 2005.

Cheng Mei 成美, Chen Daofu 陈道馥, and Xue Xiayuan 薛夏原, eds. *Ding Yilan Zhuan* 丁一岚传. Beijing: Zhongguo guoji guangbo chubanshe, 2011.

Chin, Sei-Jeong. "Print Capitalism, War, and the Remaking of the Mass Media in 1930s China." *Modern China* 30, no. 4 (2014): 393–425.

Clark, Douglas. *Gunboat Justice: British and American Law Courts in China and Japan, 1842–1943*. 3 vols. Hong Kong: Earnshaw, 2015.

Coderre, Laurence. *Newborn Socialist Things: Materiality in Maoist China*. Durham, NC: Duke University Press, 2021.

Corbin, Alain. *Village Bells: Sound and Meaning in the 19th-Century French Countryside*. New York: Columbia University Press, 1998.

Dangdai Beijing guangbo shihua 当代北京广播史话. Beijing: Dangdai Zhongguo Chubanshe, 2013.

Dangdai Zhongguo de Guangbo Dianshi Bianji Bu. *Zhongguo de Youxian Guangbo* 中国的有线广播. Beijing: Beijing Chubanshe, 1988.

Darnton, Robert. "An Early Information Society: News and Media in Mid-Eighteenth Century Paris." *American Historical Review* 105, no. 1 (Feb. 2000): 1–35.

———. *The Literary Underground of the Old Regime*. Cambridge, MA: Harvard University Press, 1982.

———. *Poetry and the Police: Communications Networks in Eighteenth-Century Paris*. Cambridge, MA: Harvard University Press, 2010.

De Giorgi, Laura. "Communication Technology and Mass Propaganda in Republican China: The Nationalist Party's Radio Broadcasting Policy and Organisation during the Nanjing Decade (1927–1937)." *European Journal of East Asian Studies* 13 (2014): 305–29.

———. "Media and Popular Education: Views and Policies on Radio Broadcasting in Republican China, 1920s–30s." *Interactions: Studies in Communication & Culture* 7, no. 3 (2016): 281–96.

Dikotter, Frank. *Mao's Great Famine*. New York: Bloomsbury, 2017.

Dillon, Nara. "New Democracy and the Demise of Private Charity in Shanghai." In *Dilemmas of Victory: The Early Years of the People's Republic of China*, edited by Jeremy Brown and Paul Pickowicz, 80–102. Cambridge, MA: Harvard University Press, 2007.

Ding Yilan 丁一岚. "Hutuo hepan ding xin meng" 滹沱河畔定心盟. *Shehui kexue zhanxian* 社会科学战线, no. 2 (1986): 129–33.

———. "Wushi nian qian de kuanghuan zhi ye" 五十年前的狂欢之夜. *Xinwen Aihaozhe* 新闻工作者, no. 9 (1995): 20.

Duara, Prasenjit. *Rescuing History from the Nation: Questioning Narratives of Modern China*. Chicago: University of Chicago Press, 1995.

———. *Sovereignty and Authenticity: Manchukuo and the East Asian Modern*. Lanham, MD: Rowman & Littlefield, 2003.

Eastman, Lloyd. *The Abortive Revolution: China under Nationalist Rule, 1927–1937*. Cambridge, MA: Harvard University Press, 1974.

Eisenstein, Elizabeth. *The Printing Revolution in Early Modern Europe*. Cambridge: Cambridge University Press, 1983.

Elleman, Bruce. "Soviet Diplomacy and the First United Front in China." *Modern China* 21, no. 4 (1995): 450–80.

Elliott, Mark. "The Limits of Tartary: Manchuria in Imperial and National Geographies." *Journal of Asian Studies* 59, no. 3 (2000): 603–46.

Elman, Benjamin A. "Naval Warfare and the Refraction of China's Self-Strengthening Reforms into Scientific and Technological Failure, 1865–1895." *Modern Asian Studies* 38, no. 2 (2004): 283–326.

———. "Rethinking the Twentieth Century Denigration of Traditional Chinese Science and Medicine in the Twenty-First Century." Prepared for the 6th International Conference on The New Significance of Chinese Culture in the Twenty-First Century, Nov. 2003.

Fang Xingdong 方兴东 and Chen Shuai 陈帅. "Zhongguo hulianwang 25 nian" 中国互联网 25 年 *Xiandai Chuanbo* 现代传播 273, no. 4 (2019): 1–10.

Foucault, Michel. *The Order of Things: An Archaeology of the Human Sciences*. New York: Routledge, 2002.

Fu Changsheng 傅畅盛. "Yi jiefang chuqi shouyinzhan gongzuo" 忆解放初期收音站工作. In *Zui shi nan wang de yi ye huiyi xin de xuan* 最是难忘的一页回忆心得选, edited by Hu Shihui 胡世徽 and Zhang Xiangqi 张湘绮, 84–88. Guizhou: Guizhou Renmin Chubanshe, 1999.

Fu Xingpei 傅兴沛. "Di-er ci Zhi-Feng zhanzheng jishi" 第二次直奉战争纪实. In *Wenshi Ziliao Jingxuan Di 4 Ce*, edited by Wenshi Ziliao Xuanji, 93–101. Beijing: Zhongguo Wenshi Chubanshe, 1990.

Garon, Sheldon. *Molding Japanese Minds: The State in Everyday Life*. Princeton, NJ: Princeton University Press, 1998.

Gatu, Dagfinn. *Village China at War: The Impact of Resistance to Japan, 1937–1945*. Copenhagen: NIAS Press, 2007.

Geng Wenbin 耿文彬. "Taixing shouyinzhan shimo" 泰兴收音站始末. In Zhengxie Taizhou Shi Xuexi Wenshi Weiyuanhui, *Taizhou Wenshi Ziliao, 1949–1952* 泰州文史资料 (1949–1952), 202–6. Jiangsu: Published by committee, 2008.

Ghosh, Arunabh. *Making It Count: Statistics and Statecraft in the Early People's Republic of China*. Princeton, NJ: Princeton University Press, 2020.

Guarneri, Julia. *Newsprint Metropolis: City Papers and the Making of Modern Americans*. Chicago: University of Chicago Press, 2017.

Guo Zhenzhi. "A Chronicle of Private Radio in Shanghai." *Journal of Broadcasting & Electronic Media* 30, no. 4 (1986): 379–92.

Habermas, Jürgen. *The Structural Transformation of the Public Sphere: An Inquiry into a Category of Bourgeois Society*. Cambridge, MA: MIT Press, 1989.

Hamm, Charles. "Music and Radio in the People's Republic of China." *Asian Music* 22, no. 2 (1991): 1–42.

Harrison, Henrietta. "Newspapers and Nationalism in Rural China, 1890–1929." *Past & Present*, no. 166 (Feb. 2000): 181–204.

Hartono, Paulina. "'A Good Communist Style': Sounding Like a Communist in Twentieth-Century China." *Representations* 151, no. 1 (2020): 26–50.

He Zhiping 贺致平. "Muping gongzuo de huiyi" 牟平工作的回忆. In Zhong-Gong

Shandong sheng wei dang shi ziliao zhengji yanjiu weiyuanwei, ed. *Shandong dang shi ziliao* 山东党史资料, no. 3 (1985): 7–17.

Headrick, Daniel. *The Invisible Weapon: Telecommunications and International Politics, 1851–1945*. Oxford: Oxford University Press, 1991.

———. *The Tentacles of Progress: Technology Transfer in the Age of Imperialism, 1850–1940*. Oxford: Oxford University Press, 1988.

———. *Tools of Empire: Technology and European Imperialism in the Nineteenth Century*. Oxford: Oxford University Press, 1981.

Herkman, Juha, Taisto Hujanen, and Paavo Oinenan, eds. *Intermediality and Media Change*. Tampere, FI: Tampere University Press, 2012.

Hershatter, Gail. *The Gender of Memory: Rural Women in China's Collective Past*. Berkeley: University of California Press, 2011.

Hilmes, Michele. *Radio Voices: American Broadcasting, 1922–1952*. Minneapolis: University of Minnesota Press, 1997.

Hirata, Koji. "Made in Manchuria: The Transnational Origins of Socialist Industrialization in Maoist China." *American Historical Review* 126, no. 3 (2021): 1072–1101.

———. "Steel Metropolis: Industrial Manchuria and the Making of Chinese Socialism." *Enterprise & Society* 21, no. 4 (2020): 875–85.

Hobsbawm, Eric. *The Age of Revolution: 1789–1848*. New York: Vintage, 1996.

Howe, Daniel Walker. *What Hath God Wrought: The Transformation of America, 1815–1848*. Oxford: Oxford University Press, 2007.

Hu Baoren 胡宝仁. "Manhua Shang-xian de diyi bu shouyinji" 漫话商县的第一部收音机. In Shaanxi Sheng Shang-Xian Weiyuanhui Wenshi Ziliao Yanjiu Weiyuanhui. *Shang-Xian Wenshi Ziliao, Di-si Ji* 商县文史资料 第4辑. 1987.

Hu Fang 胡芳. "Nanjing zhi ying—minguo boyinyuan Liu Junying" "南京之莺"——民国播音员刘俊英. *Xinwen Yanjiu Dao Kan* 新闻导刊 7, no. 24 (2016): 77, 81.

Hu Guanzhong 胡冠中, with Fang Zhenyu 方振煜, ed. "Wo suo qinli de Xiamen jiefang de san jian shi" 我所亲历的厦门解放的三件事. In *Koushu lishi: qinli Xiamen jiefang* 口述历史：亲历厦门解放, edited by Zhong-Gong Xiamen shi wei xuanchuan bu, 137–45. Xiamen: Xiamen Daxue Chubanshe, 2009.

Huang Bo 黄波. "Yong bu xiaoshi de shengyin—Qi Yue" 永不消逝的声音——齐越. *Shiji Fengcai* 世纪风采. Nov. 2019.

Huang Qixuan 黄启轩. "Shouyinzhan zai Dong xiang" 收音站在侗乡. In Zhengxie Sanjiang Dongzu Zizhi Xian Weiyuanhui, *Sanjiang Wenshi Ziliao, Di 6 Ji* 三江文史资料 第6辑, 118–120. Nanning: Guangxi Renmin Chubanshe, 2002.

Huang Xiaoying 黄小英. "Mingguo shiqi guangyin jiaoyu de lishi huigu" 民国时期播音教育的历史回顾. *Dianhua jiaoyu yanjiu* 电化教育研究 6 (2011): 110–20.

Hunan sheng aiguo weisheng yundong weiyuanhui bangongshi. *Hunan sheng chu si hai jiang weisheng yundong ziliao huibian* 湖南省除四害讲卫生运动资料汇编. 1958.

Innis, Harold. *Empire and Communications*. Oxford: Clarendon, 1950.

Jasanoff, Sheila, ed. *States of Knowledge: The Co-production of Science and Social Order*. London: Routledge, 2004.

Jiang Shiying 姜仕英. "Jiefang chuqi Chuxiong zhuanqu shouyinzhan de xuanchuan gongzuo" 解放初期楚雄专区收音站的宣传工作. In Zhongguo Renmin Zhengzhi Xieshang Huiyi Yunnan Sheng Chuxiong Yizu Zizhi Zhou Weiyuanhui Jiao Ke Wen Wei Wenshi Ziliao Weiyuanhui, *Chuxiong Zhou Wenshi Ziliao Xuanji, Di 21 Ji: Wenha tiyu shi qiu zhuanji* 楚雄州文史资料选辑, 第21辑: 文化体育史秋专辑. Chuxiong: Self-published by the Committee, 2004.

Jin Mai 靳迈. "Zai Chongqing guangbo Riben touxiang de xiaoxi" 在重庆广播日本投降的消息. *Shiji* 世纪 [Century] 5 (1997): 57.

John, Richard R. *Spreading the News: The American Postal System from Franklin to Morse*. Cambridge, MA: Harvard University Press, 1995.

John, Richard R., and Jonathan Silberstein-Loeb, eds. *Making News: The Political Economy of Journalism in Britain and America from the Glorious Revolution to the Internet*. Oxford: Oxford University Press, 2015.

Johnson, Matthew D. "Propaganda and Sovereignty in Wartime China: Morale Operations and Psychological Warfare under the Office of War Information." *Modern Asian Studies* 45, no. 2 (2011): 303–44.

Jones, Andrew F. *Circuit Listening: Chinese Popular Music in the Global 1960s*. Minneapolis: University of Minnesota Press, 2020.

Judge, Joan. *Print and Politics: Shibao and the Culture of Reform in Late Qing China*. Stanford, CA: Stanford University Press, 1996.

Kang Minzhuang 康敏庄. "Xuexi zhangwo he yunyong guangbo gongju: Huiyi Dalian diantai jian tai chuqi qingkuang" 学习和运用广播工具——回忆大连电台建台初期情况. In *Dalian guangbo huiyilu* 大连广播回忆录, edited by Zhu Qingyan 朱庆晏, 3–20. N.p. [Dalian?]: n.p., 1986.

Katz, Elihu. "Disintermediating the Parents: What Else Is New?" In *The Wired Homestead*, edited by Joseph Turow and Andrea L. Kavanaugh, 45–52. Cambridge, MA: MIT Press, 2003.

King, Rachael Scarborough. "The Manuscript Newsletter and the Rise of the Newspaper, 1665–1715." *Huntington Library Quarterly* 79, no. 3 (2016): 411–37.

Köll, Elizabeth. *Railroads and the Transformation of China*. Cambridge, MA: Harvard University Press, 2019.

Kotkin, Stephen. *Magnetic Mountain: Stalinism as a Civilization*. Berkeley: University of California Press, 1995.

———. *Stalin*. Vol. 1, *Paradoxes of Power, 1878–1928*. New York: Penguin, 2014.

Krysko, Michael. *American Radio in China: International Encounters with Technology and Communications, 1919–41*. New York: Palgrave Macmillan, 2011.

Kushner, Barak. *The Thought War: Japanese Imperial Propaganda*. Honolulu: University of Hawai'i Press, 2007.

Kwong, Chi Man. "Building a 'Total Mobilization State': Thinking about War and Society in 1920s Manchuria." *American Journal of Chinese Studies* 26, no. 1 (April 2019): 1–14.

Lacey, Kate. *Listening Publics: The Politics and Experience of Listening in the Media Age*. Cambridge, UK: Polity, 2013.

Langfur, Hal. *Adrift on an Inland Sea: Misinformation and the Limits of Empire in the Brazilian Backlands*. Stanford, CA: Stanford University Press, 2023.

Lary, Diana. *The Chinese People at War: Human Suffering and Social Transformation, 1937–1945*. Cambridge: Cambridge University Press, 2010.

Latour, Bruno. *The Pasteurization of France*. Translated by Alan Sheridan and John Law. Cambridge, MA: Harvard University Press, 1988.

Lean, Eugenia. *Public Passions: The Trial of Shi Jianqiao and the Rise of Popular Sympathy in Republican China*. Berkeley: University of California Press, 2007.

Lei, Wei. *Radio and the Social Transformation of China*. Oxford: Routledge, 2019.

Lekner, Dayton. "Echolocating the Social: Silence, Voice, and Affect in China's Hundred Flowers and Anti-Rightist Campaigns, 1956–58." *Journal of Asian Studies* 80 no. 4 (2021): 1–21.

Li Dai 李呆. *Chun feng hua kai: wo suo jingli de 1976–1985 nian* 春风花开 我所经历的 1976–1985年. Hangzhou: Zhejiang Gong Shang Daxue Chubanshe, 2018.

Li, Jie. "Revolutionary Echoes: Radios and Loudspeakers in the Mao Era." *Twentieth-Century China* 45, no. 1 (2020): 25–45.

Li Mai 李麦. "Ji-Zhong diqu de xinwen gongzuo" 冀中地区的新闻工作. In *Xinwen yanjiu ziliao congkan 1981 nian di 2 ji* 新闻研究资料丛刊 1981 年第2辑, edited by Zhongguo shehui kexue yuan xinwen yanjiusuo, 143–55. Beijing: Xinhua Chubanshe, 1981.

Li Qing 李清. "Laiyuan de di yi tai shouyinji" 涞源的第一台收音机. Zhengxie Laiyuan Xian Weiyuanhui, *Laiyuan Xian wenshi ziliao: Di-yi Ji* 涞源文史资料 第1辑 (1996): 261.

Li Yu 李煜. *Zhongguo Guangbo Xiandaixing Liubian* 中国广播现代性流变. Beijing: Zhongguo Chuanmei Daxue Chubanshe, 2016.

Liang Yi 梁毅. "Kang-zhan chuqi, Wuyang Xian dixia dang dui di douzheng de pianduan huiyi" 抗战初期, 舞阳县地下党对敌斗争的片断回忆. In *Luohe Wenshi Ziliao, Di 6 Ji: Kang-Ri Zhanzheng Zhuanji* 漯河文史资料 第6辑: 抗日战争专辑, edited by Zhongguo Renmin Zhengzhi ban shang huiyi Luohe Shi weiyuanhui wenshi ziliao weiyuanhui (n.p.: Ma Dian Yinshuachang, 1995): 196–204.

Lindsay, Hsiao Li. *Bold Plum: With the Guerillas in China's War against Japan.* Morrisville, NC: Lulu Press, 2007.

Lindsay, Michael. *The Unknown War: North China, 1937–1945.* London: Bergstrom & Boyle, 1975.

Liu, Alan P. L. *Radio Broadcasting in Communist China.* Cambridge, MA: MIT Center for International Studies, 1964.

Liu Chuansheng 刘川生. *Beijing Shifan Daxue dashi mingjia koushu shi* 北京师范大学大师名家口述史. Beijing: Guangming Ribao Chubanshe, 2012.

Liu Jiufu 刘九富, ed. *Shenyang guangbo zhi ziliao xing wengao* 沈阳广播志资料性文稿. Shenyang: Shenyang Renmin Guangbo Diantai, 1987.

Liu, Lydia. "Scripts in Motion: Writing as Imperial Technology." *Past and Present* 140, no. 2 (March 2015): 375–83.

Liu Yonghua 刘永华. "Qing dai minzhong shizi wenti de zai renshi." 清代民众识字问题的再认识. *Zhongguo Shehui Kexue Pingjia*, no. 2 (2017): 96–128.

Lovell, Stephen. *Russia in the Microphone Age: A History of Soviet Radio, 1919–1970.* Oxford: Oxford University Press, 2015.

Lu Kaitai 鲁开泰. "Huiyi Fangcheng Kang-Zhan Ribao" 回忆方城《抗战日报》. In *Fangcheng Wenshi Ziliao, Di 8 Ji* 方城文史资料选辑 第8辑, edited by Fangcheng Xian weiyuanhui wenshi ziliao yanjiu weiyuanhui, 20–22. N.p.: n.p., 1991.

Lu Tianhong 陆天虹. "Kang-Zhan shi wo zhuban de youji xiaobao." 抗战时我主办的游击小报 *Wuhan Wenshi Ziliao* 武汉文史资料, no. 4 (2012): 29–33.

Luo Dong 罗东. "Wu Yu zhi shou 'da kong jia dian'" 吴虞 只手"打孔家店." *Beijing Bao Digital* 新京报. www.bjnews.com.cn/feature/2019/04/17/568678.html.

Ma Mingyi 马明义 and Fang Chunli 方春礼. "Jiang Botong, He Wenfang zhi si" 姜伯通、何文芳之死. In Xichang Wenshi Bianshen weiyuanhui, *Xichang Wenshi Di 16 Ji* 西昌文史第16辑 (2001): 106–9.

Ma Sanli 马三立. "Xin jiu shehui liang chong tian" 新旧社会两重天. In *Chengshi jieguan qinli ji* 城市接管亲历记, 226–28. Beijing: Zhongguo Wen Shi Chubanshe, 1999.

MacFarquhar, Roderick, and Michael Schoenhals. *Mao's Last Revolution.* Cambridge, MA: Harvard University Press, 2006.

Madame Chiang Kai-shek (Song Meiling). "China's 'New Life' Movement." Originally published in *China Press*. Republished in *China's Leaders and Their Policies: Messages to the Chinese People*, by Wang Ching-Wei and General Chiang Kai-Shek. Shanghai: China United Press, 1935.

The Manchoukuo Yearbook, 1942. Hsinking, Manchoukuo: Manchoukuo Year Book Co., 1942.

Manela, Erez. *The Wilsonian Moment: Self-Determination and the International Origins of Anticolonial Nationalism.* Oxford: Oxford University Press, 2017.

Marx, Karl. *Capital*. Vol. 1. Translated by Ben Fowkes. Harmondsworth: Penguin, 1990.

McLuhan, Marshall. *The Gutenberg Galaxy: The Making of Typographic Man*. Toronto: University of Toronto Press, 1962.

———. *Understanding Media: The Extensions of Man*. Cambridge, MA: MIT Press, 1994.

Meng Yiqi 孟亦奇. "Cong zhongxue xiaoyuan toushen kangzhan de huiyi" 从中学校园投身抗战的回忆. *Jianghuai Wenshi* 江淮文史 4 (1999): 28–41.

Mimura, Janis. *Planning for Empire: Reform Bureaucrats and the Japanese Wartime State*. Ithaca, NY: Cornell University Press, 2011.

Mitchell, Richard. *Censorship in Imperial Japan*. Princeton, NJ: Princeton University Press, 1983.

Mitter, Rana. *A Bitter Revolution: China's Struggle with the Modern World*. Oxford: Oxford University Press, 2005.

———. "Classifying Citizens in Nationalist China during World War II, 1937–1941." *Modern Asian Studies* 45, no. 2 (2011): 243–75.

———. *Forgotten Ally: China's World War II, 1937–1945*. Boston: Houghton Mifflin Harcourt, 2013.

———. "The Last Warlord." *History Today*, Feb. 1, 2004.

———. *The Manchurian Myth: Nationalism, Resistance, and Collaboration in Modern China*. Berkeley: University of California Press, 2000.

Mittler, Barbara. *A Newspaper for China? Power, Identity, and Change in Shanghai's News Media, 1872–1912*. Cambridge, MA: Harvard University Asia Center, 2004.

Mokros, Emily. *The Peking Gazette in Late Imperial China: State News and Political Authority*. Seattle: University of Washington Press, 2021.

———. "Spies and Postmen: Communications Liaisons and the Evolution of the Qing Bureaucracy." *Frontiers of History in China* 14, no. 1 (2019): 19–50.

Moore, Aaron Stephen. *Constructing East Asia: Technology, Ideology, and Empire in Japan's Wartime Era, 1931–1945*. Stanford, CA: Stanford University Press, 2013.

Mullaney, Thomas. *The Chinese Typewriter: A History*. Cambridge, MA: MIT Press, 2017.

———. "Semiotic Sovereignty: The 1871 Chinese Telegraph Code in Historical Perspective." In *Science and Technology in Modern China, 1880s–1940s*, edited by Jing Tsu and Benjamin Elman, 153–83. Boston: Brill, 2014.

Murvar, Vatro. "Some Tentative Modifications of Weber's Typology: Occidental versus Oriental City." *Social Forces* 44, no. 3 (1966): 381–90.

Needham, Joseph, and Tsien Tsuen-Hsuin. *Science and Civilization in China*. Vol. 5. Cambridge: Cambridge University Press. 1987.

Negro, Gianluigi. *The Internet in China: From Infrastructure to Nascent Civil Society.* London: Palgrave Macmillan, 2017.

Netz, Reviel. *Barbed Wire: An Ecology of Modernity.* Middletown, CT: Wesleyan University Press, 2004.

Nippon Hoso Kyokai (NHK). *The History of Broadcasting in Japan.* Tokyo: History Compilation Room, Radio & TV Culture Research Institute, NHK, 1967.

Osterhammel, Jürgen. "Semi-colonialism and Informal Empire in Twentieth-Century China: Towards a Framework of Analysis." In *Imperialism and After: Continuities and Discontinuities,* edited by Wolfgang J. Mommsen and Jürgen Osterhammel, 290–314. London: Allen & Unwin, 1986.

Ou, Susan, and Heyu Xiong. "Mass Persuasion and the Ideological Origins of the Chinese Cultural Revolution." *Journal of Development Economics* 153 (2021).

Patterson, Don. "The Journalism of China." Special issue, *University of Missouri Bulletin* 23, no. 34 (1922). Journalism Series, no. 26, edited by Robert S. Mann.

Perry, Elizabeth. "Moving the Masses: Emotion Work in the Chinese Revolution." *Mobilization* 7, no. 2 (2002): 111–28.

Rawski, Evelyn. *Education and Popular Literacy in Ch'ing China.* Ann Arbor: University of Michigan Press, 1979.

Read, Donald. *The Power of News: The History of Reuters.* Oxford: Oxford University Press, 1999.

Reed, Christopher. *Gutenberg in Shanghai: Chinese Print Capitalism, 1876–1937.* Honolulu: University of Hawai'i Press, 2004.

Reinhardt, Anne. *Navigating Semi-colonialism: Shipping, Sovereignty, and Nation-Building in China, 1860–1937.* Cambridge, MA: Harvard University Press, 2018.

Rittenberg, Sidney. *The Man Who Stayed Behind.* Durham, NC: Duke University Press, 2001.

Rucheng Xianzhi Bianzuan Weiyuanhui. *Rucheng xian zhi* 汝城县志. Hunan: Hunan Renmin Chubanshe, 1997.

San Shui 三水. "Guangbo Riben touxiang xiaoxi jishi" 广播日本投降消息记事. *Shi Ting Jie* 视听界 [Broadcasting realm]. Jiangsu Sheng Guangbo Dianshi Zongtai, Jan. 1, 1988.

Sand, Jordan. *House and Home in Modern Japan: Architecture, Domestic Space, and Bourgeois Culture, 1880–1930.* Cambridge, MA: Harvard University Press, 2003.

Sato, Masumi. "Technological Development in China Viewed through the Electronics Industry: An Engineer's View." *Developing Economies* 9, no. 3 (1971): 315–31.

Scales, Rebecca. *Radio and the Politics of Sound in Interwar France, 1921–1939.* Cambridge: Cambridge University Press, 2016.

Schafer, R. Murray. *The Soundscape: Our Sonic Environment and the Tuning of the World.* Rochester, VT: Destiny, 1977.

Schoppa, R. Keith. *In a Sea of Bitterness*. Cambridge, MA: Harvard University Press, 2011.

Scott, James C. *Seeing like a State: How Certain Schemes to Improve the Human Condition Have Failed*. New Haven, CT: Yale University Press, 1998.

Seow, Victor. *Carbon Technocracy: Energy Regimes in Modern East Asia*. Chicago: University of Chicago Press, 2021.

Service, John. *The John S. and Caroline Service Oral History Project*. Vol. 1. Berkeley: Regional Oral History Office, Bancroft Library, 1978.

Shaanxi guangbo dianshi zhi 陕西广播电视志. Xi'an: 1992.

Shanghai Shi Danganguan [Shanghai Municipal Archive] et al. *Jiu Zhongguo de Shanghai Guangbo Shiye* 旧中国的上海广播事业. Beijing: Dang'an Chubanshe, 1985.

Shiquan Xian difang zhi bianzuan weiyuanhui. *Shiquan Xian Zhi* 石泉县志 (1991).

Shockey, Nathan. *The Typographic Imagination: Reading and Writing in Japan's Age of Modern Print Media*. New York: Columbia University Press, 2019.

Stamm, Michael. *Dead Tree Media: Manufacturing the Newspaper in Twentieth-Century North America*. Baltimore: Johns Hopkins University Press, 2018.

Strand, David. *An Unfinished Republic: Leading by Word and Deed in Modern China*. Berkeley: University of California Press, 2011.

Strauss, Julia. *Strong Institutions in Weak Polities: State Building in Republican China, 1927–1940*. Oxford: Oxford University Press, 1998.

"Suilu Xian Di Yi Bu Shouyinji" 绥渌县第一部收音机. In Zhengxie Fusui Xian Weiyuanhui Wenshi Ziliao Bianji Weiyuanhui, *Fusui Wenshi Ziliao, Di 5 Ji* 扶绥文史资料, 第5辑, 20–24. Guangxi: n.p., 1998.

Sun Fuyuan 孙伏园. "Huiyi Wusi Dang Nian" 回忆五四当年. In Zhongguo Shehui Kexue Yuan Jindai Yanjiusuo, *Wusi Yundong Huiyilu (Shang)* 五四运动回忆录（上）. Beijing: Zhongguo Shehui Kexue Chubanshe, 1989.

Sun Yat-sen. *Fundamentals of National Reconstruction*. Taipei: China Cultural Service, 1953.

———. *The International Development of China*. New York: Knickerbocker, 1922.

Sun Yude. "My Older Brother's First Radio." In *China in Family Photographs*, edited by Ed Krebs and Hanchao Lu, 313–17. Los Angeles: Bridge21, 2018.

Sun Zhongshan 孙中山. *Sun Zhongshan Quanji, Di 7 Juan* 孙中山全集 第7卷. Beijing: Renmin Chubanshe, 2015.

Tallon, Tina. "A Century of 'Shrill': How Bias in Technology Has Hurt Women's Voices." *New Yorker*, Sept. 3, 2019.

Tam, Gina. *Dialect and Nationalism in China, 1860–1960*. Cambridge: Cambridge University Press, 2020.

Tao Dongfeng. "Teresa Teng and the Spread of Pop Songs in Mainland China in the Early Reform Era." *Inter-Asia Cultural Studies* 23, no. 2 (2022): 269–87.

Tao Dun 陶钝. "Wusi zai Shandong nongcun" "五四"在山东农村. In Quan guo zheng xie wenshi ziliao weiyuanhui, *Wusi Yundong Qin Li Ji* 五四运动亲历记. Beijing: Zhongguo Wen Shi Chubanshe, 1999.

Teiwes, Frederick, and Warren Sun. *China's Road to Disaster: Mao, Central Politicians and Provincial Leaders in the Great Leap Forward, 1955–59*. New York: Routledge, 1999.

Thompson, Emily. *The Soundscape of Modernity: Architectural Acoustics and the Culture of Listening in America, 1900–1933*. Cambridge: Cambridge University Press, 2002.

Tian Jianping 田建平 and Zhang Jinfeng 张金凤, eds. *Jinchaji kang Ri genju di shubao chuanbo shilue, 1935–1945* 晋察冀抗日根据地书报传播史略, 1935–1945. Baoding: Hebei Daxue Chubanshe, 2010.

Tianjin Lishi Bowuguan, *Wusi Yundong zai Tianjin* 五四运动在天津. Tianjin: Tianjin Renmin Chubanshe, 1979.

Tilly, Charles. "Coercion, Capital, and European States, AD 990–1990." In *Collective Violence, Contentious Politics, and Social Change: A Charles Tilly Reader*, edited by Ernesto Castañeda and Cathy Lisa Schneider, 140–54. London: Routledge, 2017.

Tsai, Weiping. *Reading Shenbao: Nationalism, Consumerism, and Individuality in China, 1919–37*. London: Palgrave Macmillan, 2010.

Turner, Frederick. *From Counterculture to Cyberculture: Stewart Brand, the Whole Earth Network, and the Rise of Digital Utopianism*. Chicago: University of Chicago Press, 2010.

Tworek, Heidi. *News from Germany: The Competition to Control World Communications, 1900–1945*. Cambridge, MA: Harvard University Press, 2019.

US Department of State. *United States Statutes at Large*. Vol. 45, pt. 2. Washington, DC: Government Printing Office, 1929.

Van de Ven, Hans. *Breaking with the Past: The Maritime Customs Service and the Global Origins of Modernity in China*. New York: Columbia University Press, 2014.

———. *China at War: Triumph and Tragedy in the Emergence of the New China*. Cambridge, MA: Harvard University Press, 2018.

———. *From Friend to Comrade: The Founding of the Chinese Communist Party, 1920–1927*. Berkeley: University of California Press, 1992.

———. *War and Nationalism in China, 1925–1945*. London: Routledge, 2003.

Van Meer, Elizabeth. "The Transatlantic Pursuit of a World Engineering Federation: For the Profession, the Nation, and International Peace, 1918–48." *Technology and Culture* 53, no. 1 (2012): 120–45.

Wakeman, Frederick. *Policing Shanghai, 1927–1937*. Berkeley: University of California Press, 1995.

Wang Fulai 王福来. "Taizhou Di Yi Jia Guangbozhan" 台州第一家广播站. In *Taizhou Wenshi Ziliao Di 5 Ji* 台州文史资料 第5辑, edited by Wang Miaozeng 王妙增, 193–96. Zhejiang: n.p., 1998.

Wang Guoqing 王国庆. "Zhuixun WenGe de zongying zhi er" 追寻文革的踪影之二. In Sanmenxia shi weiyuanhui xuexi wenshi ziliao weiyuanhui, *Sanmenxia Wenshi Ziliao Di 17 Ji* (2007): 41–57.

Wang Helin 王鹤麟. "Yi Nanping guangbo shouyinzhan chengli hou de ji jianzhi" 忆南平广播收音站成立后的几件事. In Fujian Sheng Nanping Shi Weiyuanhui Wenshi Ziliao Weiyuanhui, *Nanping Wenshi Ziliao, Di 12 Ji*. 南平文史资料, 第12辑, 10–16. Fujian: n.p., 1991.

Wang Jiaqi 王嘉琦. "Kongbu shijie" 恐怖世界. In *Xuelei de huiyi: Lao yi bei yu dangdai shaonian tan "Jiu-Yi-Ba"* 血泪的回忆——老一辈与当代少年谈"九一八." Shenyang: Liaoning shaonian ertong chubanshe, 1991.

Wang Peijun 王培军 and Zhang Shanxiang. "Wusi Yundong zai Taizhou" 五四运动在台州. In *Taizhou Wenshi Ziliao, Di Yi Juan* 台州文史资料, 第1卷, edited by Wang Miaozeng 王妙增, 1–8. Zhejiang: n.p., 1998.

Wang Xijian 王希坚. "Huiyi Kang-Zhan shiqi Bei Jun 112 shi dixia dang de gongzuo" 回忆抗战时期东北军一一二师地下党的工作. In *Zhong-Gong Dongbei Jun dixia dang gongzuo huiyi* 中共东北军地下党工作回忆, edited by Zhong-Gong Dongbei Jun dang shi zu, 413–18. Beijing: Zhong-Gong Dang shi chubanshe, 1995.

Wang Xinghua 王星华, ed. *1958 nian de gushi* 1958 年的故事. Beijing: Zhongguo xiaonian ertong chubanshe, 2001.

Wang Xueqi 汪学起. *Di si zhanxian: Guomindang zhongyang guangbo diantai duoshi* 第四战线：国民党中央广播电台掇实. Beijing: Zhongguo Wen Shi Chubanshe, 1988.

Wang Ying 王莹. "Kang-Zhan shiqi de Yan'an Xinhua Guangbo Diantai" 抗战时期的延安新华广播电台 *Zhonghua Nuzi Xueyuan Xuebao* 中华女子学院学报, no. 5 (2015): 91–94.

Wang Yizhi 王一知. "Jieguan Changchun shi guangbo diantai de di yi ge huihe," 接管长春市广播电台的第一个回合. In *Chengshi jieguan qinli ji* 城市接管亲历记, 93–95. Beijing: Zhongguo Wen Shi Chubanshe, 1999.

Wang Yu 王雨. "Geming de fudiao: Boyinyuan yu shehuizhuiyi boyin fengge de zaidihua" 革命的复调：播音员与社会主义播音风格的在地化. In *Ting Xiandai Zhongguo*, edited by Tang Xiaobing, 58–69. Shanghai: Fudan Daxue Chubanshe.

——— [as Yu Wang]. "Listening to the Enemy: Radio Consumption and Technological Culture in Maoist China, 1949–1965." *Twentieth-Century China* 47, no. 2 (2022): 154–70.

——— [as Yu Wang]. "Listening to the State: Radio and the Technopolitics of Sound in Mao's China." PhD diss., University of Toronto, 2019.

———. "Nuxing de pinlu: Xingbie, guangbo yu 20 shiji 40 niandai zhongguo geming de tingjue wenhua" "女性的频率": 性别、广播与20世纪40年代中国革命的听觉文化. *Funu Yanjiu Luncong* 妇女研究论丛 166, no. 4 (July 2021): 5–15.

Wasserstrom, Jeffrey. *Global Shanghai, 1850–2010: A History in Fragments*. London: Routledge, 2008.

Watt, Lori. *When Empire Comes Home: Repatriation and Reintegration in Postwar Japan*. Cambridge, MA: Harvard University Press, 2009.

Wilbur, C. Martin. *The Nationalist Revolution in China, 1923–1928*. Cambridge: Cambridge University Press, 1984.

Willoughby, W. W. *Foreign Rights and Interests in China*. Vol. 2. Baltimore: Johns Hopkins University Press, 1927.

Wilson, Anne, Victoria Parker, and Matthew Feinberg. "Polarization in the Contemporary Political and Media Landscape." *Current Opinion in Behavioral Sciences* 34 (2020): 223–28.

Wrigley, Edward Anthony. *Energy and the English Industrial Revolution*. Cambridge: Cambridge University Press, 2010.

Wu, Angela Xiao. "Historicizing Internet Use in China and the Problem of the User Figure." *IEEE Annals of the History of Computing* 37, no. 4 (2015): 2–4.

Wu Daoyi 吳道一. *Zhongguang Sishi Nian* 中廣四十年. Taipei: Zhongguo Guangbo Gongsi, 1968.

Wu, Shellen Xiao. *Empires of Coal: Fueling China's Entry into the Modern World Order, 1860–1920*. Stanford, CA: Stanford University Press, 2015.

Wu, Shuling. "The Development of Poetry Helped by Ancient Postal Service in the Tang Dynasty." *Frontiers of Literary Studies in China* 4, no. 4 (2010): 553–77.

Xia, Yun. *Down with the Traitors: Justice and Nationalism in Wartime China*. Seattle: University of Washington Press, 2017.

Xibei Daxue Lishixi Zhongguo xiandai shi yanjiushi. *Xi'an shibian ziliao xuanji* 西安事变资料选辑. N.p. [Xi'an?]: n.p., 1978.

Xu, Chuan. "From Sonic Models to Sonic Hooligans: Magnetic Tape and the Unraveling of the Mao-Era Sound Regime, 1958–1983." *East Asian Science, Technology and Society* 13, no. 3 (2019): 391–412.

Xu Deheng 许德珩. "Huiyi wusi yundong" 回忆五四运动. In Quan guo zheng xie wenshi ziliao weiyuanhui, *Wusi Yundong Qin Li Ji* 五四运动亲历记. Beijing: Zhongguo Wen Shi Chubanshe, 1999.

Xu Ruizhang 徐瑞璋. "50 nian qian de hong se dianbo—yi Yan'an Xinhua Guangbo Diantai," 50年前的红色电波——忆延安新华广播电台 *Dang Shi Zong Heng* 党史纵横, no. 4 (1991): 14–16.

Xu Yi 徐毅 and Bas van Leeuwen. "19 Shiji Zhongguo Dazhong shizi lü de zai gusuan," 19 世纪中国大众识字率的再估算. In *Qing Shi Lun Cong: 2013 nian hao*

清史论丛 2013年号, 240–47. Beijing: Zhongguo Guangbo Dianshi Chubanshe, 2013.

Yan Boyuan 严博渊. "Wo dang le qi nian yeyu guangboyuan." 我当了七年业余广播员. In *Women renmin chang: Jiangxi "Xiao san xian" 9333 chang shilu(Xia)* 我们人民厂：江西"小三线"9333厂实录（下）, 730–32. Shanghai: Shanghai Renmin Chubanshe, 2015.

Yang, Daqing. *Technology of Empire: Telecommunications and Japanese Expansion in Asia*. Cambridge, MA: Harvard University Press, 2010.

Yang Jisheng. *Tombstone: The Great Chinese Famine, 1958–1962*. New York: Farrar, Straus and Giroux, 2008.

Ye Xin 叶辛. *Ye Xin Yanzhong de Shanghai* 叶辛眼中的上海. Beijing: Xinhang Chubanshe, 2018.

Ye Yonglie. "Zai Meiguo kan Jiang Jieshi Riji." *Tongzhou Gongjin*, no. 2 (2008): 46–48.

Yeh, Wen-hsin. *Shanghai Splendor: A Cultural History, 1843–1945*. Berkeley: University of California Press, 2007.

Yin Huazhen 殷华珍. "Jiyi zhong de Jun Xian renmin guangbo zhan" 记忆中的均县人民广播站. In Danjiangkou shi Zhengxie Wenshi Ziliao Weiyuanhui, *Danjiangkou Wenshi Ziliao, Di 9 Ji* 丹江口文史资料, 第9辑, 298–301. Danjiangkou: Published by committee, 2008.

Yoon, Wook. "Dashed Expectations: Limitations of the Telegraphic Service in the Late Qing." *Modern Asian Studies* 49, no. 3 (2015): 832–57.

Young, Louise. *Japan's Total Empire: Manchuria and the Culture of Wartime Imperialism*. Berkeley: University of California Press, 1998.

Yulin Shi zhi bianzuan weiyuanhui. *Yulin Shi Zhi* 榆林市志. Xi'an: Sanqing Chubanshe, 1996.

Yunnan Sheng Xundian Huizu Yizu Zizhui Xian Zhi Bianzuan Weiyuanhui. *Xundian Huizu Yizu Zizhi Xian Zhi* 寻甸回族彝族自治县志. Yunnan: Yunnan Renmin Chuban She, 1999.

Zhang Fan 张帆. "Deng Tuo furen Ding Yilan: Kai guo da dian guangboyuan" 邓拓夫人丁一岚 开国大典播音员. *Wen Shi*, no. 5 (2015): 28–29.

Zhang Jiping 张继平, ed. *Jinan xiaoqing he lishi wenhua congshu* 济南小清河历史文化丛书. Jinan: Jinan Chubanshe, 2008.

Zhang Shujun 张树军, ed. *Tu wen gongheguo nianlun 1949–1959* 图文共和国年轮 1949–1959. Shijiazhuang: Hebei Renmin Chubanshe, 2009.

Zhang Tinghao 张廷灏. "Zai Shanghai canjia wusi yundong de huiyi" 在上海参加五四运动的回忆. In *WuSi Yundong Qin Li Ji* 五四运动亲历记, edited by Quan guo zhengxie wenshi ziliao weiyuanhui, 174–77. Beijing: Zhongguo Wenshi Chubanshe, 1999.

Zhang Wenyang. "The Grammar of the Telegraph in the Late Qing: The Design and Application of Chinese Telegraphic Codebooks." *Journal of Modern Chinese History* 12, no. 2 (2018): 227–45.

Zhang Xiaohang 张小航. *Kangzhan Ba Nian Guangbo Ji* 抗战八年广播纪. Chongqing: Chongqing Chubanshe, 2015.

Zhang Xueliang 張學良 and Tang Degang 唐德剛, eds. *Zhang Xueliang Koushu Lishi* 張學良口述歷史. Taipei: Yuanliu Chuban Shiye, 2009.

Zhang Yan 张彦 and Huang Wenxuan 黄文轩. "Kangzhan shouyin yiwen" 抗战收音轶闻. *Hongyan Chunqiu* 红岩春秋 [*Hongyan Chronicle: Journal of the Chongqing Municipal Party Historical Research Institute*], Zhonggong Chongqing Shiwei Dangshi Yanjiushi 1 (2006): 39.

Zhang Yaojun 张姚俊. "Sun Zhongshan de heping tongyi xuanyan jiujing heshi shouci gongbu?" 孙中山的《和平统一宣言》究竟何时首次公布. Shanghai dang'an xinxi wang [Shanghai Archive News Web].

Zhao Hanzhou 赵汉洲. "Yiyuan Xian Shouyinzhan" 沂源县收音站. In Zhengxie Yiyuan Xian Weiyuanhui, *Yiyuan Xian Wenshi Ziliao, Di Liu Ji* 沂源县文史资料, 第6辑, 254–57. Shandong: n.p., 1997.

Zhao Jingshen 赵景深. "You guan wusi de yi dian huiyi" 有关五四的一点回忆. In *Wentan Huiyi* 文坛回忆, 12–14. Chongqing: Chongqing Chubanshe, 1988.

Zhao Yuming 赵玉明, ed. *Riben Qin Hua Guangbo Shiliao Xuanbian* 日本侵华广播史料选编. Beijing; Zhongguo Guangbo Dianshi Chubanshe, 2015.

———, ed. *Xinxiu Difang Zhi Zaoqi Guangbo Shiliao Huibian* 新修地方志早期广播史料汇编. 2 vols. Beijing: Zhongguo Guangbo Yingshi Chubanshe, 2016.

———. *Zhongguo Xiandai Guangbo Jianshi* 中国现代广播简史 [A short history of China's modern broadcasting]. Beijing: Zhongguo Guangbo Dianshi Chubanshe, 1987.

Zhengzhi Bu Wenhua Gongzuo Weiyuanhui. "Riben guanbing dui wo fang dui di xuanchuan yingxiang zuotanhui" 日本官兵對我方對敵宣傳影響座談會. *Diqing Cankao Ziliao* 敵情參考資料, no. 27 (c. 1944): 1–18.

Zhong-Gong Sichuan sheng wei dang shi gongzuo weiyuanhui. *Wusi Yundong zai Sichuan* 五四运动在四川. Chengdu: Sichuan Daxue Chubanshe, 1989.

Zhou Gucheng 周谷城. "Wusi Yundong yu Qingnian Xuesheng" 五四运动与青年学生. In Zhongguo Shehui Kexue Yuan Jindai Yanjiusuo, *Wusi Yundong Huiyilu (Xia)* 五四运动回忆录（下）. Beijing: Zhongguo Shehui Kexue Chubanshe, 1989.

Zhou Shizhao 周世钊. "Xiang Jiang de nuhou" 湘江的怒吼. In Quan guo zheng xie wenshi ziliao weiyuanhui, *Wusi Yundong Qin Li Ji* 五四运动亲历记 (Beijing: Zhongguo Wen Shi Chubanshe, 1999).

Zhou Zuguo 周祖国. "Xichang WenGe qin wen qin li" 西昌文革亲闻亲历. In Xichang Wenshi Bianshen weiyuanhui, *Xichang Wenshi Di 16 Ji* 西昌文史 第16辑 (2001): 96–105.

Index

Page numbers in *italics* refer to figures.

acoustic history, 20–21
Adorno, Theodor, 192, 194
agriculture: collectivization, 217, 218, 225, 230, 231–32; fertilizer, 229–30, 235; model village, 229; production increases, 229–30; radio programs, 160–61, 217, 219–20, 229, 235–36; water conservation campaign, 228–29, 235. See also rural areas
American Globe Wireless Company, 134, 150
Anderson, Benedict, 6, 19, 20, 27, 264
Anhui Province: amateur radio builders, 96; Nanling County, 94; rural radio listeners, 94; Shou County, 128–29; youth propaganda squadrons, 128–29

BBC, 136, 200
Beard, Mary, 171

Beijing: clandestine radio listening, 186; Cultural Revolution in, 248–49, 250, 255–56, 319n25; Japanese conquest and occupation, 113–14, 115, 126, 139, 145–46, 159; number of radio receivers, 76, 146, 147; population, 301n117; radio equipment manufacturing, 258–59; radio popularization campaign, 145–46; sparrow extermination campaign, 1–3; telegraph office, 25, 29–30; wired broadcasting network, 1–2; Yenching University, 166–68. See also Beijing radio stations; Peking University
Beijing Central Broadcasting, 251, 252
Beijing Jiaotong Daxue (Beijing Communications University), 66, 286n15
Beijing Normal University, 247

341

Beijing radio stations: Beijing Central Broadcasting, 251, 252; of CCP, 188, 203, 304n39; Japanese-controlled, 113–14, 140, 141–42; loudspeakers in public spaces, 1–2, 72, 73; municipal, 1; Nationalist, 106; New China Station, 188; of puppet government, 180, 188; under Zhang regime, 69, 71, 72–73, 75, 78, 82
Beiping New China Broadcasting Station, 188
Beiyang University, speech troupes, 39–40
Benjamin, Walter, 192
big-character posters: during civil war, 187; in Cultural Revolution, 241, 244, 247–48, 249, 250, 255, 256
Boxer Rebellion, 7, 51, 65
Britain: officials in Shanghai, 116–17; radio equipment, 70, 119; radio regulations, 74; relations with Nationalist Party, 45–46, 280n7, 281n14; spies, 61, 67; wartime morale, 136; Zhang Zuolin and, 67–68. See also Hong Kong
broadcasters (guangboyuan): influence, 196; in 1960s, 251; political messages, 230, 231–32, 234–36; tasks, 220–21, 225–26, 228, 233–34; training, 223; women, 223, 226. See also radio announcers
broadcasting, see radio; wired broadcasting networks
broadcasting posts (guangbozhan): equipment, 223, 239, 252; installation and maintenance, 222, 232–33, 268; listening meetings, 227–29, 234; local contributors, 221, 226–27, 230, 231–32, 233–34; number of, 222; operations, 223; political influence, 230–31, 252–53; programming, 223–26, 230, 233, 234–36, 251; public demand for, 222, 223–24, 233; seized in Cultural Revolution, 252–55, 256; technology, 220; terminology, 314–15n103; weather forecasts, 224, 234; in workplaces, 239, 243, 246, 251–55, 268

Cairo Conference, 142–43
Canton: radio listening in, 90; radio station, 78; refugees from, 124–25
Cao Rulin, 29, 32
Cao Zhongyuan, 58–59, 284n63, 285n70
capitalism, 15, 19, 235
Carey, James, 10
cassette tapes, music recordings, 258
CCP, see Chinese Communist Party
Central Broadcasting Administration (Chongqing), 122
Central Broadcasting Authority (Nanjing), 106
Central Broadcasting Industry Association, 121
Central Broadcasting Station, see China Central Broadcasting
Central Intelligence Agency (CIA), Foreign Broadcast Information Service, 246
Changchun radio station, 139, 181–82
Changsha: number of radio receivers, 123; radio station, 115, 118, 135
Chen Guofu, 92
Chengdu: newspapers, 25, 28–29, 31–33; radio monitor training, 102; radio station, 121
Chiang Kai-shek (Jiang Jieshi): Cairo Conference, 142–43; Japanese propaganda and, 149; Nationalist

troops of, 187; New Life Movement, 99; radio and, 84, 86, 92, 97; at war's end, 154; Xi'an incident, 106–9. See also Nationalist government; Nationalist Party

children: denunciations of parents, 211; Japanese radio popularization campaign and, 138; radio newspapers delivered by, 215; radio programming for, 204. See also schools

China Central Broadcasting (Chongqing): equipment, 119, 121; establishment, 119; female announcers, 170; information on war's end, 152–53, 179; Japanese-language propaganda, 123–24; listeners in Japanese-occupied areas, 110–11, 113, 123–25, 126–29, 133–34, 135–38, 160; news transcriptions, 119, 124–27, 160, 163; overseas listeners, 124; radio-listening network, 120–23; slowness of news reporting, 142–44

China Central Broadcasting (Nanjing): archives, 21; Chiang Kai-shek's speech on anti-Japanese resistance, 84; directors, 92, 97, 118, 172; educational programming, 120, 122; establishment, 92, 97; female announcers, 172; Japanese conquest and, 118; under Japanese control, 142; listeners in Japanese-occupied areas, 115; news transcribed by radio monitors, 97–98, 108, 114; propaganda, 92, 118

China Central Broadcasting (Nanjing; Japanese station), 139

China Maritime Customs Service, 51, 52–54, 74, 86

China News Agency, 152

China Press, 46–47, 48, 50, 54, 73–74, 75, 280n10

Chinese Communist Party (CCP): alliance with Nationalists, 47, 110; anti-Japanese resistance, 157, 158; antirightist campaign, 230–31, 234, 247; early history, 18, 44; founding of PRC, 140, 189, 190; gamification techniques, 227; Hundred Flowers Campaign, 230, 234, 242; internal enemies, 209–11, 217–18, 221; Jinchaji base area, 166, 168–69, 170, 179; Leninist organization, 4, 18, 43, 44; lower cadres, 236–37; sparrow extermination campaign, 1–5. See also Communist newsscape; Communist propaganda; socialized media newsscape

Chinese language: character encoding for Morse code, 30–31, 65–66, 246; dialects and accents, 75, 120, 161–62, 173–74, 233

Chinese opera, 72, 74, 77, 80, 89, 141

Chongqing: Japanese air raids, 119; Office of War Information branch, 130; radio meetings, 206, 208–9, 312nn70–71; topography, 119; wired broadcasting network, 204, 310–11n47. See also China Central Broadcasting (Chongqing); Nationalist government (Chongqing)

Chuan Bao, 31, 32–33

Churchill, Winston, 130, 142–43

civil war period (1945–49): CCP propaganda, 157, 185–89; CCP victory, 189; demand for news, 185, 186; PLA victories, 185, 187, 188; role of radio, 185–87, 188–89; war criminals lists, 187. See also Communist newsscape; postwar newsscape collaborationist governments: infrastructure development, 19, 111–12; propaganda, 150–51; radio stations, 112, 142, 144, 145, 146–47, 148–52, 153, 154, 180; restrictions on radio listening, 133, 138–39; at war's end, 179, 180

communal listening culture: in 1930s, 87; in 1950s, 191, 194; promoted by Nationalist Party, 92–93; in rural areas, 87, 93–97, 99–101, 102–3, 105–6, 120–21, 127, 215; in urban areas, 87, 89–91, 92–93, 105, 120–21; in wartime, 120–23, 127. See also loudspeakers; radio meetings; radio receivers

communication: evolution of meaning, 29; Sun Yat-sen on, 44–45, 195, 252; use of term, 10–11. See also newspapers; newsscapes

communication ecosystem, 9, 10

Communications Ministry: ban on imported radio equipment, 54, 58; Beijing radio station, 72; radio engineering training, 56, 66; responsibilities in 1919, 29; Zhang regime and, 71. See also Cao Rulin

communications technology: intermediality, 13–14, 27–28, 87, 220, 245–46; mass politics and, 4–5, 6–8, 248–52; in 1919, 25–26, 29–35; nationalism and, 19–20; telephone, 25–26, 220. See also mass media; radio; technopolitical process; telegraphy; wired broadcasting networks

Communist newsscape (1937–49): big-character posters, 187; clandestine radio listening, 186–87; infiltration of GMD newspapers, 156, 163, 187–88; Japanese radio engineers, 182, 183–85; Japanese radio infrastructure used, 156, 185; Japanese radio infrastructure used by, 157; materiality of news, 165–69; newspapers, 156, 157, 159–60, 163–65, 187–88, 303n5; news posted on walls and blackboards, 159, 160; radio monitors, 156, 160–61; radio newspapers, 156, 157, 158–59, 160–63, 164–66, 169, 186–87; radio stations, 156, 168–69, 180–85; role of women, 155–57; at war's end, 179–80. See also civil war period; Communist propaganda; Jinchaji Daily (Ribao); Yan'an radio station

Communist propaganda: in civil war period, 157, 185–89; Four Pests campaign, 2–5; images of communications technology, 193, 197, 201; on Japanese radio, 149; in 1950s, 195, 204–13, 217, 220, 235, 236–37; organizational structure, 157, 195, 308n11; in wartime, 149, 157, 158–64, 168–69

coproduction, 8

Crow, Carl, 34

Cultural Revolution (1966–76): aims, 240–41; beginning, 240–41, 242–48; big-character posters, 241, 244, 247–48, 249, 250, 255, 256; as communications event, 241, 255–57; denunciations, 247, 248, 249–50, 254; factions, 241, 248, 249, 250, 251, 253–55, 256–57; manifesto posted at Peking University, 242–44, 245; Red Guards, 241, 250, 251, 253–54, 255–56; roles of broadcasting, 238, 241, 245–48, 249, 250, 252–57; in rural areas, 248–50, 254–55; seizure of means of communication, 244–45,

252–55, 256; Sixteen Points Manifesto, 250; social disintegration, 241; technopolitical process, 255–57, 268; victims, 241, 247, 248, 249–50
Customs Service, see China Maritime Customs Service

Dalian: Japanese radio engineers, 182, 183–85; number of radio receivers, 185; radio audiences, 75; radio station, 68, 75, 138, 140, 182–86
Deng Tuo, 170, 177, 181, 304n39
Deng Xiaoping, 319n25
Ding Yilan: broadcasting career, 188–89, 190, 265, 304n39, 306n73; education, 155, 171; escape from Tianjin, 155; marriage and family, 170, 188; in Nanjing, 158; newspaper stories, 170; at war's end, 176, 177–79; in Zhangjiakou, 181
disintermediation, 231, 240, 252

East Asia Broadcasting Council, 140–41
East China Radio Station, 197–98
East China Shanghai Resist America Support Korea Association, 205
ecosystems, see communication ecosystem
education: radio programs, 90, 91, 120, 122; of women, 40, 171, 172–73, 174–75. See also illiteracy; People's Education Centers; schools
Eighth Route Army, 180–81
engineers, see radio engineers
Europe: radio equipment exports, 88, 202; women's political roles, 171. See also Britain; France; Germany; imperialist powers

Fang Ziwei (George T. W. Fang), 55–57, 58, 265
fascism, mass media and, 192
Fengtian (Mukden): Japanese control, 80–82, 106, 116, 117, 139; radio stations, 69–70, 71, 75, 77, 78, 79–80, 81–82, 106, 139; US consulate, 309n19; World Radio Receiving Station, 61–62
Foucault, Michel, 6
Four Olds, 250–51
Four Pests campaign, 1–5
France: ancien régime newsscape, 28; print technology and revolution, 8; radio broadcasting and nationalism, 19; unrest in 1960s, 257
Fudan University students, 40, 41, 128, 186
Fujian Province: Japanese propaganda, 143; radio monitors, 216–17

Gender: language and, 173–74; traditional roles, 170–71. See also women
Germany: Nazi propaganda, 149; Nazi radio, 145, 192; news media, 13; radio equipment, 70, 76, 77, 81; unrest in 1960s, 257; wire news services, 62. See also Telefunken
GMD, see Nationalist Party
gramophones, 41
Greater East Asia Coprosperity Sphere, 131, 148, 149, 151–52
Great Leap Forward, 194, 225–29, 232–36, 244
guangboyuan, see broadcasters
guangbozhan, see broadcasting posts
Guangdong Province: clandestine radio listening, 200; rural radio listeners, 94–95. See also Canton

Guangxi Province, radio-listening stations, 99, 219
Guizhou Province: Jun County People's Broadcasting Station, 223; wired broadcasting network, 245
Guomindang, see Nationalist Party
Guomin Gongbao, 31–32, 33

Habermas, Jürgen, 15, 27, 264
Hangzhou: clandestine radio listening, 200; radio station, 78, 139; reactions to Shanghai Incident, 105
Harbin: clandestine radio listening, 133, 200; radio station, 69–70, 75, 78, 139, 286n15; telegraph office, 66
Hebei Province: absence of information in wartime, 259–60; Ba County, 96–97; CCP radio station (Hebei People's Station), 181, 186; Laiyuan County, 101; radio listeners in wartime, 114–15, 259–60; radio-listening stations, 101; radio newspapers, 159, 160–61
Henan Province: communal radio listening in streets, 107–8; Cultural Revolution, 253; Fangchang County, 162–63; radio monitors, 164; radio newspapers, 162–63, 165; Sanmenxia, 253; underground newspaper, 164; wired broadcasting network, 253; Wuyang County, 164; Zhengzhou, 107–8
Hengyang, 123, 143
Hitler, Adolf, 192
Hong Kong: Japanese occupation, 135; political cartoons, 46; radio equipment dealers, 202; radio station, 282n21; Sun Yat-sen in, 281n14; wartime radio reception, 135; wired broadcasting network, 315n107

Hubei People's Broadcasting Station, 223
Hubei Province: Enshi, 103; Huanggang County, 227–29; radio equipment shortages, 122; radio-listening stations, 103; radios in People's Education Centers, 100; wired broadcasting network, 222, 227–29, 230–36; Wufeng County, 230–32; Xishui County, 232–36. See also Wuhan
Hunan Province: Lingling County, 3; newspaper readership, 35; radio monitors, 313n76; sparrow extermination campaign, 3
Hundred Flowers Campaign, 230, 234, 242

illiteracy: in China, 6, 12–13, 14, 16, 34; of local news reporters, 159; of radio listeners, 12–13, 90–91; of women, 278n34
imperialist powers: Boxer Rebellion and, 7, 51, 65; Japanese propaganda messages, 148; pressures on China, 7, 18. See also Britain; China Maritime Customs Service; France; Germany; semicolonialism
intermediality, 13–14, 27–28, 87, 220, 245–46
internet, 259, 261, 265–67

Japan: control of Shandong, 26, 37–40; mass media, 7, 16, 273n43; number of radio receivers, 123, 139, 144; radio equipment exports, 76, 88, 111, 112, 144, 166, 291n9; radio equipment manufacturing, 112, 145, 147; radio stations, 75, 82, 134; relations with Manchuria, 289n76; repatriation of

citizens from Manchuria, 184;
Russo-Japanese War, 64, 144, 182;
Shanghai Incident (1932), 104–6;
Sino-Japanese War (1895), 7;
surrender (1945), 152–54, 176–79;
Twenty-One Demands, 37, 38, 62;
US Armed Forces Radio, 198, 200;
wartime newsscape, 144–45. See
also Japanese propaganda; Manchuria; Sino-Japanese War (1937–45)
Japan Broadcasting Corporation, see NHK
Japanese Communist Party, 184
Japanese propaganda: anti-Soviet, 149, 301n130; ideological messages, 148; influence, 111, 142–43, 149–52; in Manchuria, 82, 138; misinformation, 143, 148–50, 151–52; in print media, 113, 149; radio popularization campaign, 138, 144, 145–46, 147
Jiangsu Province: radio listeners, 162; radio meetings, 205, 206–7; radio monitors, 217; radio newspapers, 160–61
Jiangxi Province, radio-listening stations, 100
Jiaotong, evolution of meaning, 29
Jinan: newspaper readership, 34–35; number of radio receivers, 147; radio stations, 2, 140
Jinchaji base area, 166, 168–69, 170, 179
Jinchaji Daily (Ribao), 166, 169, 170, 176–77, 181
Joffe, Adolph, 47

Kaifeng, broadcasting post, 254–55
Kellogg Switchboard & Supply Company, 55, 58–59, 73, 78
Korean War: radio meetings during, 206–13; radio monitors, 156, 216–17

Kunming newspapers, 187
Kwantung Army, 64, 81, 82, 113

land reform, 160, 213, 219
League of Nations, 104
Leninism, 4, 18, 43, 44, 47, 69
letters, 32–33, 35–36. See also postal service
Li Yuanhong, 50–51, 282n29
libraries, radios in, 90–91
Lindsay, Hsiao Li, 166, 167–68, 169, 171, 178
Lindsay, Michael, 166–69, 170, 178, 304n34
literacy, see illiteracy
lithography, 161, 165
Liu Han, 65–67, 70, 74, 80, 286n15
Liu Shaoqi, 319n25
loudspeakers: domestic manufacturing, 203, 310n43; in households, 239–40; in Nanjing decade, 87, 89–91, 92–93, 99; in rural areas, 99, 102–3, 215; in schools, 204, 205, 232, 238–40, 245; in urban areas, 17, 72, 73, 84, 90, 92–93, 104, 105, 212; in workplaces, 204, 232, 239–40, 243, 246. See also broadcasting posts; communal listening culture; wired broadcasting networks
Lyall, Leonard Arthur, 52–54

mail service, see letters; postal service
Manchuria: immigrants, 64; industrialization, 64, 67, 112; Japanese rule, 80–82, 106, 112, 133, 138, 153–54; modernization, 63–64, 67; number of radio receivers, 139, 154; railways, 64–65; Russo-Chinese War, 79, 80; Soviet troops in, 179, 181, 182; Zhang regime, 64–65, 67–68, 70–71, 82–83.

Manchuria (cont.)
  See also Fengtian; Harbin; Manchurian newsscape
Manchurian Incident, 80–82
Manchurian newsscape: foreign influences, 62; propaganda, 79–80, 138; state-building project and, 63; technopolitics, 61–63, 70–71, 76, 82–83
Manchurian newsscape, radio: Japanese control, 81–83, 138, 139–40, 151; postwar stations, 181–85; programming, 77, 80; regulations, 71, 74, 78–79, 288n52; World Radio Receiving Station, 61–62; under Zhang regime, 65, 67–68, 69–71, 74–77, 78–80
Manchurian Telegraph and Telephone Company (MTTC), 138, 139–40, 141, 142, 170
Mao Zedong: death, 238–39; founding of PRC, 140, 189, 190; Japanese propaganda and, 149; May Fourth information and, 35; mobilization goals, 221; orders in civil war, 188; at Peking University, 28; radio broadcasts, 140, 190, 245; Red Guard rally, 250; speeches read on radio, 189, 250; at war's end, 154, 176
Maoist period, see Cultural Revolution; Great Leap Forward; socialized media newsscape
Marco Polo Bridge Incident, 84, 101, 118
Maritime Customs Service, see China Maritime Customs Service
Marx, Karl, 262
mass media: emergence, 6–7, 15–16, 240; in Japan, 7, 16, 273n43; magazines, 16, 273n43; mass politics and, 7–8, 240; in socialized newsscape, 194. See also communications technology; newspapers; radio
mass politics: communications technology and, 4–5, 6–8, 248–52; in 1950s, 193, 195; radio meetings, 205–13; revolutionary parties, 44; technopolitical process and, 5–8, 9, 44, 195, 240, 257, 267; in West, 7
mass society: communal listening and, 127; development in Japanese-occupied China, 112, 154; disintermediation, 231, 240, 252; emergence, 4, 6–7, 109, 261–64; nationalism and, 11; role of state, 262, 263–64; technological construction, 4–5, 11, 15, 19, 63, 88; values, 264; in West, 6–7; wired broadcasting network and, 221, 236–37
materiality of news, 11, 165–69
May Fourth Movement: communications technology and, 25–26, 29–35; parades, 36, 37; reactions to news of, 35, 36–40; student protests, 25–26, 28–29, 31–32. See also newsscape of 1919
Maze, Frederick, 52
McLuhan, Marshall, 19, 27, 87–88
Media, use of term, 9. See also intermediality; mass media; newsscapes
Meng Yiqi, 84–85, 109, 128–29
mimeographed newspapers and fliers, 85, 103, 120, 158–60, 165–66, 214, 215, 234
mimeograph machines, 98, 108, 109, 126, 165
model workers, 221, 226, 228
MTTC, see Manchurian Telegraph and Telephone Company
Mukden, see Fengtian

Nanjing: clandestine radio listening, 200; Japanese conquest and occupation, 118, 139; loudspeakers in public spaces, 92–93; radio equipment manufacturing, 203; radio meetings, 205; radio stations, 75, 139. See also China Central Broadcasting (Nanjing)

Nanjing decade newsscape (1927–37): communal listening culture, 87; demand for news, 93–97, 100–101, 105, 107–8; female radio announcers, 170, 172–73; loudspeakers in public spaces, 87, 89–91, 92–93, 99; nationwide radio network, 85–86, 92–93; newspapers, 97–98, 99, 103; news posted on walls and blackboards, 85, 87, 101, 103; radio-listening stations, 87, 97–102, 103–4, 108, 158; radio monitors, 87, 97–99, 101–2, 108, 158; radio-news fliers, 85–86, 103, 158, 268; role of radio, 84–87, 109; in rural areas, 84–86, 87–89, 93–104, 105–6; technopolitics, 87–88, 104–9; in urban areas, 84, 88, 89–91, 92–93, 104, 105

Nanjing government, see Nationalist government (Nanjing)

nationalism: anticolonial movements, 44; communications technology and, 19–20; effects of May Fourth Movement, 26; emergence, 11, 262; in wartime, 18

Nationalist government (Chongqing): air raid shelters, 119; censorship, 163, 164, 185; Central Broadcasting Administration, 122; Central News Agency, 180, 188; effects of war on institutions, 18–19; import restrictions, 186; newspapers, 162–63, 168; postwar radio infrastructure and, 179–80; use of radio, 110–11, 113. See also China Central Broadcasting (Chongqing); Sino-Japanese War; wartime newsscape

Nationalist Government (Nanjing): Central Broadcasting Authority, 106; Ministry of Education, 99–100, 120; nationwide radio network, 85–86, 92–93, 103–4, 109; propaganda, 118; radio regulations, 78–79; retreat, 118; retreat to Chongqing, 119. See also Nanjing decade newsscape; Nationalist Party

Nationalist government (Taipei), radio broadcasts, 198, 199

Nationalist Party (Guomindang): alliance with CCP, 47, 110; early history, 18, 43, 44; Leninist organization, 18, 43, 44, 47, 69; newspapers, 46, 162–63, 187–88; relations with British, 45–46, 280n7, 281n14; Soviet Union and, 44, 45–46, 47, 182, 183, 281n13; use of radio, 92–93, 97–98, 158, 198. See also Chiang Kai-shek; civil war period; Nanjing decade newsscape; Sun Yat-sen

Nationalist propaganda: broadcast from Taiwan, 198; of Chongqing government, 110–11, 113, 119, 123, 124, 128–29, 137, 179; of Nanjing government, 92, 97–98, 118

National Radio Administration Ltd., 54–55

Nazi Germany: propaganda, 149; radio, 145, 192

New Life Movement, 99

news: Chinese terms, 13; history and use of term, 13–14; materiality, 11, 165–69; oral networks, 27, 28, 36–42, 177. See also newsscapes news, demand for: in civil war period, 185, 186; emergence of mass society and, 261–64; as human need, 259–61; in Nanjing decade, 93–97, 100–101, 105, 107–8; in socialized media newsscape, 196, 199–200; technopolitical process and, 6, 259, 262–64; tradeoffs, 266–67; in wartime, 19, 110, 114–15, 122, 124, 126–29, 135–37, 144, 148, 259–60

news media, see mass media; newspapers; newsscapes; radio

newspapers: broadcasting of contents, 245–47; as businesses, 15; circulation and readership, 14, 15–17, 33–35, 37, 98, 245–46, 278n32; coexistence with radio, 270n14; in Communist newsscape, 156, 157, 159–60, 163–65, 187–88, 303n5; history in China, 15–17, 273n41; local, 159–60; mass politics and, 6; postal distribution, 33–34, 213; radio transcriptions published by, 98, 99, 103; rural distribution problems, 16–17; secondary market, 16, 34, 37, 278n32; in Shanghai, 16, 46–47, 114, 150–51, 152, 199; in socialized media newsscape, 199, 213. See also Jinchaji Daily (Ribao); mass media; People's Daily; radio newspapers

newsscape analysis, 11–14, 264, 267

newsscape of 1919: fliers, 36–37; letters, 32–33, 35–36; natural landscape and, 42; newspapers, 26–27, 31–35, 42, 277n15; oral and auditory forms, 35, 36–42; social contexts, 41–42; speech troupes, 35, 38–40; technological infrastructure, 25–27, 29–35; telegraphy, 25–26, 28–32, 41–42. See also May Fourth Movement

newsscape of 1920s: newspapers, 46–47; radio, 45–47, 72–73, 74–77, 82–83; Sun Yat-sen and, 44–48, 73. See also Manchurian newsscape; Nanjing decade newsscape

newsscapes: acoustic history and, 20–21; authoritarian, 192–94; definition and related terms, 8–11; intermediality, 13–14, 27–28, 41–42; in post-Maoist period, 258–59, 265–66; social contexts, 26; space and, 11–12, 28; use of heuristic, 5, 9, 11–14, 267. See also Communist newsscape; Cultural Revolution; Manchurian newsscape; Nanjing decade newsscape; socialized media newsscape; wartime newsscape

NHK (Nippon Hōsō Kyōkai: Japan Broadcasting Corporation), 139, 140, 142, 144

Nie Yuanze, 242–44, 245, 248

Nippon Hōsō Kyōkai, see NHK

North China Broadcasting Association (Huabei guangbo xiehui), 139, 141, 146

Northeast China, radio stations, 74–76. See also Beijing; Manchuria; Shandong Province; Tianjin

Northeastern Army, 158

Nurhaci, 61, 67

Office of War Information (OWI), 122–23, 130, 134, 135, 141, 143

Opium Wars, 48

oral news networks, 27, 28, 36–42, 177

Osborn, Ernest George Haywood, 48, 49–51, 54–55, 280n10, 282n21

OWI, see Office of War Information

Paris Peace Conference, 26, 38
Peking University: Cultural Revolution work teams, 248–49; manifesto on cafeteria wall, 241–44, 248; students, 28, 133–34
People's Daily, 2–3, 245–48, 249, 304n39
People's Education Centers, 99–100, 102, 120
People's Liberation Army (PLA), 180–81, 185, 187, 188
People's Republic of China (PRC): founding, 140, 189, 190; post-Maoist period, 258–59, 265–66. See also Chinese Communist Party; Cultural Revolution; socialized media newsscape
Pingyuan Province, radio-listening stations, 216
PLA, see People's Liberation Army
postal services, 8, 33, 35–36, 213
posters, see big-character posters
post-Maoist period: newsscapes, 258–59, 265–66; technopolitical process, 258–59
postwar newsscape: clandestine radio listening, 185; competition for control of radio infrastructure, 179–85; number of radio receivers, 190–91. See also civil war period; Communist newsscape
PRC, see People's Republic of China
print capitalism, 19, 272n40
print technology, 6, 8, 15–16. See also newspapers
propaganda: American, 130; Chinese term, 13; in entertainment programming, 90, 230; German, 149; Soviet, 79; student teams, 109; telegraphed circulars, 17; during Xi'an incident, 107; youth squadrons in wartime, 128–29. See also Communist propaganda; Japanese propaganda; Nationalist propaganda
public sphere, 263–64

Qingdao: clandestine radio listening, 200; refugees from, 158
Qujiang, number of radio receivers, 123

radio: administrative use, 13, 14–15, 195–96, 213, 265; advantages over other media, 17, 42, 55–56, 104–5, 109, 308n11; audiences, 19, 47, 49–50, 59, 75, 77; early history in China, 14–15, 17–18, 19–21, 43–44, 45, 48–55, 86, 267–68; educational programming, 90, 91, 120, 122; military use, 51–52, 53, 67, 71; as national unifier, 12, 19–20; public interest in, 59, 75, 76–77; reception quality and state power, 241, 318n6; research, 56, 57; scholarship on, 275n65; utopian images, 265. See also radio equipment; radio receivers
Radio (Wuxiandian) magazine, 193, 197, 239, 265
radio announcers, dialects and accents, 120, 161–62, 173–74, 233. See also women radio announcers
Radio Corporation of China, 48, 49, 54. See also Osborn, Ernest George Haywood
radio engineers: American, 72; Chinese, 54, 55–57, 58, 65–67, 72, 76, 265, 283n37; Japanese, 115, 157, 182, 183–85; training, 66, 77, 286n15
radio equipment: antennas, 96, 215; demand for, 105; domestic manufacturing, 59, 120, 121, 202–3, 258–59,

radio equipment (*cont.*) 310n43; maintenance training, 164–65; of radio monitors, 165–67; repairing, 133, 144, 146–47, 199, 309n21; Robertson's demonstrations, 59, 68, 282n25; wartime shortages and black market, 121–23, 132. See also loudspeakers; radio receivers
radio equipment, amateur builders: Boy Scouts, 58; costs, 88; crystal receivers, 88–89, 95–96; illegally imported parts, 55; number of receivers, 285n70; published instructions, 57; quality of radios, 88–89, 284n54; restrictions under communist rule, 202; transmitters, 52; in wartime, 132–33, 169
radio equipment, imported: American, 49, 57, 59, 70, 76, 81, 88, 202; costs, 88; Eastern European, 202; Japanese, 76, 88, 111, 112, 166, 181; Japanese receivers, 111, 138, 144, 145, 146, 147, 291n9, 301n124; restrictions, 51, 52–54, 55, 58–59, 186; smugglers, 59, 166–69; Soviet, 174, 203, 304n40; Western European, 70, 76, 77, 81, 88, 119
radio-listening stations (shouyinzhan): equipment, 214, 215; in Nanjing decade, 87, 97–102, 103–4, 108, 158; in 1950s, 213–20; in wartime, 120–23, 127, 213–14, 313n76. See also radio monitors
radio meetings: in 1950s, 205–13, 217, 219, 312nn70–71; over wired broadcasting network, 227–29
radio-monitoring network (guangbo shouyinwang), 213–20
radio monitors (shouyinyuan): in Communist newsscape, 156, 160–61, 164; equipment, 165–67; during Korean War, 156, 216–17; linguistic challenges, 161–62; in Nanjing decade, 87, 97–99, 101–2, 108, 158; in 1950s, 196, 213–20; in 1960s, 246; succeeded by wired broadcasting network, 220–22; training, 102, 164–65, 175, 214; travel in countryside, 216, 217–18; in wartime, 120, 121, 164–66, 313n76; women, 98, 156, 175, 214–15, 313n79. See also radio-listening stations
radio-news fliers, 85–86, 103, 158, 160, 268
radio newspapers: in Communist newsscape, 156, 157, 158–59, 160–63, 164–66, 169, 186–87; in 1960s, 246; in socialized media newsscape, 2, 213–14, 215–16; in wartime, 20, 124–26, 160–63, 165, 268
radio news transcriptions: in Nanjing decade, 98, 99, 103; in wartime, 119, 120, 128–29, 143, 164–65. See also radio monitors; radio newspapers
radio receivers: costs, 86, 88, 103, 132, 166, 258–59, 291n7, 291nn9–10; crystal, 88–89, 95–96, 103, 291n10; Japanese, 111, 138, 144, 145, 146, 147, 291n9, 301n124; spread of ownership, 86, 93–97; transistor radios, 240, 258; vacuum tubes, 88, 95, 166, 167, 202. See also radio equipment
radio receivers, number of: by amateur builders, 285n70; in Beijing, 76, 146, 147; in Changsha, 123; in Dalian, 185; growth in wartime, 112, 114, 144, 147; in Japan, 123, 139, 144; in Jinan and Taiyuan, 147; in Manchuria, 139, 154; in postwar newsscape, 190–91; in Shanghai area, 59, 198, 285n70; in Tianjin, 75, 76

radio regulations: of Beijing government, 51, 52–54, 57–59, 71; British, 74; call signs, 78; in Japanese-occupied areas, 115; of Manchurian regime, 71, 74, 78–79, 288n52; of Nanjing government, 78–79; in Shanghai, 90; unification, 78–79
Radio Tokyo, 82, 134
RCA, 81, 202
receivers, see radio receivers
Red Guards, 241, 250, 251, 253–54, 255–56. See also Cultural Revolution
Reform and Opening period, 185, 258
Reformed Government of Wang Jingwei, 142, 146
Resist America and Aid Korea campaign, 207–13, 216
Resistance Daily, 162–63
Robertson, C. H., 49–50, 59, 68, 282n25
rural areas: communal listening culture, 87, 93–97, 99–101, 102–3, 105–6, 120–21, 127, 215; cooperatives, 217, 220; Cultural Revolution, 248–50, 254–55; impact of radio and wired broadcast networks, 103–4, 109, 216–20, 223–25, 227–36; land reform, 160, 213, 219; local news, 194, 195, 221, 226–27, 233–34, 316n128; loudspeakers in public spaces, 99, 102–3, 215; Nanjing decade newsscape, 84–86, 87–89, 93–104, 105–6; oral news networks, 38–41; People's Education Centers, 99–100, 102, 120; radio-listening stations, 97–102, 108, 213–20; radio newspapers in Japanese-occupied areas, 124–25; wired broadcasting networks, 2–3, 196, 216–22, 223–25, 227–36, 240, 268. See also agriculture; broadcasting posts

Russo-Chinese War, 79, 80
Russo-Japanese War, 64, 144, 182

schools: broadcasts from, 204; during Cultural Revolution, 249–50; for girls, 40; loudspeakers in, 204, 205, 232, 238–40, 245; radio-building instruction, 96; radios installed in, 76, 87, 96, 99–101, 104, 120, 147
Science (Kexue), 57
Self-Defense Daily (Zi Wei Ribao), 164
semicolonialism: financial, 51, 74; in Manchuria, 64; in Shanghai, 48–49, 50, 52–53; technopolitics and, 48–49, 50, 52–54, 57–59. See also China Maritime Customs Service; imperialist powers; newsscape of 1920s
Service, John, 107
Shaanxi Province: broadcasting posts, 226, 229–30, 316n128; Daye County, 226, 229–30; radio equipment, 121, 174; radios in schools, 100–101; Yulin County, 248–50, 251. See also Yan'an
Shandong Province: Japanese control in 1919, 26, 37–40; newspaper readership, 34–35; radio-listening stations, 215; radio newspapers, 166; radio station, 140; receiver ownership, 96, 147; rural radio listeners, 95; sparrow extermination campaign, 2; wired broadcasting network, 222; Yishui County, 95; Yiyuan County, 213, 215. See also Jinan; Qingdao
Shanghai: anti-Japanese resistance, 115–18; clandestine radio listening, 199–200; Cultural Revolution, 245,

Shanghai (cont.) 256; dismantling of shortwave radios, 198–99; foreign concessions, 43, 46, 48–49, 50, 55, 58–59, 90, 91, 116–17, 210; foreign consuls, 52–54, 55; Fudan University students, 40, 41, 128, 186; Japanese conquest (1937), 115–18, 196; Japanese invasion (1932), 104–6; Japanese occupation, 131, 132–33, 134–35, 136–37; Japanese puppet regime, 150–51, 152; loudspeakers in public spaces, 90; newspapers, 16, 46–47, 114, 150–51, 152, 199; number of radio receivers, 198; Peace Hotel, 116, 196–98; radio equipment dealers, 202; radio equipment manufacturing, 203, 258–59, 310n43; radio listeners, 75; radio meetings, 205, 207, 210–11; refugees from, 150–51, 152; regulation of public radios, 90; semicoloniality, 48–49, 50, 52–53; student speech troupes, 40; YMCA, 49–50, 59. See also Shanghai radio stations

Shanghai Datong University, 209
Shanghai Incident (1932), 104–6
Shanghai Jiaotong (Communications) University, 56, 203
Shanghai radio stations: accounts of Japanese invasion, 104–5; anti-Japanese resistance, 115–18; CCP takeover, 196–98; East China Radio Station, 197–98; German, 197–98; in International Concession, 91, 116–17, 134–35, 161, 197–98; Japanese control, 139, 142; Kellogg, 55, 58–59, 73, 78; municipal, 115–18, 135, 196, 296n22; number in 1930s, 91, 292n19; of Osborn, 48, 49–51, 54–55, 280n10; private, 91, 142, 198–99; Shenbao, 46, 55, 58–59; Soviet, 134–35, 161; Sun's speech transmitted by, 46–47; at war's end, 180

Shanxi Province: Datong, 104; Fenyang, 93, 126–27; Japanese invasion, 127; radio receivers, 93, 104; sparrow extermination campaign, 2

Shenbao, 33–34, 46, 55, 57, 58–59
shouyinyuan, see radio monitors
shouyinzhan, see radio-listening stations
Sichuan All School Student Association, 31
Sichuan Province: broadcasting posts, 255; climate, 121; Cultural Revolution in, 255; Jiangyou County, 120, 121; newspapers, 33–34; Qu County, 85–86, 108; radio listening in wartime, 120–21; radio-listening stations, 99, 100, 102, 108; radio monitors, 121; Tianquan County, 102; Xichang, 255. See also Chengdu; Chongqing
Sino-Japanese War (1895), 7
Sino-Japanese War (1937–45): aftermath, 179; anti-Japanese resistance, 82, 84, 86, 110, 113, 124, 128–29; beginning, 84–86, 113; Communist guerrillas, 157, 158; economic impact, 131–32; end of, 152–54, 176–79; influence of Japanese occupation, 111–12, 125; Japanese atrocities, 207–8; refugees, 124–25, 130, 150–51, 152, 155, 158, 167–68; state formation and, 18–19. See also collaborationist governments; Nationalist government (Chongqing); wartime newsscape

Smedley, Agnes, 107

social classes: of newspaper readers, 15–16; of radio listeners, 17, 86, 89, 91, 133–34, 236–37
socialized media newsscape (1949–58): ban on shortwave receivers, 198–99; CCP monopoly, 195, 202; clandestine radio listening, 199–202; communal listening, 191, 194, 204–13, 224, 234; demand for news, 196, 199–200; economic news, 225; in Great Leap Forward, 194, 225–29, 232–36; health information, 225; influence on daily life, 196, 225; local news in rural areas, 194, 195, 221, 226–27, 233–34, 316n128; mass politics and, 193, 195; newspapers, 199, 213; news posted on walls and blackboards, 2, 196, 203, 213–14, 215; participatory radio listening, 194, 227–29, 234–35; party directives to cadres, 203–4, 206, 212–13, 219, 220, 236–37; political impact, 191, 193; propaganda, 193, 197, 201, 204–13, 217, 220, 235, 236–37; radio as direct link between party and individual, 192–93, 207, 217–18, 231; radio as governance tool, 14–15, 195–96; radio meetings, 205–13, 217, 219, 312nn70–71; radio monitors, 196, 213–20; radio newspapers, 2, 213–14, 215–16; radio programming, 203–5, 224–25; required radio listening, 194, 204–5; takeover and rebuilding of infrastructure, 191, 196–98, 203; technopolitical process, 212, 217, 229, 236–37; trust in, 194–95. See also broadcasting posts; wired broadcasting networks
social media, 259, 261, 265–66
society, mass, see mass society

Song Meiling, 100
Song Qingling, 72
sound, see acoustic history; radio soundscape, 9, 11
South Manchuria Railway Company (Mantetsu or SMRC), 64
Soviet Union: Bolshevik Revolution, 7; conflict with Zhang regime, 79, 80; Japanese propaganda on, 149; propaganda, 79; radio equipment exports, 174, 203, 304n40; radio stations, 79, 134–35, 161; relations with CCP, 160, 174, 304n40; relations with Nationalist Party, 44, 45–46, 47, 182, 183, 281n13; Sputnik, 236, 265; TASS Agency, 79, 135; wired broadcasting network, 221; in World War II, 149, 153, 161, 179, 181, 221, 301n130
sparrow extermination campaign, 1–5
Spear, C. Ronald, 61, 67, 80
speech troupes, 35, 38–40
Stalin, Joseph, 218
state: formation, 18–19, 63; mass society and, 262, 263–64; power, 241, 318n6
Sun Yat-sen (Sun Zhongshan): on communications, 44–45, 195, 252; death and funeral, 73; final year, 69; The International Development of China, 44–45; "Peaceful Unification Manifesto," 45–47; radio broadcasts honoring, 93; use of radio, 43, 44, 45, 46–47, 50, 52, 60, 280n10; view of radio's future, 71–72, 73, 265. See also Nationalist Party

Taiwan, see Nationalist government (Taipei)
Taiyuan, number of radio receivers, 147
Taizhou, letters sent to, 35–36

Tangshan, 115
Tao Baichuan, 116–18, 196
technopolitical process: in China, 15, 18–20, 266; during Cultural Revolution, 255–57, 268; demand for news and, 6, 259, 262–64; dialectical nature, 44, 195, 217, 253; evolution after World War I, 43–44, 47–48; external competition and, 63; information saturation and, 240; introduction of radio and, 43–44; in Manchuria, 61–63, 70–71, 76, 82–83; mass politics and, 5–8, 9, 44, 195, 240, 257, 267; in Nanjing decade, 87–88, 104–9; in 1920s, 50–60, 72–74; in 1930s, 87–88; in 1950s, 195, 212, 217, 229, 236–37; political stability and, 240, 255, 257, 266; possible directions, 268; in post-Maoist period, 258–59; semicolonialism and, 48–49, 50, 52–54, 57–59; in socialized media newsscape, 212, 217, 229, 236–37; societal impact, 5–6; in urban areas, 212; utopian dreams, 265; in wartime, 110–13, 144, 145, 154
Telefunken, 62, 77, 81, 92, 119
Telegraphist's Companion (Dian You), 57, 72
telegraphy: character encoding process, 30–31, 65–66, 246; constraints and challenges, 17, 29–31, 94; in Manchuria, 82; Morse code, 30–31, 65–66, 169; news communicated in 1919, 25–26, 28–32, 39, 41–42; Shanghai operators, 296n22
telephone, 25–26, 39, 220
Teng, Theresa, 258
Tianjin: amateur radio builders, 58; communal radio listening in streets, 89–90; foreign concessions, 39, 48, 115; Japanese conquest and occupation, 114, 115, 126, 155; news of May Fourth Movement, 39; radio listeners in wartime, 115; radio meetings, 205; radios in libraries, 90–91; radio station, 71, 74–75, 78, 140, 170, 286n15; refugees from, 155; student speech troupes, 39–40; Zhang's army in, 71
transportation: boats and ships, 12, 33; communication term used for, 29; railways, 12, 17, 64–65; in remote areas, 213
Tung, Robert Ho, 80

United States: barbed wire, 12; female broadcasters, 172; media corporations, 62; military forces in China, 152; Mukden consulate, 309n19; newspapers, 13; Office of War Information, 122–23, 130, 134, 135, 141, 143; postal service, 8; radio development in, 56; radio equipment manufacture and exports, 59, 70, 76, 81, 88, 122, 202; San Francisco radio stations, 81, 130, 134, 135, 137; State Department, 202; unrest in 1960s, 257; Voice of America, 198, 199–200. *See also* Korean War
US Armed Forces Radio, Tokyo, 198, 200
urban areas: communal listening culture, 87, 89–91, 92–93, 105, 120–21; loudspeakers in public spaces, 17, 72, 73, 84, 90, 92–93, 104, 105, 212; Nanjing decade newsscape, 84, 88, 89–91, 92–93, 104, 105; technopolitical process, 212; underground radio newspapers

during Japanese occupation, 126. See also individual cities

Voice of America (VOA), 198, 199–200

Wang Jingwei, 142, 146, 151, 302n136
War against the Sparrows, 1–3
warlords: conflicts, 18; Manchurian, 61; Sun Yat-sen and, 45; Zhili Clique, 51, 52, 69. See also Zhang Xueliang; Zhang Zuolin
wartime newsscape (1937–45): Communist propaganda, 149, 157, 158–64, 168–69; demand for news, 19, 110, 114–15, 122, 124, 126–29, 135–37, 144, 148, 259–60; in Japan, 144–45; Japanese censorship and regulation, 113, 115, 123–24, 126, 132, 135, 144; Japanese soldiers and, 123–24; legacy, 154; in occupied zones, 110, 113, 114, 126, 136–38, 143; technopolitics, 110–13, 144, 145, 154. See also Communist newsscape (1937–49); Japanese propaganda; Nationalist propaganda; Sino-Japanese War
wartime newsscape (1937–45), newspapers: Communist, 159, 164, 303n5; Nationalist, 162–63, 168, 187–88; news-sheets, 114; in occupied areas, 113, 114, 126, 129, 133, 143, 149, 152, 299n72, 303n5; radio newspapers, 20, 124–26, 160–63, 165, 268
wartime newsscape (1937–45), radio: clandestine listening, 115, 130–38; classes of listeners, 133–34; under collaborationist regimes, 142, 144, 145, 146–47, 148–52, 153, 154, 180; effects on listeners in occupied areas, 136–38; infrastructure construction, 19, 111–12; Japanese, 115, 131, 134, 135, 153–54; Japanese-controlled, 138–43, 144, 147, 148–52, 153–54; Japanese-language, 151, 153–54; loudspeakers in public spaces, 114; Nationalist propaganda, 110–11; news content, 141–44; news transcriptions, 20, 119, 120, 124–26, 128–29, 143, 164–65; popularization campaign, 138, 144, 145–46, 147; radio-listening stations, 120–23, 127, 213–14, 313n76; radio newspapers, 20, 124–26, 160–63, 165, 268; in Shanghai, 115–18, 135. See also China Central Broadcasting (Chongqing)
weather forecasts, 224
Weber, Max, 263
Westinghouse, 56, 57
wired broadcasting networks: administrative use, 14–15, 223; as background of daily life, 251–52, 268; costs, 221; during Cultural Revolution, 241, 245–48, 252, 253, 257; in Hong Kong, 315n107; in Hubei, 222, 227–29, 230–36; mass society and, 221, 236–37; in Nanjing decade, 92–93; nationwide, 212, 220–22, 238–40, 245–48, 268; participatory listening, 194, 227–29, 234–35; polycentric infrastructure, 194, 195, 241, 244–45, 268; propaganda images, 201; radio meetings, 227–29; in rural areas, 2–3, 196, 216–22, 223–25, 227–36, 240, 268; Russian-language lessons, 204, 310–11n47; in Soviet Union, 221; sparrow extermination campaign, 1–3; Western perceptions of, 259; in Zhejiang, 225, 226, 238–40. See also broadcasting posts; loudspeakers
wire news services, 62

women: CCP activists, 170; dialects and accents, 173–74; education, 40, 171, 172–73, 174–75; feminism, 170–71; literacy levels, 278n34; local broadcasters, 223, 254; oral news networks, 40; political activism, 171; radio monitors, 98, 156, 175, 214–15, 313n79; roles in Communist newsscape, 155–57; traditional roles, 170–71

women radio announcers: at broadcasting posts, 223, 226; dialects and accents, 75, 161–62, 173–74; dominance in Chinese radio, 14, 170, 173; in Manchuria, 70, 170; in Nanjing, 172; Song Qingling, 72; sounds of voices, 170, 171–73, 178–79, 305n49; in Tianjin, 75; in United States, 172; at Yan'an station, 157, 170, 174–76, 305–6n57. See also Ding Yilan

workplaces: broadcasting posts in, 239, 243, 246, 251–55, 268; loudspeakers installed in, 204, 232, 239–40, 243, 246; radio listening in, 204–5, 206–7, 208–9, 210–11

World Radio Receiving Station, Mukden, 61–62

World War I: Paris Peace Conference, 26, 38; print media, 144

World War II: Cairo Conference, 142–43; end of, 152–54, 176–79; radio reports of Allied actions, 130, 135, 136–37, 161; Soviet Union and, 149, 153, 179, 181, 221, 301n130. See also Sino-Japanese War

Wu Daoyi, 97–98, 107, 109, 118, 119, 172

Wuhan: Cultural Revolution in, 253–54, 256–57; Hubei People's Broadcasting Station, 223; loudspeakers in public spaces, 90;

Nationalist retreat to, 118; newspapers, 114; radio station, 115, 139, 140

Xiamen: CCP infiltration of GMD newspapers, 188; Japanese radio station, 140; radio demonstrations, 59

Xi'an: radio equipment shortages, 121; radio stations, 82, 106–8

Xi'an incident (1936), 100–101, 104, 106–9

Xinhua News Agency, 159, 160, 161, 176, 186–87, 188. See also Yan'an radio station

Yan'an radio station: clandestine listening to, 186; competition with Chongqing station, 179–80; equipment and power, 160, 174, 180, 304n40; female announcers, 157, 170, 174–76, 305–6n57; Morse code transmissions, 159, 160, 169; news transcribed by monitors, 160–61; reports on radio newspapers, 159; working conditions, 174, 175

Yang Jisheng, 195, 232, 236

Yangtze River, post boats, 33

Yunnan Province: clandestine radio listening, 186–87; newspaper distribution, 213; radio-listening stations, 213

Zhang Xueliang: anti-Japanese resistance, 82; Manchurian Incident and, 80, 81; opium addiction, 80; radio and, 67, 78, 79, 106–7, 174; rule of Manchuria, 64, 77–78, 289n76; Xi'an incident, 106–9. See also Manchurian newsscape

Zhang Zuolin: civil war victory, 74; death, 68, 76; life of, 65; in power in

Beijing, 69; radio and, 67, 68, 69–71, 73–75; rule of Manchuria, 64–65, 67–68, 70–71, 82–83. See also Manchurian newsscape

Zhangjiakou City, 180–81

Zhangjiakou Xinhua radio station, 181

Zhejiang Province: broadcasting posts, 225, 226; mail service to, 35–36; radio-listening stations, 99; radio meetings, 208; radio stations in wartime, 148–49; wired broadcasting network, 225, 226, 238–40

Zhili Clique, 51, 52, 69

Zhou Enlai, 40, 174, 304n40

Zhu De, 176, 188

The authorized representative in the EU for product safety and compliance is:
Mare Nostrum Group
B.V Doelen 72
4831 GR Breda
The Netherlands